Tumor Suppressor Par-4

Vivek M. Rangnekar
Editor

Tumor Suppressor Par-4

Structural Features, Molecular Mechanisms and Function

Springer

Editor
Vivek M. Rangnekar
Associate Director, Markey Cancer Center
University of Kentucky
Lexington, KY, USA

ISBN 978-3-030-73571-5 ISBN 978-3-030-73572-2 (eBook)
https://doi.org/10.1007/978-3-030-73572-2

This Springer imprint is published by the registered company Springer Nature Switzerland AG
The registered company address is: Gewerbestrasse 11, 6330 Cham, Switzerland

Preface

Prostate apoptosis response-4 (Par-4, also referred to as PAWR) was first identified in prostate cancer cells. Since its discovery, the expression and function of Par-4 has been elucidated in diverse normal tissues and disease states including various cancers, neuronal degeneration, diabetes, renal, cardiac and eye diseases. This book, entitled *Tumor Suppressor Par-4,* is a timely assessment of the advances in Par-4 research over the past 25 years. Organized into two volumes, this book will provide the necessary framework to chart the next course of research investigations into the many intriguing facets of Par-4 and their application to better understand and treat relevant diseases.

The first volume of *Tumor Suppressor Par-4* comprises a series of chapters providing an overview of Par-4 structural organization and regulation that are essential for its function. Chap. 1, by Ganguly *et al.*, provides a succinct description of the discovery of Par-4, as well as the modular structural domains that regulate Par-4 function. This chapter describes the proteins that partner with Par-4 to accentuate or subdue its function. Moreover, several observations that shed light on the posttranslational regulation of human Par-4 protein by Akt1, upregulation of the Par-4 promoter by c-Myc, and downregulation of Par-4 by oncogenic Ras are also described. In Chap. 2, Xiaoqi Liu and colleagues describe the expression and role of Par-4 in the cell cycle. They provide an appraisal of Par-4 and some of its key targets in the context of cell cycle regulation, and for the first time describe Par-4 as a substrate of the polo-like kinase PLK1 in the mitotic phase of the cell cycle. Chap. 3 by Steve Pascal and colleagues provides insights into the disordered structure in full-length Par-4 but ordered structure in the cleaved carboxyl terminal sequences of Par-4. Chap. 4, by Jörg Labahn and colleagues, delivers a unique crystallographic analysis of the regulatory domain of Par-4 that is of essence in understanding its relevance in Par-4 function. The chapter streamlines the intricate process of crystallography for the uninitiated reader, yet delves into reasonable depth to excite the curiosity of experts in the field.

Three key domains of Par-4 have been identified to date. Involved in binding to diverse partner proteins, the leucine zipper domain at the carboxyl-terminus of Par-4 was first described in 1994. The centrally placed SAC domain of Par-4 is necessary for apoptosis activity and was described in 2003. In 2017, an amino-terminal domain of Par-4 called PAF was found to be secreted by therapy-sensitive cancer cells undergoing apoptosis. Secreted PAF induces paracrine apoptosis in neighboring therapy-resistant cancer cells. Chap. 5 by Hebbar *et al.* describes the PAF domain and its regulation, with emphasis on the significance of key residues within the PAF domain that are essential for regulation of Par-4 activity by ubiquitinases, such as Fbxo45, TRIM21, and perhaps COP1. In Chap. 6, Aurélie de Thonel and colleagues focus on the role of casein kinase 2 (CK2) in Par-4 phosphorylation in prostate cancer. The findings in prostate cancer may serve as a primer for CK2-Par-4 interaction in other cancers in which CK2 plays cell survival roles. Chap. 7, by Marie Cohen and colleagues, describes the co-regulation of Par-4 and GRP78 that interact in the endoplasmic reticulum (ER) and also at the plasma membrane of cancer cells, trophoblasts, and other stressed cells. As GRP78 generally functions as a chaperone, it is particularly intriguing that Par-4 is essential for translocating GRP78 from the ER to the cell surface, where extracellular (i.e., secreted) Par-4 can bind to GRP78 and effectively utilize it as a receptor to trigger apoptosis.

In Chap. 8, Bieberich and Elsherbini discuss the role of Par-4 regulation in human pluripotent stem cell death and embryo development. Their original findings were extended to human induced pluripotent stem cells (hiPSCs), which hold promise in medical innovation. They describe dissociation-induced cell membrane blebbing and apoptosis in hiPSCs occurring *via* the ROCK-actin-myosin pathway that is dependent on Par-4 function. Chap. 9 by Howard Donninger and Geoffrey Clark examines the regulation of Par-4 function by RASSF2. These facets of Par-4 in hiPSCs and embryogenesis, or in tumors where RASSF proteins serve as apoptotic effectors of the RAS oncoproteins deserve further investigation to reveal the nuances involved in the regulation of Par-4 function. Chap. 10 describes the regulation of Par-4 by ceramide in the process of cell death. Par-4 has been generally shown to promote apoptosis, and Galadari and colleagues describe an unexpected function for Par-4 in autophagy that is dependent on the activity of the guardian of the genome p53.

Although *Par-4* was first identified as a gene induced in response to intracellular Ca^{2+} elevation by ionomycin, and induction of the *Par-4* upstream promoter was later corroborated in studies that used thapsigargin, which elevates intracellular Ca^{2+} levels, most studies have focused on Par-4 protein structure and function rather than the precise mechanisms involved in *Par-4* gene induction or suppression. Chap. 11 by Leonie Young's team focuses on the suppression of the *Par-4* gene by Steroid Receptor Co-activator 1 (SRC-1) in partnership with the homeobox protein HOXC11, which is known to transcriptionally regulate pluripotency and development genes. This chapter on the suppression of the *Par-4* promoter by the SRC-1/HOXC11 interactome illustrates one of the key mechanisms involved in silencing *Par-4* leading to tumor recurrence.

The first volume provides the framework for the second volume of *Tumor Suppressor Par-4*, which describes the involvement of Par-4 in various diseases including cancer, Alzheimer's and other neurodegenerative disorders, diabetes, and renal and ocular ailments, as well as in clinically relevant processes such as inflammation, epithelial mesenchymal transition, and apoptosis, which are associated with these diseases. Moreover, Par-4 is an archetypal therapeutic target in various cancers, and Par-4 regulating drugs or Par-4 secretagogues developed for clinical trials are discussed in the second volume. Together, Volume 1 and 2 serve as a set for ready reference on basic, translational and clinical features of the fascinating tumor suppressor Par-4, along with valuable perspectives for future research investigations and clinical applications.

Lexington, KY, USA Vivek M. Rangnekar

Contents

Discovery and Overview of Par-4

Saptadwipa Ganguly, Ravshan Burikhanov, Shirley Qiu, and Vivek M. Rangnekar

Abstract Prostate apoptosis response-4 (Par-4) is a tumor suppressor that was first identified as an immediate early gene in a differential screen for genes induced during the process of apoptosis in prostate cancer cells. Par-4 is ubiquitously expressed in normal human and rodent cells and tissues but is downregulated, mutated, or inactivated in most cancer cells. Functionally, elevated levels of Par-4 induce apoptosis in cancer cells and experimental tumors or sensitize normal and cancer cells to the apoptotic action of diverse insults. Par-4 contains several functional domains, including a leucine zipper domain at the carboxy terminus, selective for apoptosis of cancer cells (SAC) domain naturally placed in the center of the molecule, and Par-4 amino-terminal fragment, which regulate its tumor suppressor and pro-apoptosis function. Par-4 is a unique tumor suppressor protein expressed both intracellularly in the cytoplasm and nucleus and extracellularly. Protein kinase A activates intracellular Par-4 by phosphorylation at Thr155 residue in rat Par-4 or Thr163 in human Par-4. Activated intracellular Par-4 induces the Fas (CD95)/FasL apoptotic pathway and also inhibits NF-κB cell survival pathway. On the other hand, Akt1 inactivates Par-4 by phosphorylation at S249 in rat Par-4 or S228 in human Par-4. Extracellular Par-4 acts via cell surface receptor glucose-regulated protein 78 (GRP78) to induce caspase-8/caspase-3 and apoptosis. Whether intracellular or extracellular, Par-4 induces cancer-selective apoptosis without harming normal cells, making Par-4 an attractive target in cancer therapy. Moreover, Par-4 plays key roles in epithelial mesenchymal transition, inflammation, neuronal degenera-

S. Ganguly
Department of Toxicology and Cancer Biology, University of Kentucky, Lexington, KY, USA

R. Burikhanov
Department of Radiation Medicine, University of Kentucky, Lexington, KY, USA

S. Qiu
Department of Microbiology, Immunology and Molecular Genetics, University of Kentucky, Lexington, KY, USA

V. M. Rangnekar (✉)
Associate Director, Markey Cancer Center, University of Kentucky, Lexington, KY, USA
e-mail: vmrang01@uky.edu

V. M. Rangnekar (ed.), *Tumor Suppressor Par-4*,
https://doi.org/10.1007/978-3-030-73572-2_1

tion, diabetes, and cardiac and retinal apoptosis. In recent years, FDA-approved drugs and new compounds that can stimulate secretion of Par-4 from normal cells of the body to induce paracrine apoptosis of cancer cells have been identified, and clinical trials are ongoing. Thus, Par-4 research has progressed from the bench to the bedside to impact the longevity of cancer patients. This chapter provides a succinct description of research on the discovery of Par-4, leading to its key domains, interactive partners, and regulation along with mechanistic evidence supporting Par-4 function.

Keywords Prostate apoptosis response-4 (Par-4) · PAWR · Nuclear factor kappa-light-chain-enhancer of activated B cells (NF-κB) · Apoptosis · Secretagogues · Selective for Apoptosis in Cancer cells (SAC) · Par-4 amino-terminal fragment (PAF) · Leucine zipper (LZ) domain · Nuclear localization signal (NLS) · Tumor suppressor · Glucose-regulated protein 78 (GRP78) · Wilm's tumor protein 1 (WT1) · Atypical protein kinase Cζ (ζPKC) · ER stress · c-Myc · Ras · Neurodegeneration · Inflammation · EMT · Pluripotent stem cells · Crystallography · PKA · GRP78 · RASS2 · CK2 · PLK1 · Ubiquitin · FBOX45

1 Discovery of Par-4

Prostate apoptosis response-4 (Par-4) was discovered in 1994 by subtractive hybridization and differential gene screen performed to identify immediate early genes responsible for inducing apoptosis as judged by DNA fragmentation in androgen-dependent and -independent cells. The initial screen performed in AT-3 rat androgen-independent rat prostate tumor cells exposed to the calcium ionophore, ionomycin, along with cycloheximide for 5 h [1]. In addition to Par-4, this screen identified four other immediate early genes that were induced by ionomycin and superinduced by ionomycin in the presence of cycloheximide, indicating their expression did not require de novo protein synthesis. The genes were designated prostate apoptosis response (Par)-1, Par-2, Par-3, Par-4, and Par-5. Northern analysis of the RNA from AT-3 rat prostate cancer cells treated with ionomycin validated the induction of the genes. Among them, Par-4 (GenBank Accession No. U63809) and Par-5 were judged to be novel according to GenBank after partial nucleotide sequencing. All the *par* genes displayed constitutive expression, but on exposure to ionomycin, they were rapidly induced as early as 30 min (except for *par-5*) in diverse cell types [1, 2].

Par-4 shows a biphasic mode of induction with an early transient peak seen around 30 min and sustained peak beginning at 3 h of the same magnitude as the first peak. The *par* genes displayed induction in ventral prostate of castrated rats, and particularly Par-4 induction reached its peak at 66 h coinciding with peak levels of apoptosis of the rat secretory epithelial cells that are known to be dependent on androgen for survival. Par-4 induction as well as apoptosis induction in the rat prostate was inhibited by pretreatment of the rats with nifedipine prior to orchiectomy [1]. As nifedipine inhibits L-type of calcium channels, we inferred that both Par-4 induction and apoptosis were dependent on intracellular calcium elevation in the prostate cells.

Early studies indicated that *Par-4* gene was induced at the RNA level in response to apoptotic insults but not growth-arresting, growth-promoting, or necrotic stress stimuli. Thus, serum-starved AT-3 cells exposed to serum or cells treated with growth-arresting agent TNF-α or oxidative stress-inducing agent H_2O_2 did not show *Par-4* gene induction as judged by Northern blot analysis. Moreover, when human prostate cancer LNCaP cells that are androgen-dependent were maintained in charcoal-stripped serum and then treated with dihydroxytestosterone (DHT, a cell survival and growth-promoting agent) for 1 week, they did not display induction of Par-4 mRNA expression, suggesting that *Par-4* is not a growth factor inducible gene [1, 3]. Par-4 induction was immediately confirmed in the luminal epithelial cells of the ventral prostate of rats that were castrated, as these cells start undergoing apoptosis within 24 h after castration [1]. On the other hand, *Par-4* was not induced in the kidneys and liver of the castrated rats that showed apoptosis in prostate, implying that Par-4 may not be involved in physiological processes where androgen withdrawal does not induce apoptosis. The deduced amino acid sequence of Par-4 predicted a 332 amino acid protein with two nuclear localization sequences (that we

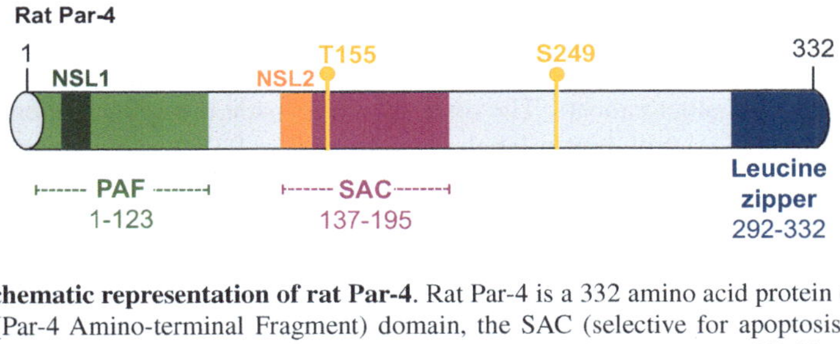

Fig. 1 Schematic representation of rat Par-4. Rat Par-4 is a 332 amino acid protein containing the PAF (Par-4 Amino-terminal Fragment) domain, the SAC (selective for apoptosis in cancer cells) domain, the leucine zipper (LZ) domain, nuclear localization sequences (NLS1 and NLS2), and functional phosphorylation sites at Thr 155 and Ser 249. At first identification of the rat cDNA, only the LZ domain (amino acids 292–332), NLS1 (amino acids 1–25), and NLS2 (amino acids 137–153) were evident in the deduced amino acid sequence. The LZ, NLS1, NLS2, and other domains such as PAF and SAC and phosphorylation sites were later functionally validated

designated as NLS1 and NLS2) and a leucine zipper (LZ) domain at the carboxyl-terminus of the protein (Fig. 1).

2 Identification and Chromosome Localization of Human Par-4/PAWR Gene

A few years after the discovery of rat Par-4, the human homolog of Par-4 was identified in studies performed using the yeast two-hybrid system in a quest to identify binding partners of atypical protein kinase C (aPKC). In these experiments, ζPKC was fused to the DNA-binding domain of the yeast LexA protein as bait and used to screen the human placenta Matchmaker cDNA library for potential binding partners. The isolated clones corresponded to the partial cDNA coding for *Par-4* indicating that Par-4 was expressed in humans as well. The full-length sequence of ζPKC as well as its regulatory domain, but not the catalytic domain of ζPKC or other serine/threonine kinases, interacted with Par-4. It was noted that the zinc finger domain of ζPKC is capable of interacting with the LZ domain at the carboxyl-terminal of Par-4 [4]. In the same year that Par-4 was identified as a partner of ζPKC, human Par-4 was identified as an interactive partner of Wilm's tumor protein WT-1 in yeast two-hybrid studies [5]. Interestingly, the LZ domain of Par-4 was found to interact with one of the four C_2-H_2 zinc finger domains of WT-1. Expression of Par-4 as a flag-tagged protein in lieu of a Par-4 antibody in these studies indicated that Par-4 protein is found in nuclear and cytosolic fractions and may play both nuclear and cytosolic roles in apoptosis.

As an interacting partner of ζPKC and WT-1 in human cells, there was significant interest in determining the chromosomal location of Par-4. In 1998, Johnstone et al. localized the human Par-4 gene to chromosome 12q21.2 [6]. Par-4 was designated as *PAWR* (PRKC apoptosis WT1 regulator) owing to redundancy of the acronym Par-4 in the literature. This region is known to be deleted in male germ cell tumors, as well as other cancers. The discovery was made using fluorescence in situ hybridization (FISH) with biotin-labeled 1.7 kb human PAWR cDNA, and the relative distance from the short arm telomere to the signals (FLpter value) was determined. A specific FISH signal was obtained near the middle of chromosome 12, and it gave a FLpter value of 0.62 ± 0.04. These results indicated a chromosomal localization at 12q21.2–q21.32 that was confirmed by PCR using a monochromosomal mapping panel. In gel electrophoresis, a band of approximately 390 bp, as predicted, was obtained only in the lane containing human genomic DNA and the cell DNA harboring human chromosome 12, implying that the *PAWR* gene is indeed localized on human chromosome 12. Radiation hybrid localization confirmed that in agreement with the FISH data, *PAWR* localized to 12q21 [6]. Many years later, studies by Zhao et al. noted that the *PAWR* gene is coded by the minus strand of chromosome 12q21.2. The gene encompasses 99.064 kb of DNA (starts at 79985745 and ends at 80084790 bp from pter) and has seven exons and six introns, with 2.2 kb nucleotides in the mRNA and 1.02 kb open reading frame. The initiation codon ATG present in Exon 2 [7].

3 Ubiquitous Expression of Par-4 in Vertebrate Tissues

The *PAWR* mRNA and protein are expressed in most normal human and rodent tissues. Intense Par-4 staining was observed in cerebellar Purkinje cells, pyramidal cells of the hypothalamus, cells of the adrenal cortex in the brain, cardiomyocytes of the heart, transitional epithelium of the distal renal tubules and the podocytes of the kidney glomeruli, centrilobular hepatocytes of the liver and bile duct epithelial cells, pseudostratified epithelium of the bronchi and columnar epithelium of bronchiole in the lung, endothelium of arterioles and venules in the duodenum, Malpighian bodies in the first concentric layer of the PALS of the spleen, and seminiferous tubules and ductular epithelial cells of the epididymis in testes. Heterogeneous staining was noted in straited muscles. Par-4 is also expressed in the hormone-sensitive tissue of the rats including the myoepithelial cells of mammary glands, granulosa cells of atretic ovarian follicles, dense mesenchyme surrounding the ventral prostate, and basal glandular epithelial cells of 6-month-old rat but not the ductal epithelial cells. Par-4 is detected in cells originating from endoderm, mesoderm, or ectoderm except for certain neurons and medulla of adrenal gland from ectodermal origin and dense connective tissue, some smooth muscle cells of mesodermal origin, glial cells, and lymphocytes. These studies indicate that the expression of Par-4 is downregulated upon differentiation. Similarly, immunohistochemical (IHC) staining for Par-4 was lower in stratum corneum cells of the skin and cells on top of the duodenal villi compared to stem cells in stratum basale and at the bottom of the crypts of Lieberkühn, respectively, which have higher staining. Par-4 is expressed in mice, horses, pigs, dogs, and cows as well as other vertebrates, fish, birds, and *Xenopus* [8]. A role for Par-4 is found in resorption of the tadpole tail via induction of apoptosis of the muscle bundle myoblasts. The induction of apoptosis in myoblasts coincides with peak levels of the thyroid hormone in the tadpole tail (Fig. 2). The expression of Par-4 is low in the pre- and pro-metamorphosis phase but is upregulated in the metamorphosis phase, suggesting a role for Par-4 in apoptosis [9].

The mouse-expressed sequence tag database was used to construct the sequence for mouse *Par-4*. The DNA analysis shows a high sequence homology in all the three species, with rat cDNA displaying 92.2% and mouse cDNA displaying 93% sequence similarity in the amino acid sequence. Human and rat Par-4 proteins have 75% identity and 84% functionally similar amino acids. The three species have 100% identical putative functional domains, suggesting a high level of protein conservation [3]. A schematic representation of the amino acid positions of the conserved domains of Par-4 in human, rat, and mouse is depicted in Fig. 3. Rat Par-4 protein contains 332 amino acids, mouse Par-4 has 333 amino acids, and human Par-4 has 340 amino acids [10].

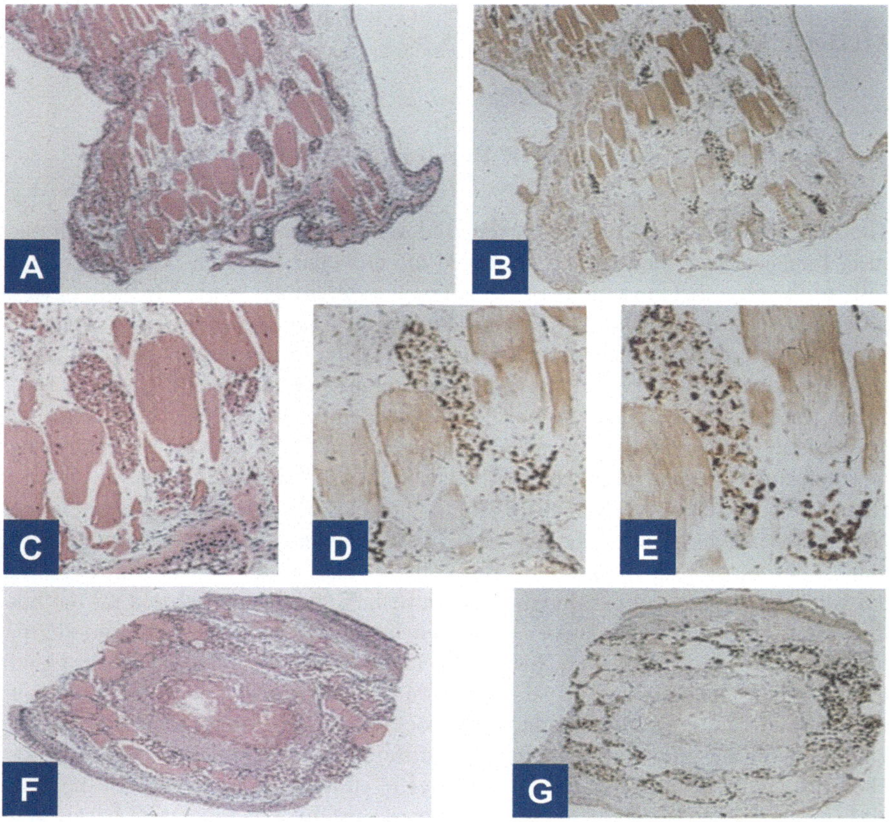

Fig. 2 Immunohistochemical staining for Par-4 in the tadpole tail undergoing resorption. Par-4 protein levels rise during the metamorphosis when the thyroid hormone levels peak in the tadpole tail, causing the myoblasts to undergo apoptosis resulting in resorption of the tail. (**a**) Hematoxylin-eosin (H&E) staining of tail (10× magnification). (**b**) Par-4 staining of the tail (10× magnification). (**c**) H&E staining of tail (40× magnification). (**d**) Par-4 staining of tail (40× magnification). (**e**) Par-4 staining of the tail (60× magnification). (**f**) H&E staining of tail (10× magnification). (**g**) Par-4 staining of the tail (10× magnification). (**a**–**e**). Longitudinal sections; (**f** and **g**). Transverse sections. Note muscle bundle myoblasts undergoing apoptosis as judged by H&E staining and Par-4 positivity. Panels **a** and **b** and **f** and **g** represent sequential sections

4 Functional Domains of Par-4

Par-4 is a multi-domain protein containing (1) two nuclear localization sequences: NLS1 (amino acids 20–25 in rat, 24–29 in human, and 21–26 in mouse) and NLS2 (amino acids 137–153 in rat, 147–163 in human, 138–154 in mouse), (2) a nuclear export signal (amino acids 291–302 in rat, 299–310 in human, 292–303 in mouse) embedded within the leucine zipper (LZ) domain and predicted to be masked upon

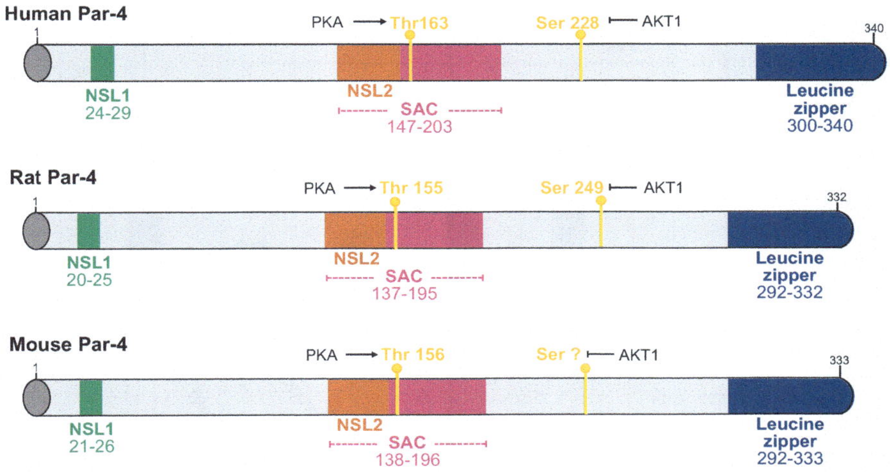

Fig. 3 Functional domains of Par-4. The functional domains of Par-4 are largely conserved across species in human, rat, and mouse Par-4. The numbers depict their relative amino acid positions. *NLS1* nuclear localization signal 1, *NLS2* nuclear localization signal 2, *SAC*- selective for apoptosis in cancer cells, *LZ* leucine zipper domain. Ser249 in rat Par-4 is functionally similar to Ser228 in human Par-4

dimerization, (3) an LZ domain (amino acids 292–332 in rat, 300–340 in human, 292–333 in mouse), (4) a partially overlapping death domain at the carboxyl-terminus region, (5) a SAC (selective for apoptosis in cancer cells) domain (amino acids 137–195 in rat, 146–203 in human, 138–196 in mouse) located in the central part of the protein, and (6) a PAF (Par-4 amino-terminal fragment) domain (amino acids 1–123 in rat, 1–124 in mouse 1–131 in human) at the amino-terminus of protein [10]. These domains are 100% conserved across the rat, mouse, and human Par-4 and play an important role in cellular localization (NLS1 and NLS2), apoptosis (SAC), and dimerization (LZ). On the other hand, the PAF domain is functionally conserved across species, but the amino acid sequence is not conserved. Moreover, the VASA domain for binding to FBXO45 is present in human Par-4 but not in rat or mouse Par-4. Par-4 is posttranslationally modified, for example, by protein kinase A (PKA) at T163 that activates Par-4 apoptotic function and casein kinase 2 (CK2) at S231 or AKT serine/threonine kinase 1 (AKT1)-mediated phosphorylation at Ser228 that inactivates Par-4 (Fig. 3). Transfection of human cancer cells with a non-phosphorylatable (S228A) mutant of human Par-4 results in loss of Par-4 inhibition by Akt1 (Fig. 4).

Rat and human Par-4 show phosphorylation at S124 and S231 by casein kinase 2 (CK2) that is responsible for inhibition of the proapoptotic function of Par-4. Phosphorylation of rat Par-4 by CK2 prevents the caspase-mediated cleavage of Par-4, thereby inhibiting its function. Such cleavage in human Par-4 is not regulated by CK2 as the phosphorylation site does not overlap the caspase cleavage site

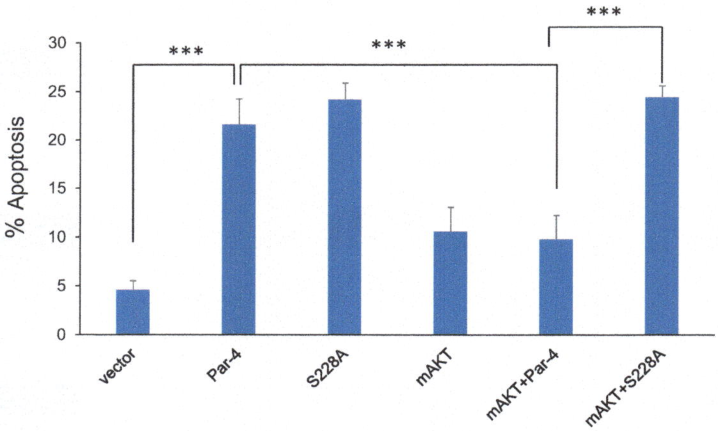

Fig. 4 Phosphorylation of Par-4 by Akt1 inhibits apoptosis. Human lung cancer cells H460 were transfected with control vector, expression constructs for Par-4, non-phosphorylatable (S228A) mutant of human Par-4, constitutively active myristoylated Akt1 (mAKT) alone and in combination with Par-4 or S228A mutant Par-4. The cells were scored for apoptosis 24 h later. Full-length Par-4 induces apoptosis that is inhibited by mAkt1. On the other hand, disruption of the S228 phosphorylation site in S228A mutant of Par-4 prevents inhibition of apoptosis by Akt phosphorylation. *** $p < 0.001$ by Student's *t*-test

(D131), but S124 phosphorylation impairs its apoptotic function. In prostate cancer, the S231 residue in human Par-4 is phosphorylated compared to its normal counterpart tissue resulting in inhibition of its apoptotic function. Accordingly, when CK2 is knocked down in prostate cancer cells, their sensitivity to Par-4 is increased, but this effect is lost if Par-4 is co-parallelly knocked down [11].

Par-4 is cleaved by caspase-3 at D131 to produce a 25 kDa C-terminal fragment that accumulates in the nucleus [11, 12]. Unlike wild-type Par-4, a D131A mutant of Par-4 does not sensitize MCF-7 breast cancer cells to cisplatin-induced apoptosis, emphasizing the need for Par-4 cleavage in its apoptosis-sensitizing action. Overexpression of the 132–340 amino acid fragment of Par-4 blocks NF-κB nuclear translocation potentially by inhibiting IκBα phosphorylation [10].

By using a domain search program, previously unrecognized domains of Par-4 were identified that may contribute to regulation of Par-4 protein function especially in cancer cells (Fig. 5). Par-4 contains an adenylyl cyclase (GGDEF) domain at the C-terminus that may convert ATP to cAMP for downstream second messenger signaling. Consistent with this potential, Par-4 has a P-loop (GXXXXGKS) sequence, which may confer ATP/GTP binding property. Moreover, Par-4 also contains the regulatory sequence DXXGQ that may stabilize the ATP/GTP attachment. Interestingly, the adenylyl cyclase domain and DXXGQ domain are evolutionarily conserved from *Xenopus* to humans, while the P-loop domain, which is absent in rodents, may have evolved later in chicken and is maintained during further evolution in rabbit, chimpanzee, rhesus monkey, and humans. Cancer cells produce

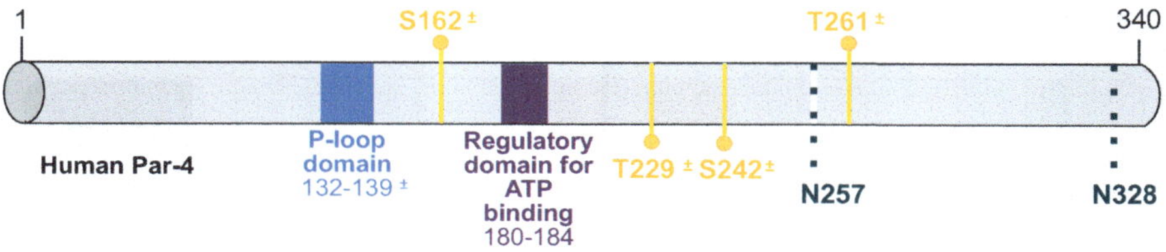

Fig. 5 Putative domains of human Par-4. Several putative domains of human Par-4 that deserve to be experimentally confirmed are indicated. ± signifies that the amino acid sequence is present in human Par-4, but not conserved in rat and mouse Par-4

excessive ATP through oxidative and glycolytic pathways, as compared to normal cells, to trigger rapid proliferation. Elevated ATP levels in cancer cells can lead to an increase in binding of ATP to the P-loop domain that is stabilized by the DXXGQ domain. This may trigger adenylyl cyclase activity of the GGDEF domain leading to conversion of ATP to cAMP, which in turn activates PKA. Activated PKA then phosphorylates Par-4 at T163 position and thereby activates the apoptotic function of Par-4. Based on this rationale, we hypothesize that (1) Par-4 may have intrinsic adenylyl cyclase activity; (2) this adenylyl cyclase activity is regulated by the P-loop and DAAGQ domains of Par-4; and (3) the adenylyl cyclase activity of Par-4 may auto-activate its apoptotic potential selectively in cancer cells. Experimental validation of this hypothesis is pending. Par-4 also contains putative protein kinase C (PKC) phosphorylation sites at T229/230 or T261/267 and sites for N-glycosylation at N257 or N328 and methylation at Lys146 or Lys227 that require functional elucidation.

Blocking the amino-terminus but not the carboxyl-terminus of Par-4 by in-frame fusion with green fluorescent protein (GFP) leads to inhibition of Par-4 secretion and apoptosis mediated by extracellular Par-4 (Fig. 6). Consistent with this observation, deletion of the first 25 amino acids that include the NLS1 sequence results in inhibition of Par-4 secretion. These observations indicate that the amino-terminus of Par-4 is essential for Par-4 secretion and that an atypical secretory peptide sequence is present at the amino terminus enabling the secretion of Par-4.

4.1 *Nuclear Localization Sequences and Nuclear Export Sequence of Par-4*

The two nuclear localization signals, NLS1 and NLS2, are localized in the amino-terminal half of Par-4 protein. Endogenous Par-4 that is not active on its own is cytoplasmic in most cell types. Deletion of NLS1 does not prevent nuclear translocation of Par-4. On the other hand, the bipartite NLS2 sequence is crucial for translocation from the cytoplasm to the nucleus and for induction of apoptosis. Par-4 can

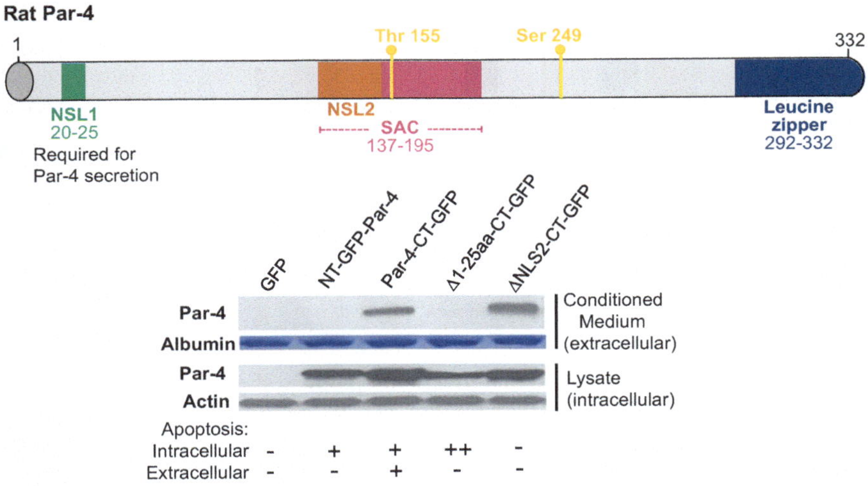

Fig. 6 NLS1 is essential for Par-4 secretion from the cells. Prostate cancer cells PC-3 were transfected with expression constructs for green florescent protein (GFP), GFP fused at the amino-terminus of Par-4 (NT-GFP-Par-4), GFP fused at the carboxyl-terminus of Par-4 (Par-4-CT-GFP), CT-GFP-Par-4 Δ1-25aa, and CT-GFP-Par-4 ΔNLS2 for 24 h, and the conditioned media and cell lysate were collected to perform Western blot analysis for Par-4. When amino acids 1–25 that include the NLS1 are deleted in ΔNLS1 construct, Par-4 is not secreted into the conditioned media and therefore does not carry out the extracellular apoptotic activity. On the other hand, this ΔNLS1 construct shows nuclear localization and induces robust apoptosis. ΔNLS2-CT-GFP fails to enter the nucleus but is secreted. However, ΔNLS2-CT-GFP fails to induce extracellular or intracellular apoptosis, implying an intact SAC domain is essential for extracellular as well as intracellular apoptosis by Par-4. When the amino-terminus of Par-4 is blocked by fusion of GFP as in NT-GFP-Par-4, this protein is not secreted. On the other hand, a GFP tag at the carboxyl-terminus as in Par-4-CT-GFP does not impede Par-4 secretion. ΔNLS2, deletion of Par-4 amino acids 137–153 (as described in Ref. 13)

activate the apoptotic pathway by Fas pro-death pathway activation and co-parallel inhibition of NF-κB transcriptional activity. Cancer cells transfected with a mutant of Par-4 lacking the NLS2 sequence did not undergo apoptosis because Par-4 was retained in the cytoplasm and neither activated the Fas pathway nor inhibited NF-κB transcriptional activity [13]. Nuclear localization is not secondary to apoptosis because after treatment with apoptosis inhibitors like zVAD-fmk, dominant-negative FADD, or the NF-κB subunit RelA, Par-4 translocated to the nucleus but did not induce apoptosis. NLS1 is not a bipartite sequence, and its deletion did not block inhibition of the NF-κB pathway nor did it impact the apoptotic function of Par-4. Thus, although NLS1 is conserved through evolution, no clear function has been assigned to it [10].

The nuclear export signal (NES) is predicted to reside at residues 299–310 (MIGKLKEEIDLL in human Par-4 [14]. The sequence is conserved in human, rat, and mouse and is masked upon dimerization, potentially enabling Par-4 to escape nuclear export and be retained in the nucleus where it can regulate topoisomerase 1

(TOP1) or NF-κB activity [14–16]. Selective inhibitors of nuclear export analogs can bind to chromosome region maintenance (CRM)-1 and inhibit its interaction with Par-4 and its nuclear export [16]. This results in inhibition of proliferation and induction of apoptosis that is inhibited by Par-4 knockdown. These observations suggest that aberrant interaction with the NES of Par-4 may promote nuclear expulsion and inhibition of apoptosis [16].

4.2 Leucine Zipper (LZ) Domain

The carboxyl-terminal of Par-4 protein that contains the LZ domain is critical for homodimeric as well as heterodimeric protein-protein interactions of Par-4. The LZ domain of Par-4 has about 40 aa found in the form of heptad repeats (A-B-C-D-E--F-G). Par-4 has five such heptad repeats with leucine at the fourth position "D." Nonpolar residues at position A and other positions are dominated by polar amino acids in each of these repeats [3]. The ultimate structure of the Par-4 LZ takes the form of scissors, consisting of parallel coiled α-helices that are held together by hydrophobic interactions between the leucine residues on one side of each helix. These helices are responsible for the formation of homodimers, as well as heterodimers, which can diverge and, depending on the interactive protein, result in binding to the major groove of the DNA with the help of the adjacent basic residues. In order for the LZ domain-containing proteins to carry out their function, homo- or heterodimer formation is critical as dimer-DNA complexes are thermodynamically more stable [17]. The type of amino acid distribution that is present in Par-4 carboxyl-terminus is linked to the phenomenon of "natively unfolded" proteins. These proteins lack stable conformational order under physiological conditions, and the changes in this domain could affect not only Par-4 folding but also its interaction with other effectors [18]. The importance of this domain is demonstrated in the Par-4 deletion mutants that lack the LZ domain and are unable to sensitize melanoma cells to apoptotic stimuli [2]. Additionally, the Par-4 LZ deletion mutants render PC12 cells insensitive to trophic factor withdrawal or exposure to Aβ-42 [19]. Some of the well-known protein partners of Par-4 that interact with its LZ domain include Wilm's tumor WT1 protein, ζPKC, DAPK3, THAP1, TOP1, AATF, UACA, and Akt1 [3, 4, 11, 13, 14, 20]. Changes in pH and temperature can induce the Par-4 carboxyl-terminus to form a self-associated coiled coil that would enable it to function as a homodimer [18].

X-ray crystallography of Par-4 revealed that it can generate homodimers by forming a parallel coiled-coil structure in the C-terminal domain [14]. A homodimerization subdomain is present in the amino-terminal region of the protein. The structure of this domain is conducive for Par-4 to form homodimers, as the 257–330 amino acid residues in this domain can form a canonical coiled-coil structure, with two right-handed helices running parallel to each other containing a left-handed super helical twist. In diffraction images, diffuse scattering is a characteristic of a coiled-coil structure displayed by Par-4. This domain displays an environment-dependent structural interconversion between a partially ordered monomer state and a predominantly coiled-coil dimer state. Charge repulsion between the two monomers of Par-4 increases instability in the LZ domain, as Par-4 prefers heterodimers to homodimers. Moreover, the homodimers are only weakly stabilized by the hydrophobic interface, whereas the heterodimeric coiled-coil interactions allowing 1–3 interhelical salt bridges and, importantly, no charge repulsions with AATF, Amida, and DAP-like kinase (DAPK3). This conformation increases the stability of the heterodimer compared to the Par-4 homodimer [14].

WT1 Wilm's tumor protein 1 (WT1) is a tumor suppressor protein whose inactivation can lead to Wilm's tumor and other cancers. The *WT1* gene, which is deleted or mutated in about 10% of sporadic Wilm's tumor cases, encodes a protein containing zinc finger domains. WT1 has four cysteine-histidine (C_2H_2)-type zinc finger domains, one of which mediates protein-protein interaction with the LZ domain of Par-4. Through this interaction, Par-4 inhibits the transcriptional activation but enhances the transcriptional repression of WT1. Thus, by binding to Par-4, WT1 brings a repressor domain to the promoter region. The LZ domain of Par-4 specifically interacts with the WT1 zinc finger domain but not with the three C_2H_2-type zinc finger domains of the Egr1 protein, which belongs to the same family of zinc finger proteins as WT1. To determine which domains of Par-4 interact with WT1, in vitro binding experiments were performed using in vitro translated WT1 and full-length rat GST-Par-4 or GST fused amino-terminal (amino acids 1 to 267) or carboxyl-terminal (amino acids 268 to 332) region of Par-4. WT1 bound both full-length GST-Par-4 and GST-Par-4 (268–332). The WT1-Par-4 interaction does not occur in the established WT1 transcriptional domain, and Par-4 acts as a critical modulator of WT1 responsible for determining if WT1 will function as an activator or repressor rather than a cofactor for WT1-mediated transcription [3, 5, 11, 20]. This interaction with WT1 suggests a role for Par-4 in the nucleus, and indeed Par-4 promotes downregulation of Bcl-2 in androgen-independent prostate cancer cells through binding with WT1 (Fig. 7). Bcl-2 contains a potential WT1 consensus binding site in the promoter region, and WT1, through its interaction with the LZ domain of Par-4, recruits Par-4 to the promoter region of Bcl-2 leading to Bcl-2 downregulation, thereby inducing apoptosis [21]. WT1 is subjected to alternative splicing, and Richard et al. [22] presented evidence that the 17AA alternative splice form of WT1 constitutes a transcriptional activation region, and its function requires direct interaction with Par-4. This was demonstrated using protein affinity chromatography of HeLa lysates using GST fusions with or without the 17AA insert, and only the GST-17AA fusion was able to bind to Par-4. Moreover, in vitro transcription assay shows that the presence of Par-4 causes transcriptional activation of the 17AA segment fused to GAL4, and this effect was inhibited by treatment with a Par-4 antibody. The ability of Par-4 to mediate transcriptional activation of WT1 17AA in cell culture was supported by chloramphenicol acetyltransferase (CAT) reporter activity assays performed in embryonic kidney 293 cells transfected with GAL4-fused WT1 17AA. These cells showed significant increase in CAT activity mediated by

GAL4-WT1 17AA in response to ultraviolet (UV) radiation but not under normal conditions. A mutant of WT1 17AA (G253A) was not able to interact with Par-4, suggesting the specificity of the interaction. The transfection of an expression construct encoding full-length WT1 or WT1 with the 17AA insertion in embryonic kidney 293 cells protected the cells from the lethal effect of UV radiation, but the vector lacking the 17AA insert or carrying the G253A mutation failed to provide this protection from UV radiation. Thus, Par-4 acts as a coactivator of WT1 17AA activation domain [22]. The interaction of the LZ domain of Par-4 with the zinc finger domain of WT1 is important for modulating WT1 activity, and consequently the EWS/WT1 fusion in desmoplastic small round cell tumor, which eliminates its ability to interact with Par-4, loses the ability to inhibit WT1-mediated transcriptional activation [23].

aPKC Atypical protein kinase C (aPKC) isoforms ζPKC and λ/ιPKC are key binding partners of Par-4. These aPKC isomers have an important role in regulation of the MAPK/ERK and NF-κB signaling pathway. Using the yeast two-hybrid system to identify the regulators of ζPKC kinase activity, Diaz-Meco et al. [4, 24] showed that using the regulatory domain of ζPKC as bait, the sequence of the interactive clones isolated corresponded to the last 272 amino acids of the rat Par-4. The regulatory domain of ζPKC or λ/ιPKC interacts with the LZ domain of Par-4 (Fig. 7). Immunoprecipitation studies using NIH 3T3 cells exposed to UV radiation for various time periods show that not only does the level of endogenous Par-4 increase in 4 h but that Par-4 co-immunoprecipitates with ζPKC, implying that Par-4 interacts with ζPKC in mammalian cells. A point mutation in Cys-167 to Ser in the ζPKC regulatory domain eliminates its binding ability to Par-4. Co-transfection of wild-type ζPKC or λ/ιPKC with Par-4 indicates that Par-4 promotes cell death primarily by inhibiting aPKCs. Transactivation of AP1 promoters is highly dependent on ζPKC or λ/ιPKC. In order to confirm that Par-4 can alter the activity of ζPKC or λ/ιPKC, NIH 3T3 cells were transfected with a Par-4 expression construct and AP1 reporter. The activity of AP1 enhancer-directed luciferase reporter plasmid was remarkably inhibited in a dose-dependent manner. On the other hand, transfection with a Par-4 mutant lacking the carboxyl-terminal or the LZ domain did not result in AP1 enhancer-directed luciferase reporter activity, suggesting that Par-4 and aPKC protein-protein interaction are mediated by the LZ domain of Par-4 [4]. As stated previously, aPKCs play an important role in the NF-κB signaling pathway. Diaz-Meco et al. [23] showed that ectopic expression of Par-4 inhibits the tumor necrosis factor-α (TNF-α)-activated nuclear translocation of p65, and about 70% of the transfected cells displayed impaired p65 nuclear translocation. Moreover, TNF-α-activated κB-dependent reporter activity was severely inhibited in a dose-dependent manner by the Par-4 expression construct, and this effect of Par-4 was abrogated completely by co-transfection of a constitutively active mutant of λ/ιPKC but not a kinase-inactive mutant of λ/ιPKC. Additionally, ectopic Par-4 expression inhibits IKKβ activation by TNF-α, with IKKβ being a target for aPKCs. Thus, the inhibition of aPKC activity by Par-4 leads to inhibition of the NF-κB survival pathway, thereby enabling the induction of apoptosis by TNF-α [24]. By phosphorylat-

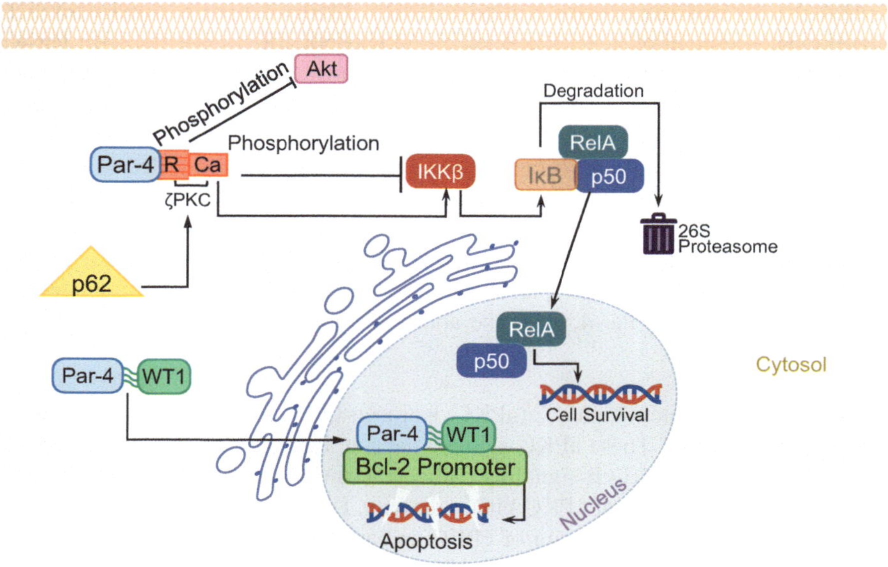

Fig. 7 Interaction of Par-4 with its partners via the leucine zipper domain. Par-4 acting through its leucine zipper (LZ) domain binds to the Wilm's tumor protein 1 (WT1). In the nucleus, the WT1-Par-4 complex binds to the promoter region of anti-apoptotic, Bcl-2 protein via the zinc finger domains of WT1. Par-4 bound to WT1 represses WT1-mediated transcription and promotes apoptosis of the cells. Thus, Par-4 acts as a transcriptional repressor when brought to the DNA by an interactive protein such as WT1. In the cytoplasm, Par-4 via its LZ domain interacts with the regulatory domain of atypical protein kinase C (ζPKC) inhibiting its kinase activity and thereby prevents the phosphorylation and activation of IKKβ. IKKβ is a positive regulator of NF-κB, and its phosphorylation results in phosphorylation of IκB and its degradation by the 26S proteasomal pathway. Degradation of IκB results in NF-κB nuclear translocation and activation of NF-κB-dependent cell survival gene transcription. However, lack of IKKβ activation and NF-κB nuclear translocation prevents transcription of NF-κB-dependent cell survival genes such as XIAP. On the other hand, when p62 binds the Par-4/ζPKC complex, it prevents inactivation of the kinase property of ζPKC and promotes phosphorylation of IKKβ and IκB degradation. This allows NF-κB nuclear translocation and transcription of genes essential for survival. Par-4/ζPKC complex can also inhibit the phosphorylation at Ser124 of Akt inhibiting activation of cell survival pathways. *R* regulatory domain of ζPKC, *Ca* catalytic domain of ζPKC

ing IKKβ, ζPKC activates the NF-κB pathway, and this pathway is inhibited by Par-4 binding to ζPKC. However, Sequestosome-1/p62, a multifunctional scaffolding protein, can bind to the Par-4-ζPKC complex to form a ternary complex that enhances the catalytic activity of ζPKC and reactivates the catalytically inactive Par-4-ζPKC complex. Formation of the ternary complex protects the cells from Par-4-mediated inhibition of NF-κB and induction of apoptosis (Fig. 7). Par-4 and p62 do not compete for binding to ζPKC but are capable of binding each other as demonstrated by pull-down and co-immunoprecipitation studies. Interestingly, p62-expressing cells are not protected from apoptosis when cellular Par-4 levels are elevated [25, 26]. It was later shown that in ζPKC-deficient cells, the NF-κB pathway is inactivated, and phosphorylation of the RelA subunit at S311 is severely impaired in response to TNF treatment [27]. Moreover, co-transfection studies with Par-4 and ζPKC indicated that ζPKC-mediated phosphorylation of Akt is inhibited by interaction of Par-4 with ζPKC. In vitro kinase assay provided evidence that ζPKC indeed phosphorylates Akt at Ser124. When ζPKC was knocked out, there was a significant decrease in phospho-Akt upon serum stimulation, and double knockout of Par-4 and ζPKC inhibited Akt activation and activity [28].

Akt1 Activation of the PI3K-Akt pathway as a result of inactivation of PTEN is one of the most well-accepted genetic lesions that leads to initiation and progression of several cancers, including prostate tumor [29]. More than 50% of primary tumors show loss or alteration in at least one of the PTEN alleles. Goswami et al. [30] demonstrated that Akt1 physically binds to rat Par-4 through interaction with the LZ domain resulting in Par-4 phosphorylation at S249 both in vitro and in vivo, sequestration in the cytosol by 14-3-3 protein, and inhibition of the proapoptotic function of Par-4. In human Par-4, Akt1 was predicted to phosphorylate the S228 residue, and phosphorylation of this residue provides a substrate site for 14-3-3 (SCANSITE 4.0). Binding and phosphorylation of human endogenous Par-4 by Akt1 were apparent when whole-cell lysates from various human cancer cell lines or cancer specimens were subjected to immunoprecipitation with Par-4 antibody, and the immunoprecipitated complexes were subjected to Western blot analysis with the Akt-substrate-specific antibody [30]. Consistently, an S228A mutant of human Par-4 induces apoptosis in cancer cells even in the presence of activated Akt1 (Fig. 4). Par-4 has been shown to function downstream of PTEN, as Par-4 is able to induce apoptosis in a PTEN-independent manner in the prostate cancer cells (Fig. 8). Wild-type and constitutively active Akt1 but not kinase-dead mutant can bind to Par-4 and inactivate the apoptotic potential of Par-4, implying that the kinase activity of Akt1 is essential for Par-4 inactivation. Inactivation of Akt1 with LY294002, a PI3K-inhibitor, results in nuclear translocation of Par-4 and inhibition of NF-κB activity. Nuclear translocation of Par-4 is also essential for binding and activating the proapoptotic molecules such as WT1, ZIPK/Daxx, or THAP1, and phosphorylation of rat Par-4 at S249 or human Par-4 at S228 prevents its nuclear translocation and apoptosis. Mutation of the phosphorylation site in rat or human Par-4 results in disruption of the phosphorylation site of Akt1 enabling rat Par-4 to induce apoptosis. Akt1 was unable to prevent the nuclear localization or the induction of apoptosis by the leucine zipperless mutant of Par-4, suggesting the importance of the LZ domain in mediating protein-protein interactions [11, 30]. Additionally, as Par-4 inhibits Akt1, the introduction of oncogenic Ras by knock-in and knockout of Par-4 in a pulmonary adenocarcinoma model resulted in Ser473 phosphorylation and activation of Akt relative to Ras-expressing Par-4 intact tumors. Par-4 knockout not only increased Akt activation but also elevated the phosphorylation levels of Akt1 substrates, GSK3β, and FOXO3 [28]. Thus, Par-4 and Akt1 regulate one another, with Akt1 binding and inhibiting human or rat Par-4 by phosphorylation [Fig. 4 and Ref. 30] and Par-4 inhibiting Akt1 indirectly by blocking Akt1 activation by ζPKC [28].

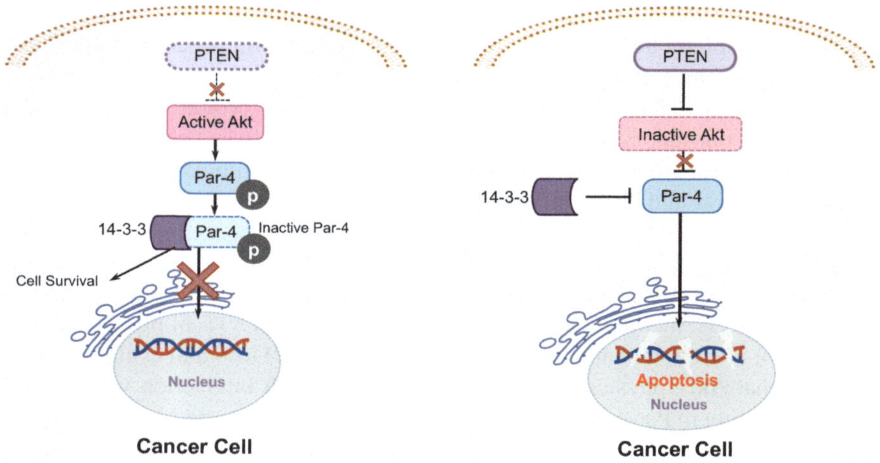

Fig. 8 Par-4 interaction via its leucine zipper domain with Akt in prostate cancer. The PI3K-Akt pathway is activated in prostate cancer often due to the loss of tumor suppressor PTEN, leading to increased levels of active Akt (left panel). Activation of Akt1 results in binding to Par-4 via the LZ domain of Par-4 leading to its phosphorylation. This phosphorylation event creates a substrate for cytosolic 14-3-3 binding to Par-4, thereby preventing Par-4 nuclear translocation that is essential for apoptosis induction. Sequestration of Par-4 by Akt1/14-3-3 complex allows prostate cancer cells to survive and proliferate. In cancer cells with active PTEN, Akt1 is inactivated so it cannot interact and phosphorylate Par-4 (right panel). This prevents binding of 14-3-3 to Par-4 allowing Par-4 nuclear translocation and apoptosis

ZIPK or Dlk/DAPK3 ZIP kinase (ZIPK) or Dlk/DAPK3 (DAP-like kinase) is a serine/threonine kinase that has a leucine zipper domain at carboxyl-terminus. DAPK3 can bind to ATF4, and its ectopic expression leads to apoptosis induction [31]. Using the mouse brain cDNA library as a screen and LZ domain of ZIPK as bait, Par-4 was identified as an interactive partner. This was confirmed by co-immunoprecipitation using epitope-tagged proteins. ZIPK interacts with Par-4 via its own LZ domain. This interaction was confirmed by direct binding assays, wherein the lack of LZ domain in Par-4 inhibited the co-immunoprecipitation of ZIPK. Co-expression of ZIPK and Par-4 promotes ZIPK relocation to the cytosol and apoptosis induction. ZIPK binds and co-localizes with Daxx in the PML (promyelocytic leukemia) oncogenic domains (POD), and this localization in POD is dependent on the ZIPK kinase activity. Recruitment of Daxx to the PODs is essential to sensitize the cells to apoptosis by IFN-Ɣ and As_2O_3. Par-4 is a substrate of ZIPK-mediated phosphorylation, and the association between ZIPK and Daxx is facilitated by Par-4, as Par-4 knockdown inhibited apoptosis induction. Overexpression of Par-4 or wild-type ZIPK enables ZIPK to phosphorylate Par-4 and facilitate the interaction between ZIPK and Daxx. Alternatively, Par-4 knockdown suppresses apoptosis by IFN-Ɣ and As_2O_3. Par-4 and ATF4 may be engaged in competitively binding to ZIPK. Under normal conditions, ZIPK is bound to ATF4 and remains in the nucleus, but in response to POD-inducing stimuli, ZIPK associates with Daxx and moves into the POD where it binds and phosphorylates Par-4. Par-4 can then compete with ATF4 to release ZIPK from ATF4 and trigger apoptosis [31–33]. Boosen et al. [34] demonstrated that Par-4 and ZIPK co-expression leads to ZIPK translocation from the nucleus to actin filaments where it results in extensive myosin light chain phosphorylation that induces apoptosis. This was prevented if either ZIPK activity was inhibited or Par-4/ZIPK interaction was disrupted. ZIPK-mediated phosphorylation of the T155 residue of Par-4 is essential for this activity, as T155A mutant did not show apoptosis induction upon co-transfection with ZIPK. This finding indicates that Par-4 is an important downstream target of ZIPK-mediated apoptosis [34].

MP Myosin phosphatase (MP) is a regulator of myosin light chain phosphorylation and is required for smooth muscle contraction and motility. This Par-4 interacting partner is a large enzyme with a catalytic subunit as well as a regulatory/targeting subunit called MYPT1. Phosphorylation of the two sites in MYTP1 (T696 and T853) leads to inhibition of MP activity. Through proximity ligation assays, surface plasmon resonance, and co-immunoprecipitation, it was demonstrated that Par-4 interacts with MP through MYPT1 and acts as an activator of MP by inhibiting MYPT1 phosphorylation [35]. The tight association of Par-4 with the contractile machinery was suggested by co-immunoprecipitation with MYPT1. Both these proteins interact via the LZ domain, and Par-4 phosphorylation leads to reduction in this interaction. Knockdown of Par-4 using small interfering RNA greatly reduces MP activity as MYPT1 phosphorylation is increased, whereas Par-4 overexpression leads to increased phosphatase activity of MP, possibly due to prevention of inhibition of MP by basal MYPT1 kinase activity. This interaction suggests that Par-4 operates through a padlock model wherein it prevents access of the inhibitory kinases to the phosphorylation sites of MYPT1 resulting in MP being active. MP inactivation requires Par-4 displacement from the complex. One such protein that inhibits MP activity by MYPT1 inhibition is ZIPK. Par-4 can target ZIPK to the MP complex and support the phosphorylation of MYPT1 inhibiting MP activity. The non-phosphorylatable T155A mutant of Par-4 prevents the ZIPK-mediated inhibition of MP, suggesting that Par-4 restricts access of ZIPK and potentially other kinases, such as ROCK, to the carboxyl-terminal inhibitory phosphorylation site(s) on MYTP1 and keeps MP in the activated state. Upon Par-4 phosphorylation at T155, its interaction with MYPT1 is disrupted exposing the inhibitory phosphorylation sites to ZIPK phosphorylation, thereby leading to MP inhibition and smooth muscle contraction [35]. Par-4 binding to MYPT1 was further confirmed to block access of ZIPK to T697 and T855 residues of MYPT1. However, MacDonald et al. [36] later demonstrated that T697 and T855 phosphorylation precedes Par-4 phosphorylation and was inhibited by nonselective kinase inhibitor but not ZIPK selective inhibitor. This observation suggests that although Par-4 binds to MYPT1, thereby preventing MP inhibition, the phosphorylation of Par-4 may not necessarily induce dissociation from MYPT1 and smooth muscle contraction [36].

THAP1 THAP domain-containing protein 1 (THAP1) is a proapoptotic factor that is associated with PML nuclear bodies (NB). It belongs to THAP family of proteins that has great similarity with the DNA-binding domain of *Drosophila melanogaster* P-element transposase. THAP1 can co-localize with Par-4 in the PML NB, thereby functioning as a proapoptotic factor that links Par-4 to PML NB [37]. The interaction of THAP1 with Par-4 was noted in yeast two-hybrid system using THAP1 as bait and confirmed using GST pull-down assay with GST-tagged Par-4 carboxyl-terminus fusion protein. This interaction is also mediated by the LZ domain of Par-4 as the deletion of this domain abolished the THAP1/Par-4 interaction. Co-transfection of GFP-THAP1 and myc-tagged Par-4 carboxyl-terminal resulted in shifting of Par-4 to the nucleus with preferential accumulation in the PML NB, confirming Par-4-THAP1 interaction in vivo. Par-4 and THAP1 interaction was only observed in PML NBs of primary human endothelial cells and fibroblasts but not human Hela cells or immortalized mouse 3T3 fibroblasts [37]. The interaction of Par-4 with THAP1 through the carboxyl-terminal domain enables THAP1 to use its zinc-dependent DNA-binding domain at the amino-terminus to bind to the promoter region of CCAR1 (cell cycle and apoptosis regulator 1). Par-4/THAP1 complex and Notch3 compete with each other to bind to the promoter of CCAR1 and competitively modulate CCAR1 pre-mRNA alternatively splicing. The two splice forms produced have opposing effect on the survival of the T cells in T-cell acute lymphoblastic leukemia (T-ALL). Par-4/THAP1 complex prevents the shift in the pre-mRNA splicing of CCAR1 induced by Notch3 from the full-length isoform to a shorter one which inhibits T-cell apoptosis leading to T-ALL. Notch3-intracellular domain (ICD) complex promotes the binding of splicing factors to 5′ and 3′ splice sites, at exon/intron boundaries leading to mis-splicing of the mRNA. This results in skipping of exon 15–22 leading to the formation of a shorter isoform that is an apoptosis inhibitor. In contrast, silencing these splicing factors prevents the Notch-ICD complex from recruiting the splicing factors, and Par-4-THAP1 complex can induce full-length CCAR1 that can promote T-ALL cellular apoptosis [38].

CHT1 This is a Na^+- and Cl^--dependent, hemicholinium-3 (HC-3)-sensitive, high affinity choline transporter which interacts with Par-4 through Par-4's LZ domain and is involved in neuronal degradation and cholinergic signaling in Alzheimer's disease [39]. Par-4 co-transfection in neural IMR-32 cells with human CHT1 eliminated the Na^+- dependent, HC-3-sensitive choline uptake that was observed with CHT1 transfection alone and was also observed in primary basal forebrain neurons. Co-immunoprecipitation studies support this interaction of CHT1 with the LZ domain of Par-4, as demonstrated by the loss of interaction between CHT1 and Par-4 carboxyl-terminal deletion mutant. The complex formation precludes the loss of CHT1-mediated choline uptake. This reduction in uptake was mediated by reducing the expression of CHT1 on the plasma membrane by Par-4 as supported by kinetic and cell surface biotinylation assays, suggesting that by lowering the incorporation of CHT1 on the cell surface, Par-4 reduces choline uptake by CHT1. Ultimately, this leads to profound loss of cholinergic function linked to memory deficit and reduction in high affinity choline transport in Alzheimer's disease [39].

AATF Apoptosis-antagonizing transcription factor (AATF) is an interacting partner of Dlk/DAPK3 that contains putative nuclear localization signals (NLS1 and NLS2), an LZ domain, and three nuclear receptor-binding motifs. Through immunoprecipitation studies, AATF was reported to antagonize apoptosis induced by Par-4 via the LZ domain and block the aberrant production of β-amyloid peptide Aβ-(1–42) in IMR-32 cells mediated by Par-4 [40]. Aβ is a heterogeneous mixture of small peptides generated by sequential cleavage of the amyloid precursor protein (APP) by β- and γ-secretase. AATF interacts directly with Par-4 in transfected cells and primary neurons and co-localizes with Par-4 in the nucleus and cytoplasm. In IMR-32 cells, withdrawal of trophic factors leads to increased production of Aβ, specifically in cells overexpressing Par-4, and this ultimately leads to death of the cells via apoptosis. Formation of the AATF/Par-4 complex was essential for AATF to exhibit its inhibitory effect on the aberrant secretion of Aβ, as Par-4 activity was inhibited upon AATF co-transfection. This observation suggests that AATF is an endogenous antagonist of Par-4 activity and inhibitor of aberrant Aβ secretion under apoptotic conditions [40].

Dopamine D2 Receptor (D2DR) The role of Par-4 in the nervous system was discovered through the interaction of the LZ domain of Par-4 with D2DR via the calmodulin binding motif in the third cytoplasmic loop [41]. The interaction was discovered in the yeast two-hybrid system using the third intracellular loop of the long isoform of human D2DR as the bait with the human fetal embryonic brain library. The LZ domain-mediated interaction was confirmed using in vitro binding assays, co-immunoprecipitation of endogenous D2DR and Par-4, and co-localization in the striatal neurons. Par-4-positive cells were also D2DR positive and co-fractionated in the synaptosomal fraction. Par-4 and calmodulin compete with each other in a Ca^{2+}-dependent manner, thereby providing a route for Ca^{2+}-mediated downregulation of D2DR efficacy. Reduced binding between Par-4 LZ and D2i3 (amino acid residues 212–373 containing fragment of D2DR) was observed with increasing calmodulin concentrations. Similarly, co-immunoprecipitation of Par-4-D2DR complexes decreased in the presence of ionomycin or thapsigargin. Par-4 siRNA resulted in enhanced cAMP accumulation in cultured rat striatal neurons suggesting that Par-4 knockdown negatively affects D2DR efficacy, essentially relieving the inhibitory tone on dopamine-mediated cAMP signaling. cAMP-responsive element-binding protein (CREB) phosphorylation at S133 is seen in response to dopamine which was severely upregulated in the Par-4ΔLZ neurons, suggesting that in the absence of Par-4/D2DR interaction, the downstream events of dopamine-mediated cAMP signaling are affected. This indicates the importance of the LZ domain in protein-protein interaction and downstream signaling. An interesting fact about the Par-4ΔLZ mice is that they display "depressive mood" in the Porsolt forced swim test and in tail suspension test. Par-4ΔLZ mice displayed increased latency to contact food, indicative of reduced motivation in an aversive environment, a feature of clinical depression in the novelty-suppressed feeding paradigm. Thus, loss of D2DR/Par-4 interaction in the Par-4ΔLZ mice disrupts dopamine signaling ultimately causing depression-like behavior [41]. In human subjects with major depression, Par-4 is decreased by 67% compared to normal controls suggesting that the D2DR/Par-4 interaction plays an important role in humans as well [42].

Amida This protein was initially discovered as a novel protein associated with Arc, a non-transcriptional immediate early gene specific to the brain. Though it is abundant in the brain, Amida is ubiquitously expressed and localized in the nucleus. Amida induces apoptosis in transfected cells after about 72 h. Arc, along with Amida, is responsible for regulating cell death in the brain [43]. The interaction of Amida and Par-4 was discovered using yeast two-hybrid screening with Par-4 as bait and then confirmed through in vitro binding assays, co-expression, and immunoprecipitation experiments [44]. Amida interacts with the LZ domain of Par-4 both in vivo and in vitro through its carboxyl-terminal domain. Deletion of the LZ domain or use of a mutant L3 that contains three alanine instead of leucine residues in the leucine zipper domain abrogates the interaction with Amida. Co-expression of Par-4 and Amida leads to translocation of Amida to the actin filaments, and this results in induction of apoptosis. This effect on the induction of apoptosis by the Amida/Par-4 complex can be inhibited by either preventing the complex formation or impairing the association of these complexes with the actin cytoskeleton, either by using a Par-4 deletion construct or co-expressing LZ mutant Par-4 L3, which still associates with the microfilament system but does not bind to Amida. Thus, relocation of Amida through the Par-4/Amida complex to the actin cytoskeleton is essential for pro-apoptotic function of this complex [44].

TOP1 Topoisomerase I (TOP1) is a monomeric enzyme in the nucleus that is responsible for relaxing the negative as well as the positive supercoils of DNA. It does so by generating single-stranded nicks, followed by rotation of the nicked strand around the intact strand and ultimately religation of the cleaved strand in an ATP-independent manner. This relaxation of the DNA is critical for DNA replication, transcription, and recombination [45, 46]. Goswami et al. [15] identified Par-4 as a binding partner of TOP1 in normal, immortalized, and cancer cells. This interaction is mediated through the LZ domain of Par-4 and was identified using GST pull-down assay with GST-Par-4 as bait and later confirmed by co-immunoprecipitation assays. Supercoiled DNA incubated with TOP1 with or without GST-Par-4 showed relaxation with TOP1 when Par-4 was absent, but not when it was present, indicating that Par-4 is able to attenuate the DNA relaxation activity of TOP1. This effect of Par-4 on DNA relaxation was not seen when the LZ domain-lacking mutant of Par-4 was used, implying that the LZ domain is critical for the interaction. Par-4 knockdown with siRNA resulted in increased TOP1 activity in the nuclear extract of non-transformed BPH-1 cells resulting in increased DNA relaxation. As transcription from TATA-box-containing promoters is selectively inhibited by TOP1, luciferase activity was measured using TATA-box-containing pTal-luc and pTal-NF-κB constructs or Inr-element containing Ras-promoter-luc construct and siRNA against Par-4 and TOP1. Increased reporter activity was observed in pTal-luc and pTal-NF-κB-luc construct but not pRas-luc construct when Par-4 was knocked down. Overexpression of Par-4 caused inhibition of these TATA-box-harboring constructs. Par-4 attenuates the activity of TOP1 but not its expression. In NIH 3T3/iRas cells, siRNA against Par-4 resulted in higher frequency of foci formation with or without IPTG induction of oncogenic Ras compared to control, whereas TOP1 knockdown inhibited foci formation. Thus, Par-4 inhibits the gene transcription activity that results in cellular transformation in a TOP1-dependent manner. Collectively these observations indicate that the interaction of Par-4 with TOP1 results in regulating TOP1 access to DNA that could otherwise contribute to aberrations in DNA replication and transcription, suggesting a role for Par-4 in safeguarding the genome [15].

AR The androgen receptor (AR) is a ligand-activated nuclear transcription factor that translocates from the cytosol to nucleus upon binding to androgen and activates cell growth and survival pathways. When the AR pathway is dysregulated in the prostate, it can contribute to prostate cancer. As a corollary, antiandrogens block the AR-signaling pathway and inhibit prostate cancer progression. Gao et al. [47] demonstrated that the DNA-binding domain of AR physically interacts with Par-4 resulting in enhanced AR DNA interaction and AR-dependent transcription. Unlike many other interacting partners of Par-4 that use the LZ domain of Par-4, the interaction with AR is not mediated through the LZ domain, and the LZ domain is not required for its cofactor, AR-dependent transcription function (Fig. 9). As c-FLIP is a direct target of AR pathway, evidence for the ability of Par-4 to influence AR-dependent transcription was provided by enhanced recruitment of AR in the presence of androgens to the c-FLIP promoter and increased c-FLIP activity. Chromatin immunoprecipitation analysis confirmed that Par-4 binds to the c-FLIP promoter in the presence of androgens. Additionally, c-FLIP expression was inhibited by amino-terminal peptide amino acid residues of Par-4 that served a dominant-negative role [47]. Collectively, the interactions of Par-4 with these different proteins suggest that Par-4 can act either as a transcription repressor or as a transcriptional coactivator depending on the binding partner, stimulus, and cellular context.

4.3 The SAC Domain of Par-4

The SAC domain is unique to Par-4 and indispensable for the pro-apoptotic function of full-length Par-4. SAC is a core domain consisting of 59 amino acids discovered by generating serial deletion mutants of Par-4 from both amino- and carboxyl-terminus. The NLS2 sequence and the Thr155 residue that are critical for Par-4 activation are located within the SAC domain. The SAC domain is responsible for the ability of Par-4 to selectively kill cancer cells [13, 48]. Upregulation of PKA activity in the normal cells using stimulants, for example, cAMP or ectopic expression of PKA, leads to sensitization of the cells to SAC-mediated apoptosis. The selectivity of SAC or Par-4 to induce apoptosis in cancer cells is due to a higher endogenous activity of PKA in cancer cells [48]. Intra-tumoral injection of an

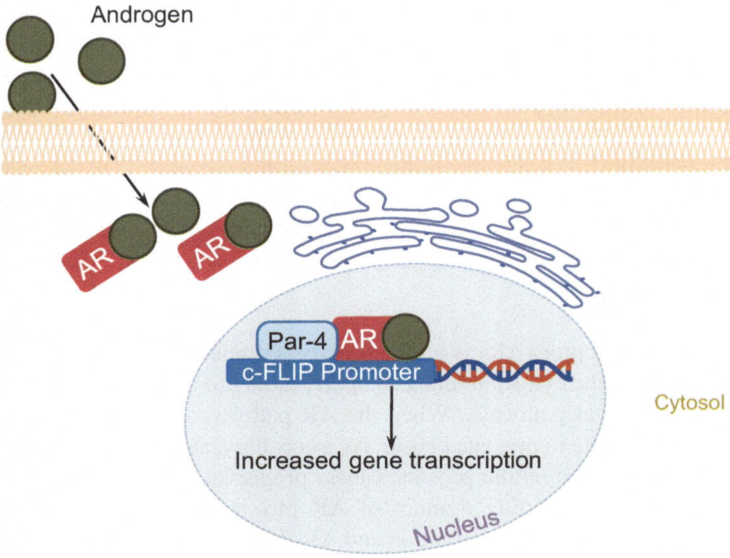

Fig. 9 Par-4 interaction with the androgen receptor (AR) via its amino-terminal domain. AR bound to androgen translocates to the nucleus, complexes with Par-4, and is recruited to the promoter region of c-FLIP, which is a direct target of AR. Par-4 acts as a transcriptional coactivator instead of co-repressor in this context, and the AR/Par-4 complex activates c-FLIP gene expression

adenoviral construct encoding the SAC domain leads to significant reduction in tumors derived from Par-4-sensitive or -resistant cells [7]. The SAC domain sequence is 100% conserved in rodent and human Par-4, and this domain is sufficient to induce Fas and FasL membrane translocation and inhibition of NF-κB activity in cancer cells similar to that noted with full-length Par-4 [13]. This observation is supported by the fact that LZ domain-deleted mutants of Par-4 can still functionally inhibit NF-κB activity and induce apoptosis. Co-expression with dominant-negative FADD inhibits SAC-mediated apoptosis similar to that seen with full-length Par-4. It is important to note that the SAC domain is unique and resembles neither the death domains nor the death effector domains of other known pro-apoptotic proteins.

An advantage of the SAC domain over full-length Par-4 is that due to the absence of the LZ domain, it cannot be bound and inactivated by Akt1 [30]. This could possibly explain the ability of SAC to induce apoptosis in Par-4-sensitive as well as Par-4-resistant cancer cells—even those with elevated Akt1 activity [30]. SAC localizes in the nucleus of normal and cancer cells, but its apoptotic function is only seen in cancer cells, because similar to Par-4, SAC activation requires Thr phosphorylation by PKA, which is elevated in cancer cells [48, 49] as seen in Fig. 10.

Glucose-regulated protein 78 (GRP78) is overexpressed on the cell surface of atherosclerotic lesions and tumor cells. On the other hand, only a few normal cells such as monocytes express background levels of GRP78 at the cell surface. GRP78

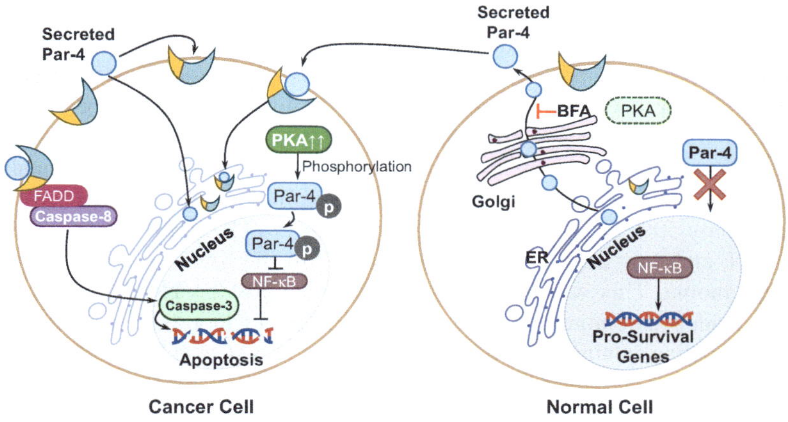

Fig. 10 Cancer cell-specific apoptosis induction by Par-4. In cancer cells, upregulation of PKA activity results in phosphorylation of Par-4 (at T155 in rat Par-4 or T163 in human Par-4). The phosphorylated form of Par-4 (Par-4-P) can then enter the nucleus and inhibit the transcriptional activity of NF-κB, a major cell survival pathway. The inhibition of NF-κB promotes induction of apoptosis in cancer cells. Moreover, Par-4 can be secreted from the ER in response to ER stress, and secreted Par-4 binds to GRP78 at the cancer cell surface. Owing to elevated ER stress in cancer cells relative to normal cells, GRP78 is upregulated on the cancer cell surface and not normal cells, thereby enabling Par-4 to specifically target cancer cells. The cell surface Par-4/GRP78 complex can activate the extrinsic apoptotic pathway through the FADD/caspase-8 pathway that leads to activation of caspase-3 and apoptosis. Alternatively, in the normal cells, the level of PKA is not high enough to cause phosphorylation, activation, and nuclear translocation of Par-4. Therefore, Par-4 does not inhibit NF-κB transcriptional activity, and there is no apoptosis in normal cells. However, Par-4 can be secreted by the normal cells, but because of the absence or downregulation of GRP78 at the cell surface of most normal cells, Par-4 does not activate the extrinsic apoptosis pathway in normal cells. Thus, intracellular Par-4 is selectively activated by PKA in cancer cells, and it selectively kills cancer cells, and as cancer cells primarily express cell surface GRP78, extracellular Par-4 selectively kills cancer cells, sparing normal cells

is therefore an attractive target for treatment of cancer and chronic vascular disease. GRP78 binding to its agonist leads to activation of various pathways that promote cell survival, for example, binding of activated alpha2-macroglobulin to GRP78 activates Akt1, which is known to suppress apoptotic pathways. Conversely, secreted Par-4 can bind to GRP78 at the surface of cancer cells near its amino-terminus to induce apoptosis [50, 51]. PC-3 cells transfected with GFP-Par-4 or GFP-SAC were unexpectedly found to induce apoptosis in the transfected cells, as well as neighboring cells indicative of a “bystander effect.” Further investigation indicated that paracrine apoptosis induction was a result of the interaction between the SAC domain of extracellular Par-4 and cell surface GRP78 that caused ER stress and cell death by activation of FADD/caspase-8/caspase-3 pathway [51].

The SAC domain is sufficient to carry out the pro-apoptotic function of Par-4 as SAC-GFP-expressing transgenic mice that ubiquitously expressed the SAC domain developed normally but lived tumor-free and a few months longer than the control mice [53]. Mouse embryonic fibroblasts (MEFs) from SAC transgenic mice upon transformation using oncogenic adenoviruses containing either oncogenic Ras or c-Myc did not show as many foci in soft agar compared to MEFs from GFP transgenic or littermate control mice, suggesting the anti-transformation effect of SAC. Because of this property, SAC-GFP mice do not develop spontaneous tumors unlike GFP transgenic mice or littermate control B6/C3H mice that grow spontaneous hepatocarcinoma and lymphoma at high frequency as they age. Additionally, in the SV40 large-T and small-t oncogene-inducible mouse prostate cancer model, TRAMP, only 50% of the offspring that were $SAC^{+/-}/TRAMP^{+/-}$ developed high-grade PIN, and only 21.4% that spontaneously lost the SAC transgene developed adenocarcinoma. This was in contrast to the 100% $SAC^{-/-}/TRAMP^{+/-}$ mice that developed adenocarcinoma. TUNEL (terminal deoxynucleotidyl transferase dUTP nick end labeling) staining of the PIN lesions from $SAC^{+/-}/TRAMP^{+/-}$ showed a higher number of apoptotic cells than the PIN lesions and adenocarcinomas from $SAC^{-/-}/TRAMP^{+/-}$, suggesting that this domain is sufficient to induce apoptosis even in vivo. The resistance to tumorigenesis in the $SAC^{+/-}/TRAMP^{+/-}$ is linked to the inhibition of NF-κB activity. As transgenic mice containing the SAC domain exhibit resistance to spontaneous and inducible tumors, the SAC domain has therapeutic potential in cancer [52]. Secreted SAC is detected in the serum of mice, and injection of Lewis lung carcinoma cells (LLC1) resulted in tumors in control mice but not in SAC transgenic mice, indicating that secreted SAC is able to retain its activity and stability even in vivo. This observation was further corroborated in experiments using bone marrow transplantation from SAC-transgenic mice to tumor-prone recipient mice that resulted in transfer of pro-apoptotic SAC activity to the recipient mice [53].

Another interacting partner of the SAC domain is cellular prion protein (PrPc) that is upregulated in temozolomide (TMZ)-induced G2/M-arrested cultures of malignant gliomas [54]. PrPc is expressed in the G2/M phase of the cell cycle and upregulated in a dose-dependent manner after TMZ treatment conferring resistance to TMZ-mediated apoptosis. Knockdown of PrPc allows TMZ to inhibit tumorigenesis of glioma cells in vivo. In order to identify the mechanism of PrPc-mediated anti-apoptotic effect in response to TMZ, a pull-down assay was performed using a soluble form of PrPc. The pull-down yielded a 38 kDa band which upon sequencing was discovered to be Par-4. The interaction of PrPc and Par-4 was confirmed using co-immunoprecipitation assays, wherein TMZ treatment increased the Par-4/PrPc association, and it was significantly higher in G2/M phase cells compared to G1/S phase. The fact that the interaction is mediated by the SAC domain of Par-4 was determined using a co-immunoprecipitation assay with SAC-GFP and ΔSAC-GFP-transfected cells. As expected, SAC-GFP but not ΔSAC-GFP yielded PrPc, indicating that the SAC domain is sufficient to mediate this interaction. In G2/M-arrested or TMZ-treated cells, PrPc interacts with the SAC domain of Par-4 and prevents its phosphorylation at T163 (located in the SAC domain of human Par-4), which is required for its activation, and inhibits its nuclear translocation, which is essential for its apoptotic function. PrPc knockdown resulted in TMZ-induced Par-4-mediated cell death with increased Bax and reduced Bcl-2 expression and significant release of cytochrome c into cytoplasm. This effect was abrogated by Par-4 knockdown or

ΔSAC-GFP transfection, suggesting that PrPc forms a complex with the SAC domain of Par-4, thereby inhibiting its phosphorylation by PKA, its activation by nuclear translocation, and induction of apoptosis [54].

4.4 The PAF Domain of Par-4

Tumors are heterogeneous in nature, consisting of therapy-sensitive and -resistant cells. Hebbar et al. [55] recently showed that treatment of therapy-sensitive cells with chemotherapeutic agents could lead to apoptosis of therapy-resistant cells in the population. This paracrine effect was attributed to the Par-4 amino-terminal fragment (PAF) that is a naturally produced fragment of Par-4 released from therapy-sensitive cancer cells following caspase-dependent cleavage of Par-4 at D131 residue. After its release from sensitive cells, PAF is capable of entering cancer cells, including the therapy-resistant cells and those resistant to secreted Par-4, to induce apoptosis. PAF is a 15 kDa caspase cleavage product that can be detected in the conditioned media in cell culture or released in the tumor microenvironment. This fragment is capable of inhibiting tumor growth in mice as well. The VASA domain (ELNNNL) of Par-4 is located within the PAF domain and enables binding to ubiquitin ligase, Fbox45, which sequesters and degrades intracellular Par-4. The VASA domain within PAF thus compromises Par-4 stability. PAF produced upon cleavage or added exogenously translocates selectively into the cancer cells where it serves as a decoy by binding via its VASA segment to Fbox45, thereby preventing Fbox45-mediated degradation of intracellular Par-4, allowing the release of intracellular Par-4 from Fbxo45 for apoptosis of cancer cells. Only 50–60% of cancer cells in culture are killed due to the action of PAF, suggesting that there must be a mechanism present in these PAF-resistant cells that allows them to escape apoptosis [55].

5 Function of Par-4 in Apoptosis

There is ample evidence in literature that intracellular Par-4 renders cells susceptible to apoptosis by diverse pro-apoptosis stimuli acting via the extrinsic caspase-8/caspase-3 pathway or the intrinsic caspase-9/caspase-3 pathway (Fig. 11). On the other hand, ectopic Par-4 can directly induce apoptosis in certain cancer cells or render them susceptible (i.e., sensitize them) to apoptosis by other pro-apoptosis stimuli. In hormone-insensitive lung cancer cells, AR-negative prostate cancer cell lines such as DU145 and PC-3, or ER-negative breast cancer cells such as MDA MB 231, Par-4 is able to induce apoptosis by facilitating translocation of Fas and FasL to the cell membrane, thereby allowing for the activation of caspase-8 and caspase-3 via the extrinsic apoptosis pathway. Par-4 also translocates to the nucleus in these cells to inhibit NF-κB cell survival activity. A single injection of the

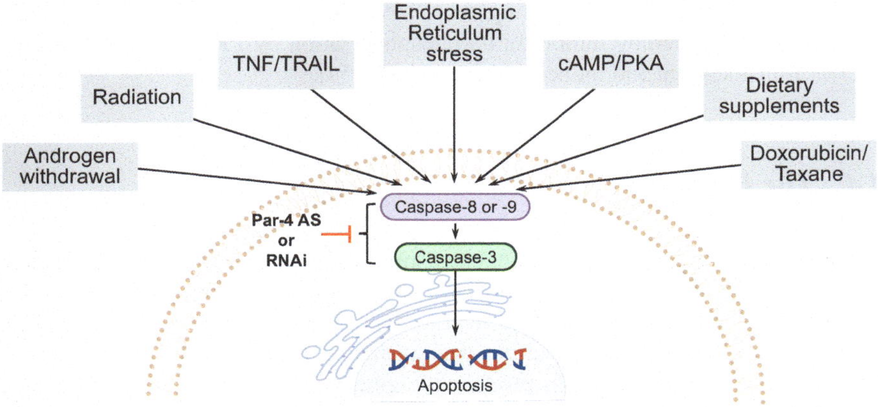

Fig. 11 Various stimuli that sensitize cells to Par-4-dependent apoptosis. Various pro-apoptotic stimuli, such as growth factor withdrawal, agents that elevate intracellular Ca^{2+}, TNF, TRAIL, UV, X-ray and gamma irradiation, or IFN-γ, sensitize mammalian cells to Par-4-dependent apoptosis via the caspase-8/caspase-3 or caspase-9/caspase-3 pathway. *AS* antisense, *RNAi* RNA interference

Par-4-producing adenovirus into PC-3 xenografts in nude mice results in enhanced FasL and Fas translocation to the cell membrane and NF-κB inhibition contributing to remarkable tumor regression [56]. Similarly, a single injection of the Par-4 adenovirus remarkably inhibits the growth of orthotopic prostate tumors from RM1 cells that are derived from urogenital sinuses and transformed by oncogenic Ras and c-Myc (Fig. 12). This signifies the immense potential of Par-4 in gene therapy. Ectopic Par-4, acting by itself, does not cause Fas or FasL membrane trafficking in AR-positive cells such as LNCaP, ER-positive cells such as MCF-7, or in normal breast or prostatic epithelial and stromal cells, and therefore, these cells are resistant to Par-4-mediated apoptosis [56].

Cellular transformation by oncogenic Ras is associated with downregulation of Par-4 and induction of NF-κB transcription activity to promote cell survival. Nalca et al. [57] showed that reintroduction of Par-4 in oncogenic Ras-transformed cells, but not in immortalized cells lacking the Ras oncogene, leads to induction of apoptosis directly without any additional apoptotic stimulus. Ectopic Par-4 inhibits NF-κB activity in the oncogenic Ras-transformed cells and induces apoptosis [57]. It was also shown by Qiu et al. [58] that downregulation of Par-4 by oncogenic Ras or c-Src is mediated by the activated Raf/MEK/ERK pathway. There was hardly any detectable Par-4 mRNA or protein in immortalized fibroblast cells transformed by H-*ras*, activated-*raf*, or v-*src*. Treatment with a MEK inhibitor PD98059 restored Par-4 mRNA and protein in IPTG-inducible oncogenic Ras-expressing cells. Moreover, Par-4 is sufficient to inhibit the cellular transformation, as judged by foci-formation ability and $[^3H]$ thymidine incorporation assays of cells expressing oncogenic Ras, indicating that Ras-mediated downregulation of Par-4 is essential for cellular transformation [58].

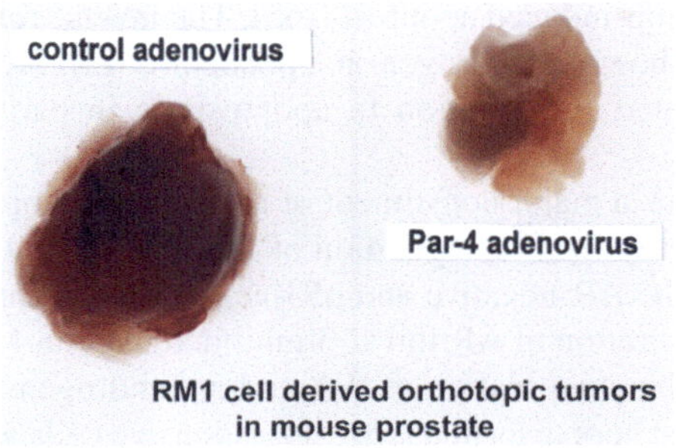

Fig. 12 Inhibition of orthotopic tumors by Par-4. About 1000 RM1 cells (derived from the mouse urogenital sinuses and transformed with oncogenic Ras and c-Myc provided by Timothy C. Thompson, Baylor Medical School) were injected into the mouse prostate, and 7 days later, the prostate tumors were injected with a Par-4 expressing adenovirus or control adenovirus. Tumor volume was measured 12 days after adenovirus injection. All the eight RM1-derived orthotopic tumors injected with Par-4 adenovirus showed >50% growth reduction relative to eight tumors injected with control adenovirus. Representative images of the tumors are shown

In the case of Jurkat T lymphocytes, overexpression of Par-4 leads to a significant decrease in Bcl-2 levels and PARP (poly ADP-ribose polymerase) cleavage, and although these events do not induce apoptosis, they increase the sensitivity of the cells to the action of different chemotherapeutic agents like arabinofuranosylcytosine and doxorubicin [59]. Treatment of the Par-4-overexpressing cells with these agents leads to increased mitochondrial membrane potential disruption, PARP cleavage, downregulation of apoptosis inhibitors cIAP1 and XIAP, and caspase-3 activation and apoptosis, which is inhibited by pretreatment with broad-spectrum caspase inhibitor suggesting the apoptosis-sensitizing function of Par-4 is caspase dependent [59]. In the case of CD95-induced apoptosis, overexpression of Par-4 increases apoptosis by upregulating CD95 receptor on cell surface, promoting cleavage of caspase-8 and caspase-10 that results in further activation of executioner caspases. Par-4 overexpression also downregulates cIAP-2 and XIAP [60]. Similarly, overexpression of Par-4 in colon cancer cell line HT29, or liposome-mediated delivery of Par-4 into colon cancer cell line-derived xenografts, increases apoptosis in response to 5-fluorouracil (5-FU) or Akt1 inhibitor phenylbutyl isoselenocyanate (ISC-4) treatment [61, 62]. In pancreatic cancer cell lines (BxPC-3, Colo-357, L3.6pl, and HPAC), treatment with nonpeptidic small-molecule inhibitors of Bcl-2 family proteins or with 3,3′-diindolylmethane (DIM) leads to Par-4 induction and Par-4-mediated apoptosis and growth inhibition. The higher the amount of Par-4, the greater the sensitivity to the inhibitor. Moreover, pretreatment with these inhibitors leads to upregulation of Par-4 and an increase in apoptosis induction upon treatment with gemcitabine [63, 64]. In PC-3 cells, ectopic Par-4 inhibits radiation-induced NF-κB transcriptional activity and causes transrepression of endogenous and radiation-induced Bcl-2, thereby altering the Bcl-2:Bax ratio and inducing apoptosis. Par-4 is responsible for inhibiting the phosphorylation and degradation of IκBα causing loss of NF-κB activity. Thus, Par-4 sensitizes prostate cancer cells to radiation-induced apoptosis [65]. The inverse relation between Par-4 and Bcl-2 has been shown in androgen-independent CWR22R xenografts, indicating that Par-4-mediated sensitization to apoptosis is in part mediated by Bcl-2 downregulation [66].

Withaferin A (WA), a major constituent of the dietary compound *Withania somnifera*, induces transcriptional upregulation of Par-4 by FOXO3a and subsequently causes apoptosis in the AR-negative and p53-negative PC-3 cells and also in PC-3 xenografts [67]. Restoration of AR in PC-3 inhibited the Par-4 induction and apoptosis. Interestingly, the combination of WA and antiandrogens leads to synergistic induction of Par-4 and apoptosis in androgen-responsive cells suggesting the ability of WA to work in androgen-dependent and -independent prostate cancer [67, 68]. The treatment of DU145 prostate cancer cells with cranberry (*Vaccinium macrocarpon*) resulted in Par-4 upregulation and caspase-8-mediated cleavage of Bid protein, followed by cytochrome-C release from the mitochondria and apoptosis induction. This observation indicates that the apoptosis induction is at least partially mediated by Par-4 via caspase-8 activation [69].

Qiu et al. [58; also reviewed in Ref. 70] noted that downregulation of Par-4 is essential for cellular transformation by oncogenic Ras through the Raf/MEK/ERK/MAPK pathway in NIH 3T3 fibroblasts. In most epithelial cells, however, this pathway was not robust enough to cause Par-4 downregulation. Similarly, Raf activation was sufficient to cause Par-4 downregulation in fibroblasts but not in epithelial cells suggesting a cell-type-dependent role for a Raf-independent pathway in Par-4 downregulation. Activation of the ERK was essential as pharmacological inhibition of ERK leads to downregulation of Par-4 and reverses the Ras-mediated transformation in either cell type. Later, Channing Der and colleagues [71] extended these findings and provided further evidence that the downregulation of Par-4 in the Ras-transformed epithelial cells was mediated by promoter silencing of Par-4 through methylation of the Par-4 promoter at the proximal CpG island. The increase in methylation of the Par-4 promoter in the Ras-transformed cells was due to the upregulation of the maintenance as well as de novo methyltransferases, Dnmt1 and Dnmt3a, respectively. Similar observations were made in cancer cells carrying endogenous Ras mutation indicating that Par-4 promoter methylation along with the Raf/MEK/ERK pathway is essential for oncogenic Ras-mediated cellular transformation in epithelial cells [71].

Studies by Qiu et al. [72] indicated that overexpression of the proto-oncogene c-Myc, which is responsible for cell cycle entry from the quiescent phase to cell proliferation, loss of differentiation, and apoptosis induction in growth factor-starved cells, can cause upregulation of Par-4 (Fig. 13). Rat-1/TGR-1 (parental) cells expressed about fourfold higher levels of Par-4 protein relative to Rat-1/myc-knockout HO15.19 cells. Also, human fibroblast HEL cells infected with the c-Myc adenovirus showed elevated expression of Par-4 compared to HEL cells infected with control adenovirus. Transient transfection of c-Myc also increased Par-4 RNA and protein levels, indicating that c-Myc upregulates Par-4 expression at the RNA and protein level (Fig. 13).

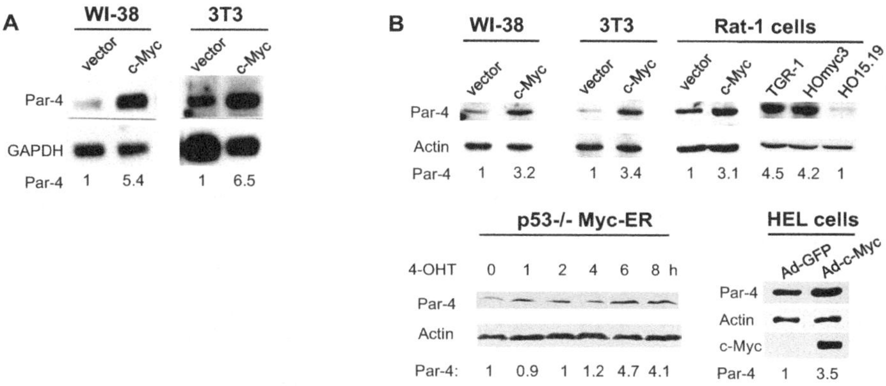

Fig. 13 Par-4 is induced by c-Myc. Human WI-38 cells or mouse NIH 3T3 cells were transiently transfected with c-Myc expression plasmid, and after 24 h, the cells were harvested and subjected to Northern blot analysis for Par-4 and GAPDH (**a**). To determine whether c-Myc caused an increase in Par-4 protein, whole-cell protein extracts were prepared from WI-38 or NIH 3T3 cells that were transfected with vector or c-Myc expression plasmid for 24 h; Rat-1A and Rat-1A/c-Myc cells, stably expressing c-Myc; Rat-1/TGR-1 (parental), Rat-1/c-Myc-knockout HO15.19, or Rat-1/c-Myc reconstituted HOmyc3 cells; p53−/− MEFs expressing myc-ER treated with 4 hydroxy tamoxifen (4-OHT) for the indicated time intervals; or human embryonic lung fibroblast cells HEL infected with c-Myc-producing adenovirus or control adenovirus for 24 h; and subjected to Western blot analysis for the indicated proteins (**b**)

Further studies indicated that c-Myc regulated Par-4 at the promoter level (Fig. 14). Ectopic c-Myc induced the expression of the Par-4 promoter-reporter construct which was abolished when E-box Myc-consensus binding site, located at position −486 in the upstream sequence, was mutated (Fig. 14). This luc activity was also inhibited by the E-box-CAT construct co-transfection, suggesting that the E-box at position −486 mediated the induction of the Par-4 gene by c-Myc. The direct binding of c-Myc to Par-4 promoter was demonstrated by chromatin immunoprecipitation (ChIP) assay (Fig. 14). The c-Myc antibody co-immunoprecipitated the DNA fragment containing the E-box. On the other hand, the c-Myc antibody did not co-immunoprecipitate the DNA fragment corresponding to the coding region of Par-4 that does not contain the E-box. Thus, c-Myc binds to an E-box sequence in the Par-4 promoter and induces Par-4 expression (Fig. 14).

Importantly, abrogation of endogenous Par-4 expression with small interfering RNA resulted in inhibition of c-Myc-inducible apoptosis (Fig. 15).

c-Myc is known to collaborate with oncogenic Ras to transform primary cells, such as MEFs. When MEFs were co-transfected with constructs expressing c-Myc and oncogenic Ras along with the Par-4 promoter-luciferase reporter plasmid, oncogenic Ras inhibited c-Myc-inducible expression of the Par-4 promoter (Fig. 16). The c-Myc adenovirus showed elevation of Par-4 RNA and protein levels, but coinfection with an oncogenic Ras virus suppressed Par-4. Additionally, co-expression of oncogenic Ras abrogated c-Myc-inducible apoptosis, but co-expression of c-Myc, oncogenic Ras and Par-4, resulted in apoptosis (Fig. 16). Thus, c-Myc and

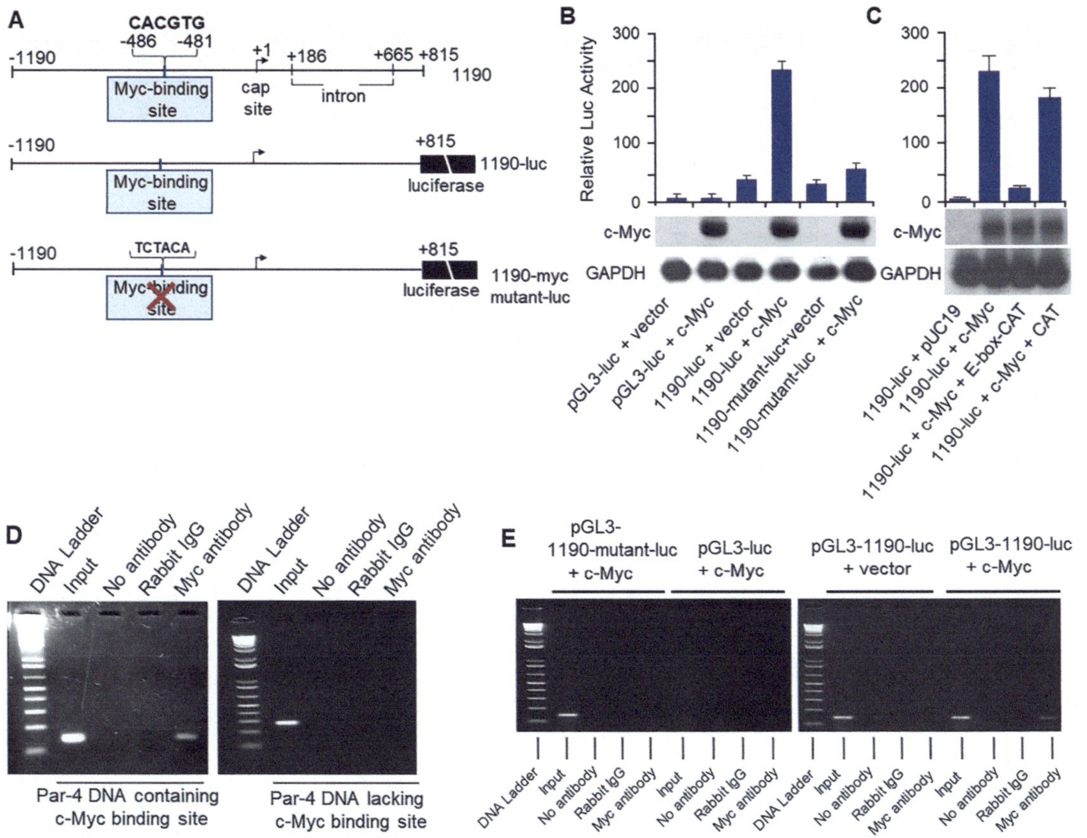

Fig. 14 c-Myc induces the Par-4 promoter via an upstream E-box element. NIH 3T3 cells were co-transfected with the 1190-luc reporter construct containing the Par-4 genomic sequence from −1190 to +815 with the consensus E-box c-Myc binding site CACGTG, or the 1190-mutant myc-luc construct in which the c-Myc-binding site was scrambled (**a**) and with c-Myc expression construct or control vector as a driver. Relative luciferase (luc) activity normalized to β-galactosidase activity is shown in panel B. Expression of the constructs was verified by Northern blot analysis (**b**, lower panel). In order to ascertain c-Myc binding to the E-box in the Par-4 promoter, NIH 3T3 cells were co-transfected with pGL3-1190-luc reporter plasmid and c-Myc expression construct, along with either a CAT construct containing an upstream E-box sequence or a CAT construct without an E-box as control, and whole-cell lysates were processed for luc activity (**c**). Expression of the constructs was verified by Northern blot analysis (**c**, lower panel)

To determine whether c-Myc was bound to the endogenous Par-4 promoter, sheared genomic DNA from PC-3 cells transfected with c-Myc expression construct was subjected to immunoprecipitation with the c-Myc antibody. The DNA in the immunoprecipitated complex was subjected to PCR by using primers flanking either the E-box in the Par-4 promoter (**d**, left panel) or a sequence within the coding region of Par-4 that lacks an E-box as a control (**d**, right panel)

To determine the specificity of c-Myc binding to the E-box in the Par-4 promoter, NIH 3T3 cells were transfected with pGL3-1190-luc, pGL3-1190-mutant-luc, or pGL3-luc with c-Myc expression plasmid or control vector. Sheared total DNA was subjected to immunoprecipitation with c-Myc antibody, control rabbit IgG antibody, or no antibody, followed by PCR with primers specific for the ectopically expressed human Par-4 promoter sequence (**e**). The PCR products were resolved on agarose gels and identified based on expected size relative to the 1 kb-Plus (Invitrogen) molecular weight marker and by nucleotide sequencing. In each set of ChIP experiments, immunoprecipitation with normal rabbit IgG antibody or with no antibody added or sheared DNA subjected directly to PCR without immunoprecipitation (input) were used as controls

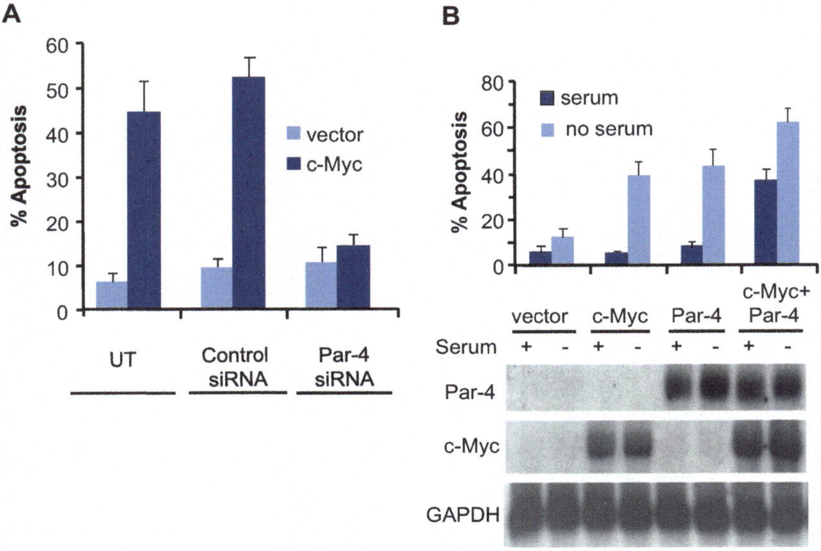

Fig. 15 Par-4 mediates c-Myc-inducible apoptosis. MEFs (p53$^{-/-}$) were transfected with siRNA for Par-4, control siRNA, or left untransfected (UT) and 24 h later transfected with vector or c-Myc expression constructs (**a**). The cells were scored after 24 h for apoptosis by Annexin V staining and confocal microscopy (**a**)

MEFs were transfected with vector, c-Myc, or Par-4, and the cells were maintained in the presence or absence of serum for 24 h (**b**). Apoptotic cells were scored by Annexin V staining and confocal microscopy (**b**, upper panel). Expression of the constructs was verified by Northern blot analysis (**b**, lower panel)

oncogenic Ras inversely regulate Par-4 expression, and oncogenic Ras abrogates c-Myc-inducible apoptosis by inhibiting Par-4 expression, thereby promoting cellular transformation. These findings suggested that cellular transformation by the coupled expression of c-Myc and oncogenic Ras was dependent on suppression, by oncogenic Ras, of c-Myc-inducible Par-4 expression and apoptosis (Fig. 16).

Endogenous Par-4 has both intracellular and extracellular activity, and almost all mammalian cells secrete Par-4. Over the years, studies have shown that Par-4 can be secreted by both normal/immortalized cells, including human embryonic kidney epithelial cells (HEK 293), immortalized human embryonic lung fibroblasts (HEL), and mouse fibroblast (NIH 3T3) cells, and cancer cells and is able to cause apoptosis specifically in the cancer cells. Endogenous Par-4 is secreted from cells via the classical endoplasmic reticulum-Golgi (ER-Golgi) pathway and pretreatment with brefeldin A (BFA) inhibits secretion of Par-4 [51]. Par-4 secretion occurs independent of apoptosis but is dependent on induction of ER stress. Treatment with ER stress inducers (thapsigargin and tunicamycin) for a short time period leads to enhanced Par-4 secretion and can be inhibited by blocking the ER stress pathway. Apoptosis by secreted Par-4 requires the SAC domain that interacts with cell surface receptor GRP78 [51]. Normally, GRP78 is an ER-resident protein that binds to

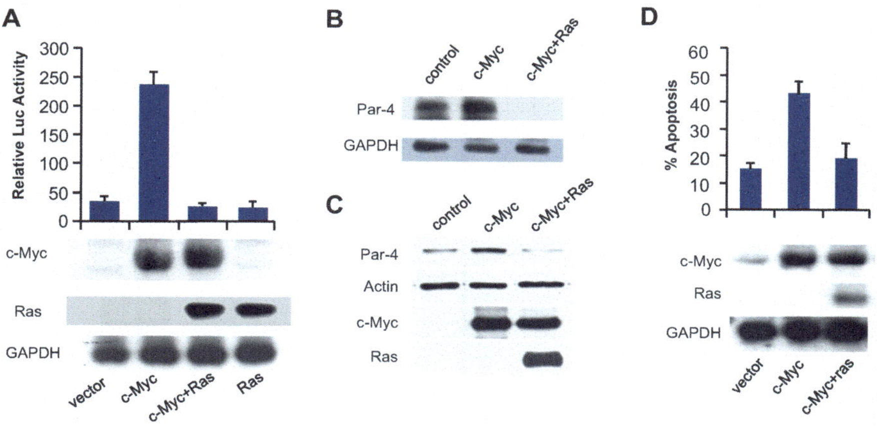

Fig. 16 Oncogenic Ras downregulates Par-4 to block the c-Myc-inducible apoptotic pathway. MEFs ($p53^{-/-}$) were co-transfected with the c-Myc expression construct in the presence or absence of the oncogenic Ras expression construct and the Par-4 promoter-reporter construct 1190-luc, and luc activity was determined (**a**, upper panel). Expression of the constructs was verified by Northern blot analysis (**a**, lower panel)
To test the effect of c-Myc and Ras on endogenous Par-4, the cells were infected with adenoviral constructs expressing c-Myc, oncogenic Ras, or GFP (control), and the effect on endogenous Par-4 RNA or protein was determined by Northern blot analysis (**b**) or Western blot analysis (**c**), respectively
To quantify apoptosis, MEFs ($p53^{-/-}$) were co-transfected with vector, Par-4, c-Myc, or oncogenic Ras expression constructs as indicated and scored for apoptosis by Annexin V staining (**d**, upper panel). Expression of the constructs was verified by Northern blot analysis (**d**, lower panel)

ER stress sensors, such as ATF-6, PERK, and IRE1, and inhibits their activity. GRP78 is highly expressed on cancer cells as these cells have inherently higher ER stress [51, 73]. Co-immunoprecipitation and pull-down studies have shown that endogenous Par-4 is bound to GRP78 protein [51]. Induction of ER stress leads to an increase in binding of GRP78 to intracellular Par-4 that enables its translocation from the ER to the plasma membrane. Once it is displayed at the cell surface, GRP78 can bind extracellular Par-4 and use FADD as an adaptor to recruit caspase-8 to the membrane leading to its activation and onset of apoptotic cascade [51] (Fig. 17).

GRP78 is a popular target for cancer therapy as it is primarily expressed on the surface of cancer cells and only a few normal cells, such as peripheral blood mononuclear cells [73]. Treatment of Wilm's tumors, which express low levels of Par-4 and are cisplatin resistant, with ectopic Par-4 and cisplatin results in co-localization of Par-4 with its receptor GRP78, followed by induction of high expression of ER proteins ATF4 and Bax and activation of ER stress apoptosis pathway [74]. Par-4 and cisplatin combination treatment also inhibited xenograft tumor growth in nude mice by co-localizing Par-4 and GRP78. This observation corroborates the finding that GRP78 translocation is essential for Par-4-mediated apoptosis in tumors [74].

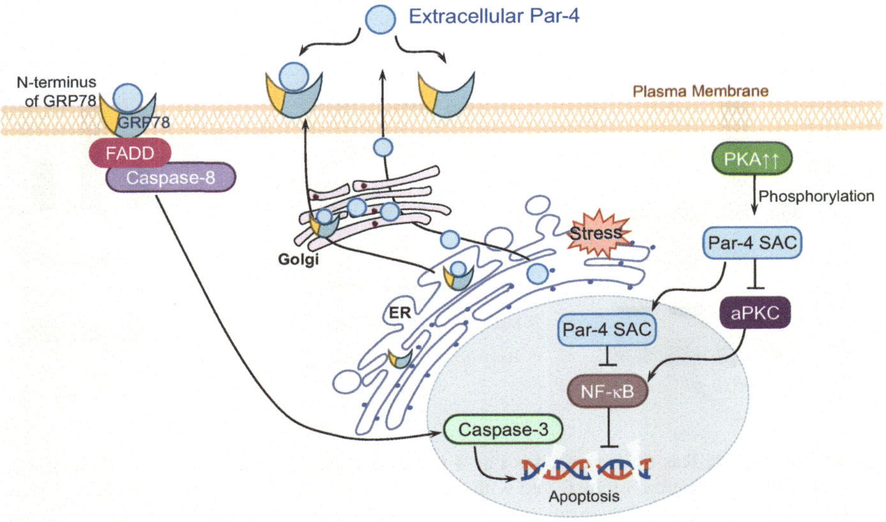

Fig. 17 Key components of apoptosis mediated by intracellular and extracellular Par-4. Cytoplasmic Par-4 inhibits NF-κB activation upon binding aPKC in the cytoplasm, resulting in apoptosis induction. PKA phosphorylates Par-4 at T155 in the SAC domain to activate its nuclear translocation. Several apoptotic stimuli trigger Par-4 translocation to the nuclear where Par-4 can inhibit NF-κB-mediated activation of pro-survival signals and promote induction of apoptosis. This nuclear action of Par-4 is mediated by its SAC domain. On the other hand, intracellular Par-4 is present in the ER as well, where it either complexes with GRP78 or exists in unbound state. Upon generation of ER stress, the Par-4/GRP78 complex translocates to the cell surface. Par-4 secretion may happen after Par-4 dissociation from GRP78 either before its translocation to the plasma membrane or after translocation to cell surface. Unbound Par-4 may also be secreted via GRP78-independent pathways. Secreted Par-4 binds to GRP78 on the cell surface to amplify the ER stress response and FADD/caspase-8-dependent apoptosis. The extrinsic apoptosis pathway activated by secreted Par-4 needs the function of intracellular Par-4 for the translocation of GRP78 to membrane from ER. The exact role that nuclear translocation of Par-4 plays in the action of extracellular Par-4 is under investigation

Apoptosis is a favored mechanism for killing cancer cells as it minimizes an inflammatory response or activation of secondary pathways that could be potentially harmful to the patient. On the other hand, cells that die by necrosis release pro-inflammatory cytokines that attract immune cells, leading to a continuous cycle of cell death and sustained inflammation [75]. There are two pathways for apoptosis, extrinsic and intrinsic, depending upon the nature of stimulus. The extrinsic pathway is activated upon binding of FasL to Fas/CD95, or TNF-α, or TRAIL to their corresponding death receptors such as TNF-R1 or DR4/DR5, respectively. This leads to recruitment of death domain protein FADD and procaspase-8. Trimeric death receptor, FADD, and procaspase-8 together form the death-inducing signaling complex (DISC). This complex allows activation of procaspase-8 to caspase-8 that further activates other downstream effectors. The intrinsic pathway, on the other hand, is activated by intracellular stimuli such as ER stress, oxidative stress, hypoxia, or DNA damage. Although these two pathways start separately, they are interlinked by common signaling factors downstream such as the executioner caspases [11]. As shown by several groups over the years, Par-4 has the ability to induce apoptosis directly on its own as demonstrated by the fact that overexpression of Par-4 can lead to induction of apoptosis in many cancer cell types. Endogenous Par-4 sensitizes cells to a wide range of apoptotic stimuli such as increased intracellular Ca^{2+}, growth factor withdrawal, TNF-α, or UV radiation, and in cells that are resistant to direct apoptosis by overexpressed Par-4, ectopic Par-4 can sensitize the cells to the action of these apoptotic stimuli (Fig. 11). Additionally, Par-4 can overcome protective anti-apoptotic mechanisms conferred by the absence of other major tumor suppressor proteins such as wild-type p53 or PTEN or the presence of anti-apoptotic proteins such as Bcl-2. However, there is a difference in the "apoptosis-inducing" and "apoptosis-sensitizing" function of Par-4 that is dependent on the ability or inability, respectively, of Par-4 to translocate to the nucleus in the absence of another apoptotic stimulus [13, 48]. The ability of Par-4 to induce apoptosis in vivo has been demonstrated using a number of models [28, 52, 53, 56, 76–79]. In the melanoma xenograft model generated in SCID (severe combined immunodeficiency) mice using the A375-C6 melanoma cells, overexpression of Par-4 leads to reduced tumor development and increased apoptosis [80].

Apoptosis mediated by Par-4 involves co-parallel inhibition of cell survival pathway and activation of the apoptotic pathway. Neither pathway on its own is sufficient for Par-4-induced apoptosis. It was initially shown by Chakraborty et al. [56] that in androgen-independent cells like PC-3 and DU-145, but not androgen-dependent cells LNCaP or normal prostatic epithelial cells, Par-4 is able to induce apoptosis. Par-4 induces cell membrane trafficking of Fas and FasL that leads to the Fas-FADD interaction resulting in formation of DISC. This was demonstrated by direct immunofluorescent analysis of PC-3 cells transfected with Par-4 or control vector. PC-3 cells transfected with Par-4 show increased Fas expression or FasL trafficking at the cell surface but without change in localization of Fas or FasL in LNCaP cells. Moreover, co-transfection of Par-4 and dominant-negative or dominant-interfering mutants of FADD, Fas, and caspase-8 or use of FasL decoy Fas-Fc or Fas antagonist antibody ZB4 blocked the Par-4-induced apoptosis. On the other hand, dominant-negative-caspase-9 did not block apoptosis by Par-4, implying that the interaction of Fas to FasL and Fas/FADD/caspase-8 pathway is critical for Par-4-mediated apoptosis. In human tumors, it is not sufficient to activate the apoptotic cascade to cause cell death as it leads to aberrant activation of other cell survival pathways. NF-κB transcriptional activity is often significantly elevated in cancer cells, leading to expression of anti-apoptotic genes like XIAP, c-IAP1, and c-IAP2, which then block apoptosis progression. However, Par-4 can translocate to the nucleus and inhibit NF-κB transcriptional activity to overcome apoptosis resistance. Also, several groups have shown that Par-4 in the cytosol represses NF-κB-dependent gene transcription by inhibiting the TNF-α-induced translocation of RelA, a NF-κB subunit to the nucleus, by atypical PKC inhibition. Inhibition of the NF-κB pathway alone using IκB super-repressor or dominant-negative IKKβ did not induce apoptosis in PC-3. Par-4 induces apoptosis via activation of both the Fas/

FasL pathway and NF-κB inhibition not only in cell culture studies but also in vivo in mouse tumors [56].

The trafficking of Fas/FasL to the membrane and NF-κB inhibition is dependent on the phosphorylation of Par-4 at T155 by PKA whose enzymatic activity is upregulated in cancer cells compared to normal cells, thereby giving Par-4 the selectivity to induce apoptosis only in cancer cells [48]. RNAi-mediated knockdown of PKA or inhibition of PKA function by PKI eliminates apoptosis induction by ectopic Par-4 or the SAC domain [48]. The PKA phosphorylation site in Par-4 (T155 in rat Par-4 or T163 in human Par-4) is also critical as an alanine mutation of this residue abrogates the ability of ectopic Par-4 to induce apoptosis (Fig. 17). Interestingly, although phosphorylated by elevated PKA levels in cancer cells, endogenous Par-4 may not induce apoptosis owing to an active intracellular mechanism most likely involving inhibitory phosphorylation of Par-4 by cell survival kinases such as Akt, PLK1, or CK2. On the other hand, endogenous Par-4 can induce apoptosis when cells are subjected to exogenous treatments that elevate PKA activity and T155 (or T163) phosphorylation [48].

Secretion of endogenous Par-4 can be induced using a number of exogenous stimuli that can cause ER stress without inducing apoptosis. In fact, the first clues to the secretion and cancer-specific killing potential of Par-4 were provided by the bystander effects noted in xenografts with a single intra-tumoral injection of Par-4 adenovirus that resulted in >80% of the cells undergoing apoptosis in the tumor, when parallel injection of a GFP-expressing adenovirus into xenografts indicated that the much lower transduction frequency could not explain the remarkable tumor shrinkage [48]. Cancer-specific apoptosis by secreted Par-4 was first formally demonstrated with conditioned media obtained from Par-4-GFP and SAC-GFP-transfected cancer cells that induced paracrine apoptosis in cancer cells but not in non-transformed cells as judged by immunocytochemistry for active caspase-3 [51]. The secreted form of Par-4 is systemic and detected in their serum of mice and humans.

The interaction of secreted Par-4 with GRP78 was identified using a pull-down assay with GST-Par-4 as bait and confirmed using co-immunoprecipitation with Par-4 or GRP78 antibody. Co-immunoprecipitation experiments showed that the SAC domain of Par-4 is sufficient to interact with GRP78. Through immunoprecipitation of the membrane fraction, colocalization, and biotinylation studies, it was demonstrated that treatment with TRAIL or recombinant TRX-Par-4 or TRX-SAC protein led to increased GRP78 expression of the cell surface, suggesting that Par-4 can translocate it to the cell membrane and co-localizes with GRP78 in cancer cells but not in normal cells. Interaction with GRP78 is essential for apoptosis induction by extracellular Par-4, as a neutralizing antibody directed toward the amino-terminal region of GRP78 blocked the induction of apoptosis induced by TRX-Par-4 or TRX-SAC. An antibody against the carboxyl-terminus of GRP78 did not block apoptosis by extracellular Par-4, suggesting that the interaction of Par-4 with the amino-terminus of GRP78 is essential for induction of apoptosis. Normal cells did not express adequate cell surface levels of GRP78, but transfection of normal cells with a membrane-directed form of full-length GRP78, not an amino-terminus lacking mutant of GRP78, reconstituted the exogenous apoptotic pathway of extracellular Par-4 in normal cells. Alternatively, the knockdown of GRP78 resulted in reduced expression of GRP78 at the cell surface and resistance to apoptosis induced by TRX-Par-4 in cancer cells. The importance of extracellular Par-4 and GRP78 interaction was further elucidated by the fact that upon treatment with neutralizing antibody against Par-4 or GRP78, the cells were resistant to even TRAIL-induced apoptosis. Interestingly, the knockdown of endogenous Par-4 resulted in downregulation of recombinant Par-4-induced translocation of GRP78 to the plasma membrane suggesting that endogenous Par-4 is critical for ER stress-dependent GRP78 induction at the cell surface [51].

The mechanism involving Par-4-bound cell surface GRP78-mediated apoptosis involves activation of the extrinsic apoptosis pathway. Specifically, the Par-4-GRP78 complex activates FADD, which leads to caspase-8 co-localization with GRP78 at the cell surface, resulting in activated caspase-8 and caspase-3. This pathway is inhibited by dominant-negative FADD, implying that FADD is essential to recruit caspase-8 at the plasma membrane and activate the caspase-8/caspase-3 apoptotic cascade (Fig. 17). ER stress-activated protein, PKR-like ER Kinase (PERK), can promote FADD- and caspase-8-dependent apoptosis suggesting that it could be a potential link between ER stress and extracellular Par-4-induced apoptosis. Consistent with this reasoning, treatment with recombinant Par-4 or TRAIL significantly increased phospho-PERK levels, whereas knockdown of PERK resulted in inhibition of caspase-8-dependent apoptosis [51].

6 Mechanism of Par-4 Secretion and Apoptosis of Cancer Cells

Secretion of Par-4 via the ER-Golgi pathway is dependent upon p53 function, as treatment with Nutlin-3a, which inhibits the ubiquitin ligase MDM2 and activates p53 function, leads to increased secretion of Par-4 from the $p53^{+/+}$ but not $p53^{-/-}$ fibroblasts [81]. Par-4 secretion is not inhibited by pan-caspase inhibitor z-VAD-fmk, which blocks apoptosis, suggesting that it is not a post-apoptosis event. In addition, the serum from $p53^{+/+}$ mice treated with Nutlin-3a which had elevated Par-4 levels caused apoptosis ex vivo in cancer cells [81]. The secretion of Par-4 is dependent upon inhibition of UACA, an inflammation-associated protein that sequesters Par-4 in ER and inhibits its secretion. UACA was identified as a top binding partner of Par-4 using yeast two-hybrid system (Table 1). The interaction of Par-4 and UACA was confirmed by co-immunoprecipitation and co-localization studies [82]. Interestingly, UACA is a novel target of p53. Transcriptionally active p53 binds to its consensus binding site in the promoter region of UACA as judged by chromatin-immunoprecipitation assays and inhibits UACA expression [81]. UACA expression is induced by NF-κB activity, and inhibition of NF-κB activity

Table 1 Interactive partners of Par-4. The yeast two-hybrid system (Hybrigenics, France) was used to determine the interacting partners of Par-4 with rat Par-4 (1–332) as bait in the N-LexA-bait-C fusion vector pB27 and human normal lung-derived library as prey. A total of 359 clones were processed to determine interacting partners

Interacting partner	Description	Strength of interaction	References
LOC441734	It is similar to hypothetical protein DKFZp434I1020 (discontinued by NCBI)	A	Gene ID: 441734
UACA (uveal autoantigen with coiled-coil domains and ankyrin repeats)	Encodes ankyrin repeats and coiled-coil domains containing protein which may play a role in apoptosis	A	Gene ID: 55075
ZNF460 (zinc finger protein 460)	Zinc finger proteins are known to interact with nucleic acids and have diverse function including acting as transcription factors	A	Gene ID: 10794
SMARCA4 (SWI/SNF-related, matrix-associated, actin-dependent regulator of chromatin, subfamily a, member 4)	Part of the SWI/SNF family of proteins that regulate the gene transcription by altering the chromatin structure around the genes in an ATP-dependent manner	B	Gene ID: 6597
STX11 (syntaxin 11)	Gene belongs to a family that is responsible for regulation of protein transport among late endosomes and the trans-Golgi network	B	Gene ID: 8676
ZNF134 (zinc finger protein 134)	Belongs to the zinc finger protein family	B	Gene ID: 7693
CTAGE5 (MIA SH3 domain ER export factor 2)	The receptor encoded by the gene is responsible for large pre-chylomicrons and pre-very-low-density lipoproteins exported in the endoplasmic reticulum	C	Gene ID: 4253
CCDC57 (coiled-coil domain containing 57)	Exact function not known. Mutation can cause immune system phenotype, homeostasis/metabolism phenotype	C	Gene ID: 284001
GOLGA3 (golgin A3)	The gene encodes a protein that is hypothesized to play a role in in nuclear transport and Golgi apparatus localization	C	Gene ID: 2802
SFRS11 (serine- and arginine-rich splicing factor 11)	Exact function not known but based on structure and localization studies may play a role in mRNA splicing	C	Gene ID: 9295
ZNF165 (zinc finger protein 165)	Belongs to the Kruppel family of zinc finger proteins and acts as transcriptional regulators. May play a role in spermatogenesis	C	Gene ID: 7718

(continued)

Table 1 (continued)

Interacting partner	Description	Strength of interaction	References
ZNF563 (zinc finger protein 563)	Exact function not known	C	Gene ID: 147837
DKFZp434I1610 (zinc finger protein 700)	Maybe involved in nucleic acid binding and transcriptional regulation	D	Gene ID: 90592
HKR1 (zinc finger protein 875)	Perhaps involved in transcriptional regulation	D	Gene ID: 284459
LOC344065 (zinc finger protein 570 pseudogene)	Function not known	D	Gene ID: 344065
PDCD7 (programmed cell death 7)	Protein encoded is associated with the U11 small nuclear ribonucleoprotein (snRNP), a component of minor U12-type spliceosome	D	Gene ID: 10081

A, very high confidence in the interaction; B, high confidence in the interaction; C, good confidence in the interaction; D, moderate confidence in the interaction. This category is the most difficult to interpret because it mixes two classes of interactions: *false*-positive interactions and interactions hardly detectable by the Y2H technique (low representation of the mRNA in the library, prey folding, prey toxicity in yeast)

results in downregulation of UACA expression and release of Par-4 for secretion in normal cells [81]. In cancer cells, UACA is upregulated relative to normal cells, and UACA prevents intracellular Par-4 from translocating GRP78 to the cell surface in cancer cells [82]. UACA knockdown causes Par-4-resistant A549 cells to increase GRP78 cell surface translocation and become sensitive to apoptosis induced by recombinant Par-4 [82]. The antimalarial drug chloroquine (CQ) and its derivative hydroxychloroquine (HCQ) are FDA-approved drugs that serve as secretagogue of Par-4 [83]. CQ inhibits UACA expression and promotes the secretion of Par-4 via the classical secretory pathway in cells with wild-type p53, but not in cells lacking wild-type p53. CQ also causes robust secretion of Par-4 in mice, and secreted Par-4 is able to induce cancer cell apoptosis in a paracrine manner and inhibit metastatic lung tumor nodules in mice in a Par-4-dendent manner [83]. In fact, p53 regulates other components of the secretory pathway including Rab8b, a member of the Rab family of GTPases, which plays a role in vesicle transport of proteins from the Golgi to the plasma membrane for secretion [83]. CQ induced Rab8b expression in a p53-dependent manner, and Rab8b knockdown or Rab8b null MEFs did not show CQ-induced Par-4 secretion. Par-4 secretion was restored upon reintroducing GFP-Rab8b in Rab8b knocked-down cells. The role for Rab8b in Par-4 secretion was further supported by co-localization studies, wherein CQ increased Par-4 co-localization in Rab8b-containing vesicle. Par-4 entry into the Rab8b vesicles was inhibited by BFA. Rab8b contains p53 consensus-binding sites in its promoter region, and therefore CQ-induced p53 activation leads to Rab8b induction, thereby promoting the secretion of Par-4 by the classical pathway. Thus, CQ-induced Par-4 is secreted from the normal cells in a p53- and Rab8b-dependent manner [83].

Another potent secretagogue of Par-4 is Arylquin-1 (Aq-1), a 3-arylquinoline which causes a dose-dependent secretion of Par-4 from normal and immortalized human cells but not from cancer cells [84]. Aq-1-induced secretion is blocked by BFA, indicating that induction of Par-4 secretion by Aq-1 occurs by the classical secretory pathway. Studies with Aq-1 indicated that vimentin, a type III intermediate filament protein, normally sequesters Par-4 preventing its secretion and that treatment with Aq-1 results in binding of Aq-1 to vimentin, thereby displacing Par-4 and facilitating Par-4 secretion [84]. Through biotinylation, co-immunoprecipitation studies, and in silico molecular modeling studies, it was determined that Aq-1 induces a conformational change in vimentin that allows for the secretion of Par-4. The Par-4 secreted in response to Aq-1 retains its ability to induce apoptosis in a paracrine manner by binding to cell surface GRP78 selectively in cancer cells but not in normal cells [84].

7 Par-4 as a Tumor Suppressor

Intracellular Par-4 functions as a proapoptotic tumor suppressor protein by inducing apoptosis selectively in cancer cells in response to various apoptotic stimuli. Par-4 is phosphorylated at the T155 residue (in rat Par-4), and its apoptotic potential is activated by PKA and DAPK3 [34, 48]. Par-4's apoptotic ability is often associated with its nuclear translocation, and elevated levels of nuclear Par-4 can overcome the cell survival mechanisms contributed by Bcl-xL, Bcl-2, or absence of other major tumor suppressors such as p53 and PTEN [20].

Par-4 is downregulated in >70% clear cell type of renal cell carcinoma [85]. Moreover, acute lymphoblastic leukemia (ALL), neuroblastoma, hypopharyngeal carcinoma (HPC), oral cavity tumors, endometrial cancer, pancreatic cancer, glioblastoma, and breast cancer show downregulation of Par-4. An example of Par-4 loss in breast cancer in shown in Fig. 18. In ~30% of endometrial tumors and in ALL, downregulation of Par-4 is caused by promoter methylation [86, 87]. Chromosome 12q21 is often deleted or unstable in gastric cancer, pancreatic cancer, and male germ cell tumors [6]. Also, K-Ras point mutation, which serves as the driver mutation in pancreatic cancer, downregulates Par-4, and consequently a larger majority of pancreatic tumors display significant loss of Par-4. In endometrial cancer, Par-4 exhibits a single base mutation in exon 3 that converts arginine residue to a stop codon resulting in truncated Par-4 protein [86]. In prostate cancer, endogenous Par-4 is inactivated via binding and phosphorylation (at S228 in human Par-4 and S249 in rat Par-4) by Akt1. This phosphorylation event promotes Par-4 binding to the chaperone protein 14-3-3 and sequestration of Par-4 in the cytoplasm, thereby preventing apoptosis [30]. High levels of Par-4 lead to better response in patients with hypopharyngeal carcinoma undergoing concurrent chemoradiotherapy or radiotherapy alone, by inducing apoptosis in the cells posttreatment [88]. In addition, Par-4 increase contributed to enhanced autophagic flux that also resulted in cell death [88]. TRIM21 is a partner of Par-4 in colon cancer cells, and the interaction is

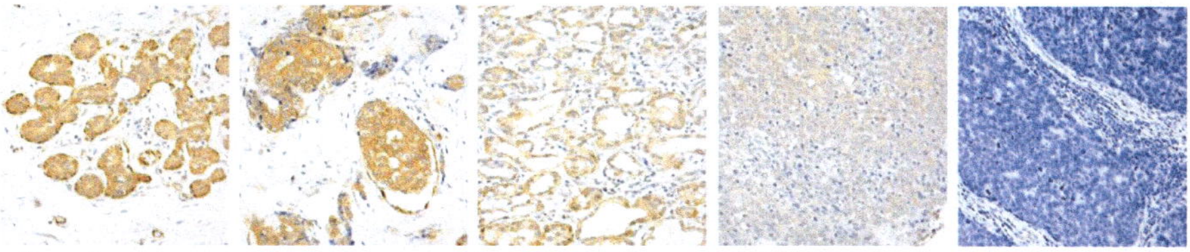

Status	Par-4 levels	3	2	1	0
	Triple Negative	50%	25%	17%	8%
	ER/PR Negative	40%	20%	40%	
	HER2 Negative	28%	44%	22%	6%
	Normal	100%			

Stage	Par-4 levels	3	2	1	0
	Any T, Tis, N0-N1	52%	44%	4%	
	Any T, N2	11%	72%	17%	
	Any T, N3		43%	29%	28%
	Normal	100%			

Fig. 18 Par-4 expression is downregulated in breast cancer. Immunohistochemical staining of Par-4 (brown) of untreated breast tumors shows that the expression of Par-4 is progressively downregulated as the cancer progresses to advanced-stage invasive carcinoma, irrespective of the estrogen (ER), progesterone (PR) receptor, and human epidermal growth factor receptor 2 (HER2) status of the tumor. The breast cancer tissue microarrays (TMAs) were prepared and processed for IHC by pathologist Rachel Stewart and Dana Napier, Biospecimen Procurement and Translational Pathology Shared Resource Facility, MCC; TMAs were scored by pathologist Craig Horbinski and all the data were analyzed by Nikhil Hebbar

mediated by PRY-SPRY domain of TRIM21 [89]. TRIM21 increases the resistance of colon cancer cells to cisplatin and can downregulate Par-4 in response to cisplatin treatment. Par-4 forced expression can increase the sensitivity of colon and pancreatic cancer cells to cisplatin [89].

Additional genetic evidence for the tumor suppressor role of Par-4 is provided by studies in Par-4 knockout mice. Homozygous deletion of Par-4 in mice results in spontaneous tumors in various organs such as the lungs, liver, urinary bladder, and endometrium and prostatic intraepithelial neoplasia (PIN) lesions in the prostate [78]. Moreover, Par-4 knockout mice develop inducible tumors in the endometrium with estrogen and in the bladder with the chemical carcinogen, N-butyl-N-(4-hydroxybutyl) nitrosamine (BBN) [78]. Interestingly, heterozygous deletion of Par-4 results in the same frequency of tumors as that noted with homozygous deletion of Par-4, indicating haploinsufficiency in the tumor suppressor function of Par-4 [78]. Simultaneous Par-4 ablation and PTEN heterozygosity lead to invasive prostate adenocarcinoma in mice [90, 91]. PTEN haploinsufficiency cooperates with Par-4 deficiency in prostate cancer initiation and progression, and their simultaneous inactivation leads to enhanced Akt activation followed by activation of NF-κB pathway leading to invasive prostate carcinomas [90, 91]. Ectopic Par-4 expression results in less efficient tumor formation when the cells are injected into nude mice intradermally [92]. Moreover, stable transfection of Par-4 in Caki cells makes them more sensitive to apoptosis by TRAIL and other ER stress-inducing agents by decreasing the level of an anti-apoptotic protein, XIAP, and inactivation of Akt by reducing its phosphorylation [92].

Relative to normal rat and human cholangiocytes, Par-4 expression is downregulated in cholangiocarcinoma (CCA) patient biopsies and cell lines [93]. Reduced Par-4 expression is associated with reduced Bax/Bcl-2 ratio and inversely correlated with proliferation markers. The induction of Par-4 by transfection or WA leads to apoptosis induction in the CCA cell lines, whereas Par-4 silencing promotes cellular proliferation [93]. The tumor suppressor role of Par-4 is further strengthened by the fact that oncogenic Ras, -Raf, and -Src can downregulate Par-4, and restoration of its levels abrogates the oncogene-induced transformation by inhibiting extracellular signal-regulated kinase (ERK1/ERK2) expression and activity [58]. A mouse lung cancer model that contains oncogenic Ras introduced by knock-in strategy and doxycycline-inducible lung cancer model in which the target cells are type II pneumocyte display enhanced lung carcinogenesis when Par-4 is lost compared to when Par-4 is present [28]. The tumor burden in the lungs of these mice was about 18% of total volume with intact Par-4, but this number significantly increased when Par-4 was knocked down. Par-4 knockout mice have higher nuclear staining for RelA in the lungs suggesting that Par-4 loss increases basal NF-κB activity that leads to enhanced XIAP levels and protection from apoptosis [28]. Alternatively, long-term reduction of Par-4 expression in Ras-transformed epithelial cells is carried out by MEK-dependent hypermethylation of Par-4 promoter [20, 71]. Overexpression of Par-4 enhanced the sensitivity of the Bcr-Abl-positive myeloid cells K562 to imatinib and inhibitors of histone deacetylases, resulting in higher apoptotic rate in doxorubicin and TRAIL and Fas-resistant cells [94]. In nasopharyngeal carcinoma cell lines, serum withdrawal or Par-4 overexpression leads to induction of apoptosis, and this involves dissociation of functional Par-4 from the Par-4/Akt complex [95]. Recently, the importance of Par-4 in drug responsiveness and overall survival of ovarian cancer patients was demonstrated by Tan et al. [96]. Using online databases, general expression profile, and target-disease associations, they showed that high *PAWR* expression is associated with better patient survival (OS: HR = 0.84, $P = 0.0077$; PFS: HR = 0.86, $P = 0.049$). In addition, Par-4 expression could predict responsiveness to platinum drugs, and paclitaxel sensitivity was higher in patients with higher *PAWR* expression. Par-4 is downregulated in ovarian cancer relative to matched normal fallopian tubes [96]. It is particularly interesting that endogenous Par-4 is unable on its own to induce robust apoptosis in cancer or normal cells, but Par-4 inhibition results in apoptosis inhibition.

Cancer deaths occur predominantly due to tumor recurrence and metastasis, and Par-4 plays an important role in breast cancer recurrence [97, 98]. Par-4 downregulation is necessary and sufficient to promote breast cancer recurrence. Par-4 is downregulated in residual breast tumor cells that survive neoadjuvant chemotherapy, thereby increasing the risk of relapse in patients. Breast tumors in mouse models consist of a heterogeneous population of cells. Those that express high levels of Par-4 are eliminated by chemotherapy, but the ones that have low Par-4 expression survive and increase the probability of local and distant metastasis. Overexpression of Par-4 prevents orthotopic tumor recurrence in mice, and this observation is corroborated by meta-analysis data on human breast cancer gene expression. Breast tumor cells that have low Par-4 are therapy resistant and responsible for tumor recurrence independent of SNAIL [97, 98]. Loss of nuclear staining of Par-4 protein has been associated with poor prognosis for breast cancer patients including those with luminal A subtype breast cancer. The prognosis is particularly worse in breast cancer patients who have tumors that are HER2-positive, EGFR-positive, or pAKT-positive and negative for nuclear Par-4 [99]. Docetaxel is used to effectively treat metastatic breast cancer, but patients may soon develop resistance. Experimental evidence supports the possibility that Par-4 may play a role in overcoming this resistance to chemotherapy by modulating the canonical and noncanonical Wnt/β-catenin pathways [100].

TCGA and REMBRANT data analysis of glioblastoma (GBM) patients shows that low Par-4 levels correlate with low survival period, but this is not applicable to astrocytomas and oligodendrogliomas, indicating that low *PAWR* expression is a predictive risk factor for GBM [101]. Par-4 acts as a negative regulator of gastric cancer development and progression. Accordingly, inhibition of miR-199a-3p, which targets Par-4, or overexpression of Par-4 decreases the progression of gastric cancer [102].

8 Role of Par-4 in Cancer Stem Cells

Cancer stem cells are a rare subpopulation of cells that have been hypothesized to promote tumor formation and growth. They are quiescent in nature and are extremely resistant to conventional radiation and chemotherapy. Their existence has been recorded in a number of solid tumors including breast, colon, brain, lung, and potentially melanoma [103]. Tamoxifen induces Par-4 upregulation that leads to breakdown of mitochondrial membrane potential, downregulation of Bcl-2, and reduced activation of Akt and 42/44 kDa ERK in human glioblastoma stem cells, ultimately killing these cells. Knockdown of Par-4 in these cells inhibits tamoxifen-induced cell death. Secreted Par-4 and GRP78 were highly expressed upon tamoxifen treatment in the glioma stem cells, and treatment with neutralizing antibody inhibited cell death, suggesting that secreted Par-4 is involved in killing them. Par-4 can be projected as a stemness inhibiting drug as upregulation of Par-4 upon tamoxifen treatment leads to downregulation of neural stem cell markers, Nestin, Bmi1, Vimentin, Sox2, and Musashi [104]. Similarly, multicellular spheroids consisting of GBM stem cells were sensitive to TAM-induced cell death upon exposure to Par-4 containing conditioned media from TAM-treated monolayer cells. Upon combining TAM and ζPKC inhibition, Par-4 secretion was induced and resulted in killing the multicellular spheroids possibly through binding with membrane GRP78 [101].

9 Role of Par-4 in Epithelial Mesenchymal Transition and Inflammation

The process of metastasis is facilitated by reactivation of an evolutionary conserved developmental program called epithelial to mesenchymal transition (EMT). During this process, differentiated epithelial cells undergo a series of dramatic changes ranging from alterations in their morphology with loss of cell to cell contact and matrix remodeling to more invasive and less differentiated mesenchymal cells. Though the role of Par-4 in apoptosis has been studied extensively, its potential role in EMT has not been investigated to that depth. In recent years, there has been growing interest in the role of Par-4 in EMT. To this effect, Nguyen et al. [105] noted that in colon cancer cell lines, SW480 and SW620, overexpression of Par-4 inhibited cell migration and invasion, whereas Par-4 knockdown enhanced these processes as seen in scratch assays and Matrigel invasion assays. The increase in Par-4 levels altered the morphology of SW620 cells in a manner that is characteristic of mesenchymal to epithelial transition (MET). This was potentiated by the upregulation of E-cadherin and downregulation of vimentin as judged by RT-PCR and Western blot analysis. Moreover, Par-4 increases the expression of other tight junction proteins like claudin, occludin, and ZO-1, suggesting that Par-4 is a promoter of the epithelial-like phenotype [105].

Induction of Par-4 by either ectopic cDNA expression or the pharmacological-inducer NGD16 leads to downregulation of major mesenchymal markers vimentin and Twist-1, and upregulation of E-cadherin and MET-associated marker ALK2, in a dose- and time-dependent manner in pancreatic cancer cells [106]. Boyden chamber assays, cell scattering, and wound healing assays demonstrated that these cells had reduced motility and invasion capability post-NGD16 treatment, in agreement with corresponding changes in their gene expression profile. Par-4 expression also promotes the MET phenotype in pancreatic cancer as evidenced by increased Smad4 protein levels and characteristic morphological changes. The induction of MET is dependent on intact ALK2/Smad4 signaling pathway, as Smad4 mutant cells did not show an increase in E-cadherin expression post-NGD16 treatment or GFP-Par-4 transfection. Alternatively, when endogenous Par-4 is knocked down, ALK2 and Smad4 expression completely disappears, and the MET phenotype is reversed along with reduced E-cadherin expression. As E-cadherin expression is controlled by binding of Twist-1 to the promoter region and Par-4 increases E-cadherin expression, it is possible that Par-4 negatively controls the transcriptional activity of Twist-1. This was demonstrated by using a Twist-1 luciferase reporter construct co-transfected into cells with GFP-Par-4 or treatment of the transfectants with NGD16. NGD16-treated syngeneic metastatic pancreatic cancer mouse models had reduced metastatic burden compared to control treatment, suggesting that Par-4 can inhibit metastasis in vivo [106]. However, in recurrent breast tumors, Twist-1 can cause epigenetic silencing of Par-4 by binding to the Par-4 promoter where Twist-1 induces a unique bivalent chromatin domain, and this bivalent configuration is able to confer plasticity to the Par-4 promoter [107]. By reversing such silencing of Par-4 or by restoring histone conformation, recurrent breast tumor cells can be made sensitive to chemotherapy both in vitro and in vivo [107].

When pancreatic cancer (Panc-1, MiaPaca-2) and breast cancer (MDA-MB-231) cells are treated with gemcitabine, there is upregulation of MDM-2 and EMT properties [108]. However, treatment of these cells with NGD16 leads to inhibition of invasion and EMT properties. In addition to diminishing pro-EMT markers and upregulating epithelial markers, Par-4 induction also diminishes MDM-2, a downstream effector of Twist-1. Par-4 activation leads to inhibition of MDM-2 that causes upregulation of p53, followed by inhibition of the EMT factors. Additionally, partial knockdown of MDM-2 bolstered the antiproliferative and anti-invasive effects of NGD16. In mouse mammary carcinoma model, NGD16 inhibited tumor growth and lung metastasis, supporting the metastasis inhibitory potential of Par-4 [108].

In Panc-1 cells, Par-4 overexpression led to a significantly high level of microRNA, miR-200c [109]. Interestingly, in the GFP-Par-4-transfected cells, TGF-β1, TGF-β2, ZEB-1, and Twist-1 mesenchymal markers were downregulated as evidenced by Western blot and RT-PCR analysis; with a concurrent increase in E-cadherin and miR-200c. Contrarily, knockdown of Par-4 significantly upregulated the EMT markers and decreased the levels of miR-200c. ZEB-1 was halted at the transcriptional level when Par-4 was overexpressed. The results were similar when Par-4 NGD16 was used. The wound healing assay also demonstrated reduced ability of the Panc-1 cells to seal the gap when they were treated with NGD16 or miR-200c. Successful metastasis requires the cells to have good motility, which can be measured by scattering assay. Par-4 transfection or treatment with NGD16 hindered the scattering phenomenon of these cells compared to vehicle or mock transfection. Even in the syngeneic mouse pancreatic cancer model, treatment of NGD16 elevated levels of miR-200c and E-cadherin with a simultaneous decrease in ZEB-1 and vimentin, suggesting that Par-4 is able to inhibit EMT even in in vivo [109].

Tumor-associated macrophages (TAMs) contribute to cancer progression and metastasis through induction of inflammation. It is known that nonsteroidal anti-inflammatory drugs (NSAID) can prevent the progression of several cancers by inhibiting inflammation. Recently, the conditioned medium from human monocytes led to proliferation of ER-dependent MCF-7 and ER-independent MDA-MB-231 cells. This proliferative phenotype was attributed to upregulation of Ras and downregulation of Par-4 [110]. Ras upregulation leads to upregulation of COX-2, PGE_2, BCL-2, *IL-6*, *IL-1β*, VEGF-A, and MMP-9. On the contrary, NSAID treatment led to Ras downregulation that inhibited growth. It also causes upregulation of Par-4 and caspase-3, which activates the apoptotic machinery that leads to death of these breast cancer cells. This observation suggests that NSAIDs inhibit inflammation-driven progression of breast cancer by downregulating Ras and upregulating Par-4 [110].

10 Role of Par-4 in the Neurons, Heart, and Eye

In neurodegenerative diseases such as Alzheimer's, Parkinson's, Huntington's, and stroke, death of one or more population of neurons—through apoptosis initiated by oxidative stress and related DNA damage, disruption of cellular calcium homeostasis, or ER stress—is considered to be one of the underlying mechanisms of disease. Though the process of apoptosis at the time of nervous system development is essential, its reactivation later in life in these neurons can have detrimental effects [3]. Par-4 is upregulated in neurons in patients diagnosed with Alzheimer's, Huntington's, or Parkinson's disease, amyotrophic lateral sclerosis, or HIV encephalitis [3]. In Alzheimer's disease, Par-4 expression is associated with hippocampal and cortical neurons undergoing cell death and with cholinergic signaling under non-apoptotic conditions [19]. Neuronal cells can be protected against the apoptotic effect of Par-4 by either inhibition of Par-4 using an antisense oligodeoxynucleotide or a dominant-negative mutant in cell culture and animal models of neurodegenerative diseases [111, 112]. The blockage of apoptosis using a broad-spectrum caspase inhibitor or co-expression of Par-4 leucine zipper domain, which serves as a dominant-negative regulator of Par-4 function, significantly reduced Aβ secretion in human neuroblastoma IMR-32 cells, suggesting that the Par-4-mediated increased Aβ secretion is through a caspase-dependent pathway [113].

Moreover, Par-4 has a role in neuronal cell death induced by environmental and genetic factors. The withdrawal of trophic factors in cultured embryonic rat hippocampal neurons leads to an increase in Par-4 mRNA and protein levels. This is followed by an increase in reactive oxygen species, mitochondrial membrane depolarization, caspase activation, nuclear chromatin condensation, and activation of apoptosis [114]. Par-4 expression in the synaptosomes after treatment with apoptotic or oxidative insults leads to local induction of cell death in the synaptic compartments. Par-4 antisense oligonucleotides can suppress mitochondrial dysfunction and caspase activation in dendritic compartments of cultured hippocampal neurons following glutamate treatment. Upregulation of Par-4 protein is completely blocked by antisense oligonucleotides in response to the various insults resulting in inhibition of neuronal cell death [112].

The brain of Alzheimer's disease patients shows a significant increase in the mRNA and protein of Par-4 in the hippocampal tissue compared to their age-matched controls. In the two brain regions affected in Alzheimer's, hippocampus, and inferior parietal cortex, Par-4 protein is upregulated to 15–20-fold and 8-fold higher, respectively. Treatment of the primary rat hippocampal neurons with Aβ 1-42 and staurosporine or overexpression of Par-4 in PC12 cells leads to significant induction of apoptosis that can be inhibited by Par-4 antisense oligonucleotides, implying that Par-4 is causally linked with apoptosis in these neurons. Cells lacking the LZ domain show a reduced mitochondrial reactive oxygen species accumulation and lower induction of apoptosis and caspase-3 activation compared to full-length Par-4-transfected cells indicating the importance of the LZ domain in mediating neuronal apoptosis. Moreover, Par-4 enhances the induction of apoptosis in neurons transfected with L286V and Ml46V PS1 mutations (linked to autosomal dominant, early-onset, inherited Alzheimer's) compared to the wild-type PS1-expressing cells [19, 115]. An increase in the oxidative stress levels in the brain is a key feature of Alzheimer's that leads to upregulation of pro-apoptotic factors such as Bax, caspase-3, and Par-4. The treatment of transgenic mouse models of Alzheimer's with melatonin leads to significant downregulation of Bax, caspase-3, and Par-4 with concomitant decrease in neuronal apoptosis, corroborating the importance of Par-4-mediated apoptosis in the disease. Par-4 increases neurotoxic Aβ production in the transfected neural cells, indicating a role for Par-4 in apoptosis and amyloidogenic processing of amyloid precursor protein (APP). Par-4 can also increase the Aβ production by interacting with cytosolic tail of BACE1 (β-secretase is membrane-bound aspartyl protease), an enzyme that catalyzes the first step of Aβ generation as evidenced by co-immunoprecipitation and in vitro pull-down assays, and co-transfection experiments with Par-4 and BACE1. On the contrary, knockdown of Par-4 leads to significant decrease in β-secretase cleavage of APP, resulting in less Aβ production, suggesting that Par-4 accelerates this process. In time course experiments, overexpression of Par-4 increased Aβ production after activating the apoptotic cascade, indicating that Par-4 plays an important in the pathogenesis of Alzheimer's disease [116].

Focal ischemic stroke results in dysfunction and death of neurons in the ischemic brain tissue. Par-4-mediated apoptosis occurs in cultured hippocampal neurons upon trophic factor withdrawal and exposure to glutamate. There is sustained increase in Par-4 levels in neurons undergoing apoptosis after transient forebrain ischemia in the hippocampal and striatal neurons in rats, and ischemia is reversed upon Par-4 knockdown [111].

Parkinson's disease is characterized by the progressive loss of dopaminergic neurons in the substantia nigra, a midbrain dopaminergic nucleus with a critical role in modulating motor movement and reward functions. Par-4 rapidly increases in the neuronal cell bodies in the substantia nigra and their axon terminals in the striatum in monkeys and mice after treatment with MPTP (1-methyl-4-phenyl-1,2,3,6-tetrahydropyridine), and functional studies indicate that an increase in Par-4 contributes to loss of tyrosine hydroxylase immunoreactivity and apoptosis [117]. Apoptosis of dopaminergic neurons with oxidative stress-generating agents (mitochondrial complex I inhibitor rotenone or Fe^{2+}) is inhibited by Par-4 antisense oligonucleotides [117]. However, Moos et al. [118] found that Par-4 immunoreactivity was absent in the substantia nigra and putamen of some Parkinson's disease patients, suggesting that death of the dopaminergic neurons may follow a degenerate pathway that circumvents the induction of Par-4.

Par-4 levels are higher in the lumbar spinal cord of amyotrophic lateral sclerosis (ALS) patients compared to neurologically normal patients [119]. Par-4 levels are higher in ALS-linked Cu/Zn superoxide dismutase (Cu/Zn-SOD) transgenic mice relative to wild-type mice, when their hind limb has already undergone paralysis suggesting that Par-4 may play a role in loss of the motor neurons in this region. Indeed, exposure of primary mouse spinal cord motor neurons to oxidative insult not only resulted in a rapid and significant increase in Par-4, but this increase preceded apoptosis in the neurons. Alternatively, pretreatment of the motor neurons with Par-4 antisense oligonucleotides not only prevented upregulation of Par-4 but also inhibited oxidative stress-induced mitochondrial dysfunction and apoptosis. These and other supporting observations indicate that Par-4 plays a role in death of the motor neurons in the ventral horn region of ALS patients [119]. In ALS-linked Cu/Zn-SOD transgenic mice bearing the G93A mutation, Par-4 knockdown protects the spinal motor neurons from apoptosis posttreatment with excitotoxic and apoptotic insults [120]. Thus, Par-4 plays a role in the pathogenesis of various neurodegenerative diseases, and the underlying mechanism requires further investigation.

In the myocardium, the microRNA miR-17 attenuates cardiac aging by targeting Par-4 [121]. Downregulation of Par-4 by miR-17-3p results in increased transcription of CEBPB, FKA, Oct4, Sca-1, vimentin, and N-cadherin and reduced levels of E-cadherin. This results in EMT and self-renewal in the mouse cardiac fibroblast (MCF) ultimately repressing senescence. A similar effect is seen when Par-4 is knocked down. Transfection with miR-17-3p or Par-4 knockdown in MCF results in decreased intensity of β-galactosidase staining (a marker for senescence), lower apoptosis rate when treated with H_2O_2, and extended survival. The opposite effect was seen with miR-17-3p inhibitor or Par-4 overexpression as these transfected cells had increased proliferation rates compared to the corresponding control cells. Par-4 was confirmed to be the target for miR-17-3p-mediated downregulation because there was a reduction in Par-4 luciferase reporter signal as well as the amount of Par-4 protein expressed upon co-transfection of miR-17-3p and a fragment of Par-4. Par-4 reduction in these experiments was abrogated when the miR-17-3p target site in Par-4 was mutated. Par-4-mediated EMT and self-renewal repression are attributed to downregulation of FAK and CEBPB. Par-4 inhibits CEBPB transcription by binding to the promoter of CEBPB, and this in turn results in a decreased amount of CEBPB binding to the FAK promoter and repression of FAK transcription. Repression of CEBPB and FAK can be reversed either by knocking down Par-4 or transfecting with miR-17-3p. ChIP assays confirmed Par-4 binding to CEBPB and CEBPB binding to FAK. Thus, Par-4 acts as a positive modulator of cardiac aging and cardiac fibroblast cellular senescence by repressing the expression of CEBPB and FAK [121]. The expression of circular RNAs (circRNAs), circ_0068655, and Par-4 was increased in myocardial infarction patients and hypoxia-subjected human-induced pluripotent stem-derived cardiomyocytes (HCMs) [122]. Conversely, expression of miR-498, which is known to inhibit the upregulation of Par-4 in the cardiomyocytes, is downregulated in this model. Thus, hypoxia caused upregulation of circ_0068655 in HCMs, allowing it to sequester and inhibit the activity of miR-498. This in turn allows for Par-4 upregulation in the cardiomyocytes leading to their apoptosis and preventing cardiac remodeling after myocardial infarction [122].

Par-4 was identified as a novel interacting partner of Paired-like homeodomain transcription factor 2 (PITX2*)* using PITX2 as bait in the yeast two-hybrid assay [123]. Three transcript variants encoding distinct isoforms have been identified for this gene [123]. Pitx2 is responsible for the establishment of the left-right axis and asymmetry; asymmetrical development of the heart, lungs, and spleen; hindlimb buds; twisting of the gut and stomach; and also the development of the eyes [124, 125]. Pitx2 deletion results in irregular morphogenesis of organs [126]. Mutations in *PITX2* result in Axenfeld-Rieger syndrome, which is a group of disorders that mainly affects the development of the eye and is associated with defective cornea and iris. Affected individuals may have an off-center pupil or extra holes in the eyes that can look like multiple pupils. The interaction of PITX2 with Par-4 is mediated by its homeodomain and adjacent carboxyl-terminal inhibitory domain, but the mutations causing Axenfeld-Rieger syndrome in this domain did not affect the PITX2-Par-4 interaction. Co-immunoprecipitation using deletion mutants of Par-4 revealed that the carboxyl-terminal portion of Par-4 harboring the LZ domain is sufficient to mediate this interaction with PITX2. PITX2-regulated luciferase reporter activity was reduced by 50% upon co-transfection of Par-4, suggesting that Par-4 inhibits PITX2 transcriptional activity. The LZ domain of Par-4 is sufficient to mediate the transcriptional inhibition activity of PITX2. Immunohistochemistry studies co-localized Par-4 and PITX2 in developing murine eye. PITX2 and Par-4 are ubiquitously expressed in periocular mesenchyme, in particular, within the mesenchyme of the presumptive iridocorneal angle, and continue into merging anterior segment structures, including the iridocorneal angle, corneal stroma, and corneal endothelium in subsequent stages. This co-expression was also seen in mesenchyme-derived corneal stroma and corneal endothelium [127]. Interestingly, Par-4 forms an interactive network with PITX2, FOXC1, and FOXC2 that may be associated with ocular phenotypes and glaucoma pathology caused by mutations in PITX2, FOXC1, and FOXC2. Consistently, both FOXC1 and FOXC2 interact with the LZ domain of Par-4. They also interact with each other in vitro and with recombinant PITX2. FOXC1 and FOXC2 are able to activate transcription by binding to *FGF19* and *FOXO1A* regulatory elements as seen in nonpigmented ciliary epithelial (NPCE) cells, and their transcriptional activity was enhanced upon Par-4 co-transfection. However, unlike the interaction of Par-4 with PIXT2 that leads to reduced transcriptional activity, FOXC1 interaction with Par-4 synergistically increases PIXT2 activity. Thus, these proteins form a regulatory network that plays a role in development or maintenance of ocular function [128].

Moreover, Par-4 expression is associated with pancreatic β cell apoptosis, which is a critical process underlying type 2 diabetes [129]. Par-4/NF-κB signaling induced apoptosis in the islet β cells as seen by the high glucose/palmitate intervention that mimics type 2 diabetes in vitro. High-glucose/palmitate intervention induced expression and secretion of Par-4 as well as increased expression and activation of NF-κB which lead to cell apoptosis. Downregulating Par-4 reduced this effect, and its overexpression potentiated this effect [129]. Also, high-glucose/fatty acid levels lead to induction of Par-4 that interacts with telomerase reverse transcriptase (TERT) and inhibits TERT leading to β cell apoptosis. Inhibition of Par-4 induced by high-glucose/fatty acid levels or knocking down Par-4 in diabetic mice leads to upregulation of TERT, reducing the apoptotic rate [130].

Par-4 also plays an important role in ER stress-induced apoptosis in host defense against *Mycobacterium tuberculosis* (Mtb) [131]. In response to infection by Mtb strain H37Ra, Par-4 and GRP78 levels are increased in macrophages allowing

GRP78 surface translocation and activation of the apoptotic cascade. Additionally, Mtb growth in the macrophages is inhibited by Par-4 overexpression and increased by Par-4 knockdown [131]. Increased Par-4, GRP78, and Par-4-dependent caspase activation was also noted in Bacillus Calmette-Guérin (BCG)-infected prostate cancer cells suggesting that ER stress-induced Par-4 acts as an important defense mechanism against mycobacterial infection and cancer. Collectively, these observations indicate that Par-4 is ubiquitously expressed in diverse normal cells and tissues, aberrant expression of Par-4 in humans causes various diseases, and induction of Par-4 induces caspase-mediated apoptosis.

11 Future Perspectives and Conclusion

Since its discovery about 25 years ago, Par-4 has made its journey from the bench to bedside and is currently in clinical trials involving compounds that stimulate the secretion of Par-4 from normal cells of the body to selectively kill cancer cells. However, a number of areas related to Par-4 have not been fully explored. (1) Par-4 is a tumor suppressor and induces apoptosis in cancer cells, but its regulation at transcriptional and translational levels is not fully understood. Although the initial studies suggested induction of Par-4 at the RNA level by apoptotic agents, only a few agents that induce Par-4 protein levels have been identified. One caveat is that the detection of elevated levels of intact Par-4 protein may be time-sensitive as induction of Par-4 protein will activate caspase-3 that can cause caspase-3-mediated cleavage of Par-4 at the D131 residue (or D123 in rat Par-4 or D124 in mouse Par-4). Therefore, it may be easier to detect loss of full-length Par-4 protein following treatment with agents that induce Par-4 RNA but cause cleavage of Par-4 due to caspase-3 activation. (2) The precise role of Par-4 in transcriptional repression needs to be carefully investigated, so as to enable the use of Par-4 to kill Par-4-resistant cells. E2F1 has been shown to partner with Par-4 to regulate the expression of the Smac promoter and contribute to glutamate-induced apoptosis in human bone mesenchymal stem cells [132]. This interaction with E2F1 needs to be investigated in other apoptotic situations to determine whether other E2F1 targets are also regulated by the Par-4-E2F1 interaction [132]. (3) Par-4 has a putative P-loop domain that is typically responsible for binding nucleotides and regulating protein activity as in the case of Ras, yet its functional role in Par-4 is unknown. It is possible that binding of ATP/GTP with the P-loop domain may regulate the activity of Par-4 in cancer cells relative to normal cells. There is also a putative "GGDEF" motif homologous to the adenylyl cyclase domain. Adenylyl cyclase is known to generate the second messenger cAMP from hydrolysis of ATP and regulate a number of signaling pathways through activation of PKA, which is a Par-4 activator, and the putative adenylyl cyclase domain of Par-4 deserves further investigation. (4) As oncogenic Ras suppresses Par-4, c-Myc induces Par-4, and the combination of oncogenic Ras and c-Myc induces cellular transformation by a pathway that requires Par-4 downregulation and inhibition of apoptosis, it would be logical to test whether overexpression of c-Myc in cells where endogenous Par-4 expression is inhibited will be sufficient to induce cellular transformation. (5) The mechanism by which Par-4 is secreted by the BFA-sensitive ER-Golgi-plasma membrane pathway despite lacking the canonical secretory peptide sequence needs further investigation. It is known that the SAC domain of Par-4 is secreted, that it then interacts with GRP78 leading to apoptosis, but the precise conditions and mechanism under which Par-4 and GRP78 are present together on the cell surface are unresolved. The pathway determining that both of them are present simultaneously on the cell surface needs to be studied in depth as it provides an attractive model to design monoclonal antibodies or small molecule agents to either mimic the Par-4-like killing ability or to target GRP78. Thus, a better understanding of factors regulating Par-4 secretion and subsequent interaction with GRP78 will enable the design of strategies for robust secretion of Par-4 and killing tumors. (6) Several Par-4 secretagogues have been identified, but some cancer cells exhibit resistance to secreted (extracellular) Par-4 due to low levels cell surface GRP78. As an example, lung cancer cells A549 are typically resistant to apoptosis by extracellular Par-4 but can be sensitized and rendered sensitive to Par-4 by agents that inhibit NF-κB activity or UACA resulting in elevated cell surface GRP78 levels. Our initial screen has identified compounds that elevate cell surface GRP78 and sensitize cancer cells to apoptosis by recombinant Par-4. One example of such a compound is curcumin, a dietary compound that increases GRP78 levels at the plasma membrane and renders A549 lung cancer cells sensitive to Par-4-mediated apoptosis (Fig. 19). The action of FDA-approved drugs that elevate cell surface GRP78 selectively in cancer cells could be complemented by recombinant Par-4, Par-4 secretagogues, or GRP78 targeting antibodies to kill tumor cells.

(7) Par-4 has a number of interacting partners, and most of its interactions are mediated through the LZ domain. An interacting partner that binds Par-4 and sequesters it efficiently is UACA. Future studies designed to suppress UACA in normal and cancer cells, thereby enabling elevated levels of secreted Par-4, are required to take advantage of the selective cancer killing ability of Par-4. Alternatively, it is possible that the mRNA encoding Par-4 protein can be packaged using liposomes or suitable nanoparticles to target normal and tumor cells, anticipating that the normal cells will synthesize and secrete Par-4 protein that will then kill tumor cells in a paracrine manner, and the tumor cells that express the mRNA will have elevated levels of intracellular Par-4 to induce apoptosis. (8) Finally, as remarkable tumor shrinkage was achieved by adenoviral delivery of Par-4, effective strategies to build high local levels of Par-4 in the tumor and its microenvironment should be explored to provide promising outcomes. Concomitantly, Par-4 levels need to be monitored in neuronal and other disorders where Par-4 may aggravate the disease.

Acknowledgments We thank former members of the Rangnekar laboratory who contributed to the findings cited in this article. These members include Stephen Sells, Swati Joshi-Barve, Steve Jensen, Patrick Collini, Jason Cook, Steve Monnig, Suzanne Humphreys, Sumathi Krishnan, Scott Crist, Prakash Nair, Aysegul Nalca, Mala Chakraborty, Nadia El-Guendy, Krishna M. Vasudevan, Ping Du, Sushma Gurumurthy, Mamta Goswami, Seong-Su Han, Anindya Goswami, Pam

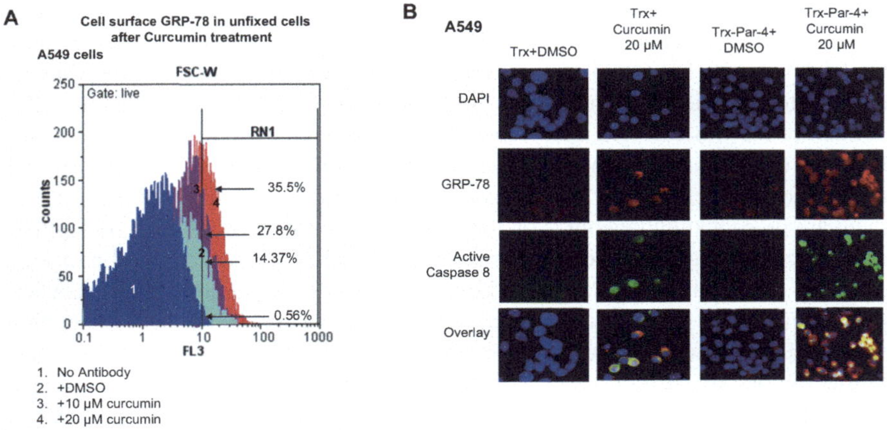

Fig. 19 Curcumin increases GRP78 translocation to the cell surface and increases Par-4-mediated apoptosis in human lung cancer cells. (**a**) **Increase in GRP78 translocation to cell surface following curcumin treatment**. A549 were treated with either 10 or 20 μM of curcumin or DMSO vehicle for 24 h, and the unfixed cells were subjected to flow cytometry for GRP78 in order to determine the percentage of cells with GRP78 cell surface translocation. Note the GRP78 translocation increases in a dose-dependent manner with curcumin compared to control vehicle. (**b**) **Curcumin increases Par-4-mediated apoptosis**. A549 were treated with TRX (thioredoxin control) + DMSO, TRX-Par-4 + DMSO, TRX+ 20 μM curcumin, and TRX-Par-4 + 20 μM curcumin for 24 h and subjected to immunocytochemistry against GRP78 (red) (goat antibody) and caspase-8 (green) (rabbit antibody). TRX and TRX-Par-4 used at 100 nM amounts. Curcumin increases GRP78 translocation to the cell surface consequently leading to increased binding of recombinant Par-4 to cell surface GRP78 and activation of the caspase-8 pathway compared to just recombinant Par-4 or curcumin alone

Ranganathan, Jaya Singh, Yanming Zhao, Graham Warren, Sunil Nooti, Tripti Bhattarai, Nikhil Hebbar, Nidhi Shukla, James Sledziona, Nathalia Araujo, and Tingfei Chen. We also thank our collaborators, David Wood, David Watt, Chunming Liu, Subbarao Bondada, Chang-Guo Zhan, and Peng Wang, at the University of Kentucky.

Funding for these studies was provided by NIH/NCI R01 and R21 grants to VMR, the Prostate Cancer Foundation, the University of Kentucky Markey Cancer Center, and the University of Kentucky Center for Clinical and Translational Science. We thank the Biospecimen Procurement and Translational Pathology Shared Resource Facility (SRF), Cancer Research Informatics SRF, Flow Cytometry and Immune Monitoring SRF, Redox Metabolism SRF, Oncogenomics SRF, and Biostatistics and Bioinformatics SRF of the University of Kentucky Markey Cancer Center (P30 CA177558) for supporting Par-4 research projects in the Rangnekar laboratory. We thank Donna Gilbreath, Elise Wright and Danielle Story of the Research Communications Office, Markey Cancer Center, for their support in preparing this book chapter.

References

1. Sells SF, Wood DP Jr, Joshi-Barve SS, Muthukumar S, Jacob RJ, Crist SA, Humphreys S, Rangnekar VM (1994) Commonality of the gene programs induced by effectors of apoptosis in androgen-dependent and -independent prostate cells. Cell Growth Differ 5(4):457–466

2. Sells SF, Han SS, Muthukkumar S, Maddiwar N, Johnstone R, Boghaert E, Gillis D, Liu G, Nair P, Monnig S, Collini P, Mattson MP, Sukhatme VP, Zimmer SG, Wood DP Jr, McRoberts JW, Shi Y, Rangnekar VM (1997) Expression and function of the leucine zipper protein Par-4 in apoptosis. Mol Cell Biol 17(7):3823–3832
3. El-Guendy N, Rangnekar VM (2003) Apoptosis by Par-4 in cancer and neurodegenerative diseases. Exp Cell Res 283(1):51–66
4. Díaz-Meco MT, Municio MM, Frutos S, Sanchez P, Lozano J, Sanz L, Moscat J (1996) The product of par-4, a gene induced during apoptosis, interacts selectively with the atypical isoforms of protein kinase C. Cell 86(5):777–786
5. Johnstone RW, See RH, Sells SF, Wang J, Muthukkumar S, Englert C, Haber DA, Licht JD, Sugrue SP, Roberts T, Rangnekar VM, Shi Y (1996) A novel repressor, par-4, modulates transcription and growth suppression functions of the Wilms' tumor suppressor WT1. Mol Cell Biol 16(12):6945–6956
6. Johnstone RW, Tommerup N, Hansen C, Vissing H, Shi Y (1998) Mapping of the human PAWR (par-4) gene to chromosome 12q21. Genomics 53(2):241–243
7. Zhao Y, Rangnekar VM (2008) Apoptosis and tumor resistance conferred by Par-4. Cancer Biol Ther 7(12):1867–1874
8. Boghaert ER, Sells SF, Walid AJ, Malone P, Williams NM, Weinstein MH, Strange R, Rangnekar VM (1997) Immunohistochemical analysis of the proapoptotic protein Par-4 in normal rat tissues. Cell Growth Differ 8(8):881–890
9. Rangnekar VM (2001) Apoptosis by par-4 protein. In: Advances in cell aging and gerontology, vol 5. Elsevier, pp 215–236
10. Hebbar N, Wang C, Rangnekar VM (2012) Mechanisms of apoptosis by the tumor suppressor Par-4. J Cell Physiol 227(12):3715–3721
11. de Thonel A, Hazoumé A, Kochin V, Isoniemi K, Jego G, Fourmaux E, Hammann A, Mjahed H, Filhol O, Micheau O, Rocchi P, Mezger V, Eriksson JE, Rangnekar VM, Garrido C (2014) Regulation of the proapoptotic functions of prostate apoptosis response-4 (Par-4) by casein kinase 2 in prostate cancer cells. Cell Death Dis 5(1):e1016–e1016
12. Chaudhry P, Singh M, Parent S, Asselin E (2012) Prostate apoptosis response 4 (Par-4), a novel substrate of caspase-3 during apoptosis activation. Mol Cell Biol 32(4):826–839
13. El-Guendy N, Zhao Y, Gurumurthy S, Burikhanov R, Rangnekar VM (2003) Identification of a unique core domain of par-4 sufficient for selective apoptosis induction in cancer cells. Mol Cell Biol 23(16):5516–5525
14. Tiruttani Subhramanyam UK, Kubicek J, Eidhoff UB, Labahn J (2017) Structural basis for the regulatory interactions of proapoptotic Par-4. Cell Death Differ 24(9):1540–1547
15. Goswami A, Qiu S, Dexheimer TS, Ranganathan P, Burikhanov R, Pommier Y, Rangnekar VM (2008) Par-4 binds to topoisomerase 1 and attenuates its DNA relaxation activity. Cancer Res 68(15):6190–6198
16. Azmi AS, Aboukameel A, Bao B, Sarkar FH, Philip PA, Kauffman M, Shacham S, Mohammad RM (2013) Selective inhibitors of nuclear export block pancreatic cancer cell proliferation and reduce tumor growth in mice. Gastroenterology 144(2):447–456
17. Miller M (2009) The importance of being flexible: the case of basic region leucine zipper transcriptional regulators. Curr Protein Pept Sci 10(3):244–269
18. Dutta K, Alexandrov A, Huang H, Pascal SM (2001) pH-induced folding of an apoptotic coiled coil. Protein Sci 10(12):2531–2540
19. Guo Q, Fu W, Xie J, Luo H, Sells SF, Geddes JW, Bondada V, Rangnekar VM, Mattson MP (1998) Par-4 is a mediator of neuronal degeneration associated with the pathogenesis of Alzheimer disease. Nat Med 4(8):957–962
20. Shrestha-Bhattarai T, Rangnekar VM (2010) Cancer-selective apoptotic effects of extracellular and intracellular Par-4. Oncogene 29(27):3873–3880
21. Cheema SK, Mishra SK, Rangnekar VM, Tari AM, Kumar R, Lopez-Berestein G (2003) Par-4 transcriptionally regulates Bcl-2 through a WT1-binding site on the bcl-2 promoter. J Biol Chem 278(22):19995–20005

22. Richard DJ, Schumacher V, Royer-Pokora B, Roberts SG (2001) Par4 is a coactivator for a splice isoform-specific transcriptional activation domain in WT1. Genes Dev 15(3):328–339
23. Kim J, Lee K, Pelletier J (1998) The DNA binding domains of the WT1 tumor suppressor gene product and chimeric EWS/WT1 oncoprotein are functionally distinct. Oncogene 16(8):1021–1030
24. Diaz-Meco MT, Lallena MJ, Monjas A, Frutos S, Moscat J (1999) Inactivation of the inhibitory kappaB protein kinase/nuclear factor kappaB pathway by Par-4 expression potentiates tumor necrosis factor alpha-induced apoptosis. J Biol Chem 274(28):19606–19612
25. Chang S, Kim JH, Shin J (2002) p62 forms a ternary complex with PKCzeta and PAR-4 and antagonizes PAR-4-induced PKCzeta inhibition. FEBS Lett 510(1–2):57–61
26. Sanchez P, De Carcer G, Sandoval IV, Moscat J, Diaz-Meco MT (1998) Localization of atypical protein kinase C isoforms into lysosome-targeted endosomes through interaction with p62. Mol Cell Biol 18(5):3069–3080
27. Duran A, Diaz-Meco MT, Moscat J (2003) Essential role of RelA Ser311 phosphorylation by zetaPKC in NF-kappaB transcriptional activation. EMBO J 22(15):3910–3918
28. Joshi J, Fernandez-Marcos PJ, Galvez A, Amanchy R, Linares JF, Duran A, Pathrose P, Leitges M, Cañamero M, Collado M, Salas C, Serrano M, Moscat J, Diaz-Meco MT (2008) Par-4 inhibits Akt and suppresses Ras-induced lung tumorigenesis. EMBO J 27(16):2181–2193
29. Vivanco I, Sawyers CL (2002) The phosphatidylinositol 3-Kinase–AKT pathway in human cancer. Nat Rev Cancer 2(7):489–501
30. Goswami A, Burikhanov R, de Thonel A, Fujita N, Goswami M, Zhao Y, Eriksson JE, Tsuruo T, Rangnekar VM (2005) Binding and phosphorylation of par-4 by akt is essential for cancer cell survival. Mol Cell 20(1):33–44
31. Kawai T, Matsumoto M, Takeda K, Sanjo H, Akira S (1998) ZIP kinase, a novel serine/threonine kinase which mediates apoptosis. Mol Cell Biol 18(3):1642–1651
32. Kawai T, Akira S, Reed JC (2003) ZIP kinase triggers apoptosis from nuclear PML oncogenic domains. Mol Cell Biol 23(17):6174–6186
33. Kögel D, Reimertz C, Mech P, Poppe M, Frühwald MC, Engemann H, Scheidtmann KH, Prehn JH (2001) Dlk/ZIP kinase-induced apoptosis in human medulloblastoma cells: requirement of the mitochondrial apoptosis pathway. Br J Cancer 85(11):1801–1808
34. Boosen M, Vetterkind S, Kubicek J, Scheidtmann K-H, Illenberger S, Preuss U (2009) Par-4 is an essential downstream target of DAP-like kinase (Dlk) in Dlk/Par-4–mediated apoptosis. Mol Biol Cell 20(18):4010–4020
35. Vetterkind S, Lee E, Sundberg E, Poythress RH, Tao TC, Preuss U, Morgan KG (2010) Par-4: a new activator of myosin phosphatase. Mol Biol Cell 21(7):1214–1224
36. MacDonald JA, Moffat LD, Al-Ghabkari A, Sutherland C, Walsh MP (2013) Prostate-apoptosis response-4 phosphorylation in vascular smooth muscle. Arch Biochem Biophys 535(1):84–90
37. Roussigne M, Cayrol C, Clouaire T, Amalric F, Girard J-P (2003) THAP1 is a nuclear pro-apoptotic factor that links prostate-apoptosis-response-4 (Par-4) to PML nuclear bodies. Oncogene 22(16):2432–2442
38. Lu C, Li JY, Ge Z, Zhang L, Zhou GP (2013) Par-4/THAP1 complex and Notch3 competitively regulated pre-mRNA splicing of CCAR1 and affected inversely the survival of T-cell acute lymphoblastic leukemia cells. Oncogene 32(50):5602–5613
39. Xie J, Guo Q (2004) Par-4 inhibits choline uptake by interacting with CHT1 and reducing its incorporation on the plasma membrane. J Biol Chem 279(27):28266–28275
40. Guo Q, Xie J (2004) AATF inhibits aberrant production of amyloid beta peptide 1-42 by interacting directly with Par-4. J Biol Chem 279(6):4596–4603
41. Park SK, Nguyen MD, Fischer A, Luke MP-S, Affar EB, Dieffenbach PB, Tseng H-C, Shi Y, Tsai L-H (2005) Par-4 links dopamine signaling and depression. Cell 122(2):275–287
42. Glantz LA, Gilmore JH, Overstreet DH, Salimi K, Lieberman JA, Jarskog LF (2010) Pro-apoptotic Par-4 and dopamine D2 receptor in temporal cortex in schizophrenia, bipolar disorder and major depression. Schizophr Res 118(1–3):292–299

43. Irie Y, Yamagata K, Gan Y, Miyamoto K, Do E, Kuo CH, Taira E, Miki N (2000) Molecular cloning and characterization of Amida, a novel protein which interacts with a neuron-specific immediate early gene product arc, contains novel nuclear localization signals, and causes cell death in cultured cells. J Biol Chem 275(4):2647–2653
44. Boosen M, Vetterkind S, Koplin A, Illenberger S, Preuss U (2005) Par-4-mediated recruitment of Amida to the actin cytoskeleton leads to the induction of apoptosis. Exp Cell Res 311(2):177–191
45. Champoux JJ (2001) DNA topoisomerases: structure, function, and mechanism. Annu Rev Biochem 70(1):369–413
46. Li M, Liu Y (2016) Topoisomerase I in human disease pathogenesis and treatments. Genomics Proteomics Bioinformatics 14(3):166–171
47. Gao S, Wang H, Lee P, Melamed J, Li CX, Zhang F, Wu H, Zhou L, Wang Z (2006) Androgen receptor and prostate apoptosis response factor-4 target the c-FLIP gene to determine survival and apoptosis in the prostate gland. J Mol Endocrinol 36(3):463–483
48. Gurumurthy S, Goswami A, Vasudevan KM, Rangnekar VM (2005) Phosphorylation of Par-4 by protein kinase A is critical for apoptosis. Mol Cell Biol 25(3):1146–1161
49. Ranganathan P, Rangnekar VM (2005) Regulation of cancer cell survival by Par-4. Ann N Y Acad Sci 1059:76–85
50. Sato M, Yao VJ, Arap W, Pasqualini R (2010) GRP78 signaling hub a receptor for targeted tumor therapy. Adv Genet 69:97–114
51. Burikhanov R, Zhao Y, Goswami A, Qiu S, Schwarze SR, Rangnekar VM (2009) The tumor suppressor Par-4 activates an extrinsic pathway for apoptosis. Cell 138(2):377–388
52. Zhao Y, Burikhanov R, Qiu S, Lele SM, Jennings CD, Bondada S, Spear B, Rangnekar VM (2007) Cancer resistance in transgenic mice expressing the SAC module of Par-4. Cancer Res 67(19):9276–9285
53. Zhao Y, Burikhanov R, Brandon J, Qiu S, Shelton BJ, Spear B, Bondada S, Bryson S, Rangnekar VM (2011) Systemic Par-4 inhibits non-autochthonous tumor growth. Cancer Biol Ther 12(2):152–157
54. Zhuang D, Liu Y, Mao Y, Gao L, Zhang H, Luan S, Huang F, Li Q (2012) TMZ-induced PrPc/par-4 interaction promotes the survival of human glioma cells. Int J Cancer 130(2):309–318
55. Hebbar N, Burikhanov R, Shukla N, Qiu S, Zhao Y, Elenitoba-Johnson KSJ, Rangnekar VM (2017) A naturally generated decoy of the prostate apoptosis response-4 protein overcomes therapy resistance in tumors. Cancer Res 77(15):4039–4050
56. Chakraborty M, Qiu SG, Vasudevan KM, Rangnekar VM (2001) Par-4 drives trafficking and activation of Fas and Fasl to induce prostate cancer cell apoptosis and tumor regression. Cancer Res 61(19):7255–7263
57. Nalca A, Qiu SG, El-Guendy N, Krishnan S, Rangnekar VM (1999) Oncogenic Ras sensitizes cells to apoptosis by Par-4. J Biol Chem 274(42):29976–29983
58. Qiu SG, Krishnan S, el-Guendy N, Rangnekar VM (1999) Negative regulation of Par-4 by oncogenic Ras is essential for cellular transformation. Oncogene 18(50):7115–7123
59. Boehrer S, Chow KU, Beske F, Kukoc-Zivojnov N, Puccetti E, Ruthardt M, Baum C, Rangnekar VM, Hoelzer D, Mitrou PS, Weidmann E (2002) In lymphatic cells par-4 sensitizes to apoptosis by down-regulating bcl-2 and promoting disruption of mitochondrial membrane potential and caspase activation. Cancer Res 62(6):1768–1775
60. Bergmann M, Kukoc-Zivojnov N, Chow KU, Trepohl B, Hoelzer D, Weidmann E, Mitrou PS, Boehrer S (2004) Prostate apoptosis response gene-4 sensitizes neoplastic lymphocytes to CD95-induced apoptosis. Ann Hematol 83(10):646–653
61. Kline CL, Shanmugavelandy SS, Kester M, Irby RB (2009) Delivery of PAR-4 plasmid in vivo via nanoliposomes sensitizes colon tumor cells subcutaneously implanted into nude mice to 5-FU. Cancer Biol Ther 8(19):1831–1837
62. Sharma AK, Kline CL, Berg A, Amin S, Irby RB (2011) The Akt inhibitor ISC-4 activates prostate apoptosis response protein-4 and reduces colon tumor growth in a nude mouse model. Clin Cancer Res 17(13):4474–4483

63. Azmi AS, Ahmad A, Banerjee S, Rangnekar VM, Mohammad RM, Sarkar FH (2008) Chemoprevention of pancreatic cancer: characterization of Par-4 and its modulation by 3,3′ diindolylmethane (DIM). Pharm Res 25(9):2117–2124
64. Azmi AS, Wang Z, Burikhanov R, Rangnekar VM, Wang G, Chen J, Wang S, Sarkar FH, Mohammad RM (2008) Critical role of prostate apoptosis response-4 in determining the sensitivity of pancreatic cancer cells to small-molecule inhibitor-induced apoptosis. Mol Cancer Ther 7(9):2884–2893
65. Chendil D, Das A, Dey S, Mohiuddin M, Ahmed MM (2002) Par-4, a pro-apoptotic gene, inhibits radiation-induced NF kappa B activity and Bcl-2 expression leading to induction of radiosensitivity in human prostate cancer cells PC-3. Cancer Biol Ther 1(2):152–160
66. Qiu G, Ahmed M, Sells SF, Mohiuddin M, Weinstein MH, Rangnekar VM (1999) Mutually exclusive expression patterns of Bcl-2 and Par-4 in human prostate tumors consistent with down-regulation of Bcl-2 by Par-4. Oncogene 18(3):623–631
67. Srinivasan S, Ranga RS, Burikhanov R, Han SS, Chendil D (2007) Par-4-dependent apoptosis by the dietary compound withaferin A in prostate cancer cells. Cancer Res 67(1):246–253
68. Das TP, Suman S, Alatassi H, Ankem MK, Damodaran C (2016) Inhibition of AKT promotes FOXO3a-dependent apoptosis in prostate cancer. Cell Death Dis 7(2):e2111
69. MacLean MA, Scott BE, Deziel BA, Nunnelley MC, Liberty AM, Gottschall-Pass KT, Neto CC, Hurta RA (2011) North American cranberry (Vaccinium macrocarpon) stimulates apoptotic pathways in DU145 human prostate cancer cells in vitro. Nutr Cancer 63(1):109–120
70. Vasudevan KM, Ranganathan P, Rangnekar VM (2005) Regulation of Par-4 by oncogenic Ras. In: Balch B, Der C, Hall A (eds) Methods in enzymology, vol 407. Elsevier Science, Inc., New York, pp 422–442
71. Pruitt K, Ulkü AS, Frantz K, Rojas RJ, Muniz-Medina VM, Rangnekar VM, Der CJ, Shields JM (2005) Ras-mediated loss of the pro-apoptotic response protein Par-4 is mediated by DNA hypermethylation through Raf-independent and Raf-dependent signaling cascades in epithelial cells. J Biol Chem 280(24):23363–23370
72. Qiu SGS, Zhao Y, Vasudevan KM, Hann S, Rangnekar VM (2007) Par-4 function is essential for c-Myc-inducible apoptosis. Unpublished
73. Araujo N, Hebbar N, Rangnekar VM (2018) GRP78 is a targetable receptor on cancer and stromal cells. EBioMedicine 33:2–3
74. Wang J, Li Y, Ma F, Zhou H, Ding R, Lu B, Zou L, Li J, Lu R (2017) Inhibitory effect of Par-4 combined with cisplatin on human Wilms' tumor cells. Tumour Biol 39(7):1010428317716689
75. Karch J, Molkentin Jeffery D (2015) Regulated necrotic cell death. Circ Res 116(11):1800–1809
76. Barradas M, Monjas A, Diaz-Meco MT, Serrano M, Moscat J (1999) The downregulation of the pro-apoptotic protein Par-4 is critical for Ras-induced survival and tumor progression. EMBO J 18(22):6362–6369
77. Lafuente MJ, Martin P, Garcia-Cao I, Diaz-Meco MT, Serrano M, Moscat J (2003) Regulation of mature T lymphocyte proliferation and differentiation by Par-4. EMBO J 22(18):4689–4698
78. García-Cao I, Duran A, Collado M, Carrascosa MJ, Martín-Caballero J, Flores JM, Diaz-Meco MT, Moscat J, Serrano M (2005) Tumour-suppression activity of the proapoptotic regulator Par4. EMBO Rep 6(6):577–583
79. Garcia-Cao I, Lafuente MJ, Criado LM, Diaz-Meco MT, Serrano M, Moscat J (2003) Genetic inactivation of Par4 results in hyperactivation of NF-kappaB and impairment of JNK and p38. EMBO Rep 4(3):307–312
80. Lucas T, Pratscher B, Krishnan S, Fink D, Günsberg P, Wolschek M, Wacheck V, Muster T, Romirer I, Wolff K, Pehamberger H, Eichler HG, Rangnekar VM, Jansen B (2001) Differential expression levels of Par-4 in melanoma. Melanoma Res 11(4):379–383
81. Burikhanov R, Shrestha-Bhattarai T, Hebbar N, Qiu S, Zhao Y, Zambetti GP, Rangnekar VM (2014) Paracrine apoptotic effect of p53 mediated by tumor suppressor Par-4. Cell Rep 6(2):271–277

82. Burikhanov R, Shrestha-Bhattarai T, Qiu S, Shukla N, Hebbar N, Lele SM, Horbinski C, Rangnekar VM (2013) Novel mechanism of apoptosis resistance in cancer mediated by extracellular PAR-4. Cancer Res 73(2):1011–1019
83. Burikhanov R, Hebbar N, Noothi SK, Shukla N, Sledziona J, Araujo N, Kudrimoti M, Wang QJ, Watt DS, Welch DR, Maranchie J, Harada A, Rangnekar VM (2017) Chloroquine-inducible Par-4 secretion is essential for tumor cell apoptosis and inhibition of metastasis. Cell Rep 18(2):508–519
84. Burikhanov R, Sviripa VM, Hebbar N, Zhang W, Layton WJ, Hamza A, Zhan CG, Watt DS, Liu C, Rangnekar VM (2014) Arylquins target vimentin to trigger Par-4 secretion for tumor cell apoptosis. Nat Chem Biol 10(11):924–926
85. Cook J, Krishnan S, Ananth S, Sells SF, Shi Y, Walther MM, Linehan WM, Sukhatme VP, Weinstein MH, Rangnekar VM (1999) Decreased expression of the pro-apoptotic protein Par-4 in renal cell carcinoma. Oncogene 18(5):1205–1208
86. Moreno-Bueno G, Fernandez-Marcos PJ, Collado M, Tendero MJ, Rodriguez-Pinilla SM, Garcia-Cao I, Hardisson D, Diaz-Meco MT, Moscat J, Serrano M, Palacios J (2007) Inactivation of the candidate tumor suppressor par-4 in endometrial cancer. Cancer Res 67(5):1927–1934
87. Chatterton Z, Morenos L, Mechinaud F, Ashley DM, Craig JM, Sexton-Oates A, Halemba MS, Parkinson-Bates M, Ng J, Morrison D, Carroll WL, Saffery R, Wong NC (2014) Epigenetic deregulation in pediatric acute lymphoblastic leukemia. Epigenetics 9(3):459–467
88. Wang LJ, Chen PR, Hsu LP, Hsu WL, Liu DW, Chang CH, Hsu YC, Lee JW (2014) Concomitant induction of apoptosis and autophagy by prostate apoptosis response-4 in hypopharyngeal carcinoma cells. Am J Pathol 184(2):418–430
89. Nguyen JQ, Irby RB (2017) TRIM21 is a novel regulator of Par-4 in colon and pancreatic cancer cells. Cancer Biol Ther 18(1):16–25
90. Moscat J, Diaz-Meco MT, Wooten MW (2009) Of the atypical PKCs, Par-4 and p62: recent understandings of the biology and pathology of a PB1-dominated complex. Cell Death Differ 16(11):1426–1437
91. Diaz-Meco MT, Abu-Baker S (2009) The Par-4/PTEN connection in tumor suppression. Cell Cycle 8(16):2518–2522
92. Lee TJ, Lee JT, Kim SH, Choi YH, Song KS, Park JW, Kwon TK (2008) Overexpression of Par-4 enhances thapsigargin-induced apoptosis via down-regulation of XIAP and inactivation of Akt in human renal cancer cells. J Cell Biochem 103(2):358–368
93. Franchitto A, Torrice A, Semeraro R, Napoli C, Nuzzo G, Giuliante F, Alpini G, Carpino G, Berloco PB, Izzo L, Bolognese A, Onori P, Renzi A, Cantafora A, Gaudio E, Alvaro D (2010) Prostate apoptosis response-4 is expressed in normal cholangiocytes, is down-regulated in human cholangiocarcinoma, and promotes apoptosis of neoplastic cholangiocytes when induced pharmacologically. Am J Pathol 177(4):1779–1790
94. Brieger A, Boehrer S, Schaaf S, Nowak D, Ruthardt M, Kim SZ, Atadja P, Hoelzer D, Mitrou PS, Weidmann E, Chow KU (2004) In bcr-abl-positive myeloid cells resistant to conventional chemotherapeutic agents, expression of Par-4 increases sensitivity to imatinib (STI571) and histone deacetylase-inhibitors. Biochem Pharmacol 68(1):85–93
95. Lee JW, Lee KF, Hsu HY, Hsu LP, Shih WL, Chu YC, Hsiao WT, Liu PF (2007) Protein expression and intracellular localization of prostate apoptosis response-4 (Par-4) are associated with apoptosis induction in nasopharyngeal carcinoma cell lines. Cancer Lett 257(2):252–262
96. Tan J, Tao K, Zheng X, Liu D, Ma D, Gao Q (2020) Expression of PAWR predicts prognosis of ovarian cancer. Cancer Cell Int 20(1):598
97. Alvarez JV, Pan TC, Ruth J, Feng Y, Zhou A, Pant D, Grimley JS, Wandless TJ, Demichele A, Chodosh LA (2013) Par-4 downregulation promotes breast cancer recurrence by preventing multinucleation following targeted therapy. Cancer Cell 24(1):30–44
98. Shrestha-Bhattarai T, Hebbar N, Rangnekar VM (2013) Par(−4)oxysm in breast cancer. Cancer Cell 24(1):3–5

99. Nagai MA, Gerhard R, Salaorni S, Fregnani JH, Nonogaki S, Netto MM, Soares FA (2010) Down-regulation of the candidate tumor suppressor gene PAR-4 is associated with poor prognosis in breast cancer. Int J Oncol 37(1):41–49
100. de Bessa Garcia SA, Pavanelli AC, Cruz EMN, Nagai MA (2017) Prostate apoptosis response 4 (PAR4) expression modulates WNT signaling pathways in MCF7 breast cancer cells: a possible mechanism underlying PAR4-mediated docetaxel chemosensitivity. Int J Mol Med 39(4):809–818
101. Jagtap JC, Parveen D, Shah RD, Desai A, Bhosale D, Chugh A, Ranade D, Karnik S, Khedkar B, Mathur A, Natesh K, Chandrika G, Shastry P (2015) Secretory prostate apoptosis response (Par)-4 sensitizes multicellular spheroids (MCS) of glioblastoma multiforme cells to tamoxifen-induced cell death. FEBS Open Biol 5:8–19
102. Song R, Li Y, Hao W, Yang L, Chen B, Zhao Y, Sun B, Xu F (2020) Circular RNA MTO1 inhibits gastric cancer progression by elevating PAWR via sponging miR-199a-3p. Cell Cycle 19(22):3127–3139
103. Yoo MH, Hatfield DL (2008) The cancer stem cell theory: is it correct? Mol Cells 26(5):514–516
104. Jagtap JC, Dawood P, Shah RD, Chandrika G, Natesh K, Shiras A, Hegde AS, Ranade D, Shastry P (2014) Expression and regulation of Prostate Apoptosis Response-4 (Par-4) in human glioma stem cells in drug-induced apoptosis. PLoS One 9(2):e88505
105. Nguyen J, Kline C, Caballero N, Irby R (2015) Overexpression of prostate apoptosis response protein-4 in colon cancer cells can inhibit metastasis by upregulating E-cadherin expression. J Colon Rectal Cancer 1:20–34
106. Katoch A, Suklabaidya S, Chakraborty S, Nayak D, Rasool RU, Sharma D, Mukherjee D, Faheem MM, Kumar A, Sharma PR, Senapati S, Kumar LD, Goswami A (2018) Dual role of Par-4 in abrogation of EMT and switching on mesenchymal to epithelial transition (MET) in metastatic pancreatic cancer cells. Mol Carcinog 57(9):1102–1115
107. Mabe NW, Fox DB, Lupo R, Decker AE, Phelps SN, Thompson JW, Alvarez JV (2018) Epigenetic silencing of tumor suppressor Par-4 promotes chemoresistance in recurrent breast cancer. J Clin Invest 128(10):4413–4428
108. Ahmad SM, Nayak D, Mir KB, Faheem MM, Nawaz S, Yadav G, Goswami A (2020) Par-4 activation restrains EMT-induced chemoresistance in PDAC by attenuating MDM-2. Pancreatology 20(8):1698–1710
109. Katoch A, Jamwal VL, Faheem MM, Kumar S, Senapati S, Yadav G, Gandhi SG, Goswami A (2021) Overlapping targets exist between the Par-4 and miR-200c axis which regulate EMT and proliferation of pancreatic cancer cells. Transl Oncol 14(1):100879
110. Thabet NA, El-Guendy N, Mohamed MM, Shouman SA (2019) Suppression of macrophages- induced inflammation via targeting RAS and PAR-4 signaling in breast cancer cell lines. Toxicol Appl Pharmacol 385:114773
111. Culmsee C, Zhu Y, Krieglstein J, Mattson MP (2001) Evidence for the involvement of Par-4 in ischemic neuron cell death. J Cereb Blood Flow Metab 21(4):334–343
112. Duan W, Rangnekar VM, Mattson MP (1999) Prostate apoptosis response-4 production in synaptic compartments following apoptotic and excitotoxic insults: evidence for a pivotal role in mitochondrial dysfunction and neuronal degeneration. J Neurochem 72(6):2312–2322
113. Guo Q, Xie J, Chang X, Du H (2001) Prostate apoptosis response-4 enhances secretion of amyloid beta peptide 1-42 in human neuroblastoma IMR-32 cells by a caspase-dependent pathway. J Biol Chem 276(19):16040–16044
114. Chan SL, Tammariello SP, Estus S, Mattson MP (1999) Prostate apoptosis response-4 mediates trophic factor withdrawal-induced apoptosis of hippocampal neurons: actions prior to mitochondrial dysfunction and caspase activation. J Neurochem 73(2):502–512
115. Feng Z, Qin C, Chang Y, Zhang JT (2006) Early melatonin supplementation alleviates oxidative stress in a transgenic mouse model of Alzheimer's disease. Free Radic Biol Med 40(1):101–109

116. Xie J, Guo Q (2005) PAR-4 is involved in regulation of beta-secretase cleavage of the Alzheimer amyloid precursor protein. J Biol Chem 280(14):13824–13832
117. Duan W, Zhang Z, Gash DM, Mattson MP (1999) Participation of prostate apoptosis response-4 in degeneration of dopaminergic neurons in models of Parkinson's disease. Ann Neurol 46(4):587–597
118. Moos T, Jensen PH (2004) Absence of prostate apoptosis response-4 protein in substantia nigra of Parkinson's disease autopsies. Acta Neuropathol 107(1):23–26
119. Pedersen WA, Luo H, Kruman I, Kasarskis E, Mattson MP (2000) The prostate apoptosis response-4 protein participates in motor neuron degeneration in amyotrophic lateral sclerosis. FASEB J 14(7):913–924
120. Xie J, Awad KS, Guo Q (2005) RNAi knockdown of Par-4 inhibits neurosynaptic degeneration in ALS-linked mice. J Neurochem 92(1):59–71.126
121. Du WW, Li X, Li T, Li H, Khorshidi A, Liu F, Yang BB (2015) The microRNA miR-17-3p inhibits mouse cardiac fibroblast senescence by targeting Par4. J Cell Sci 128(2):293
122. Chai Q, Zheng M, Wang L, Wei M, Yin Y, Ma F, Li X, Zhang H, Liu G (2020) Circ_0068655 promotes cardiomyocyte apoptosis via miR-498/PAWR axis. Tissue Eng Regener Med 17(5):659–670
123. Szeto DP, Rodriguez-Esteban C, Ryan AK, O'Connell SM, Liu F, Kioussi C, Gleiberman AS, Izpisúa-Belmonte JC, Rosenfeld MG (1999) Role of the Bicoid-related homeodomain factor Pitx1 in specifying hindlimb morphogenesis and pituitary development. Genes Dev 13(4):484–494
124. Marcil A, Dumontier É, Chamberland M, Camper SA, Drouin J (2003) Pitx1 & Pitx2 are required for development of hindlimb buds. Development 130(1):45
125. Semina EV, Reiter R, Leysens NJ, Alward WLM, Small KW, Datson NA, Siegel-Bartelt J, Bierke-Nelson D, Bitoun P, Zabel BU, Carey JC, Murray JC (1996) Cloning and characterization of a novel bicoid-related homeobox transcription factor gene, RIEG, involved in Rieger syndrome. Nat Genet 14(4):392–399
126. Gage PJ, Camper SA (1997) Pituitary homeobox 2, a novel member of the bicoid-related family of homeobox genes, is a potential regulator of anterior structure formation. Hum Mol Genet 6(3):457–464
127. Acharya M, Lingenfelter DJ, Huang L, Gage PJ, Walter MA (2009) Human PRKC apoptosis WT1 regulator is a novel PITX2-interacting protein that regulates PITX2 transcriptional activity in ocular cells. J Biol Chem 284(50):34829–34838
128. Acharya M, Huang L, Fleisch VC, Allison WT, Walter MA (2011) A complex regulatory network of transcription factors critical for ocular development and disease. Hum Mol Genet 20(8):1610–1624
129. QiNan W, XiaGuang G, XiaoTian L, WuQuan D, Ling Z, Bing C (2016) Par-4/NF-κB mediates the apoptosis of islet β cells induced by glucolipotoxicity. J Diabetes Res 2016:4692478
130. Liu C, QiNan W, XiaoTian L, MengLiu Y, XiaGuang G, WeiLing L, ZiWen L, Ling Z, GangYi Y, Bing C (2018) TERT and Akt are involved in the Par-4-dependent apoptosis of islet *β* cells in Type 2 diabetes. J Diabetes Res 2018:7653904
131. Han JY, Lim YJ, Choi JA, Lee JH, Jo SH, Oh SM, Song CH (2016) The role of Prostate Apoptosis Response-4 (Par-4) in Mycobacterium tuberculosis infected macrophages. Sci Rep 6:32079
132. Lu C, Chen JQ, Zhou GP, Wu SH, Guan YF, Yuan CS (2008) Multimolecular complex of Par-4 and E2F1 binding to Smac promoter contributes to glutamate-induced apoptosis in human- bone mesenchymal stem cells. Nucleic Acids Res 36(15):5021–5032

Par-4 in Cell Cycle Regulation

Yanquan Zhang, Fengyi Mao, and Xiaoqi Liu

Abstract Cell cycle must be carefully controlled to ensure genetic materials are replicated once and only once per cycle and equal sister chromatid segregation into two daughter cells. While cyclin-dependent kinases are master regulators of cell cycle, p53 and NF-κB pathways are also critical for the regulation of the cell cycle and especially the targets modulated by Par-4. Par-4 can be both cytosolic and nuclear. Cytosolic Par-4 inhibits nuclear localization of NF-κB and decreases Bcl-2 expression, leading to G1 arrest. Mechanistically, Par-4 can directly interact with NF-κB or represses ζPKC activity thus inhibits NF-κB indirectly, eventually resulting in inhibition of DROSHA expression and regulation of miRNA. Nuclear Par-4 functions as co-transcriptional factor of THAP for CCAR1, which subsequently activates the p53/p21 axis, resulting in G1 arrest. While Par-4-associated inhibition of DNA topoisomerase activity results in S phase arrest, the distribution of Par-4 in mitotic cells determines the fate of daughter cells. PLK1 is a serine/threonine kinase involved in many cell cycle-related events, including Cdk1 activation, centrosome maturation, microtubule-kinetochore attachment, bipolar spindle formation, sister chromatid segregation, and cytokinesis. PLK1 also has well-established roles in DNA damage response, especially p53 regulation, via direct phosphorylation of multiple p53 regulators such as Topors, GTSE1, and Numb. Further, PLK1 is transcriptionally regulated by NF-κB. Based on all the research performed so far, we are proposing an intriguing working model in which NF-κB, PLK1, and Par-4 are well integrated in the context of acquisition of radioresistance of prostate cancer.

Keywords Cell cycle · NF-κB · p53 · p21 · Bcl-2 · DROSHA · THAP1 · CCAR1 · PLK1 · Par-4

Authors Yanquan Zhang and Fengyi Mao have equally contributed to this chapter.

Y. Zhang · F. Mao · X. Liu (✉)
Department of Toxicology and Cancer Biology, University of Kentucky, Lexington, KY, USA
e-mail: xiaoqi.liu@uky.edu

V. M. Rangnekar (ed.), *Tumor Suppressor Par-4*,
https://doi.org/10.1007/978-3-030-73572-2_2

1 Introduction

Par-4 gene was first identified as an immediate early apoptotic gene in prostate cancer cells. Par-4 protein functions as a pro-apoptotic tumor suppressor that induces apoptosis selectively in cancer cells. However, genes that regulate apoptosis are often involved in cell cycle progression. Accumulating evidence shows that manipulation of the cell cycle may prevent or induce an apoptotic response depending on the cellular context [1, 2]. Therefore, it is reasonable that Par-4 also participates in the regulation of cell cycle.

2 Cell Cycle

2.1 Cellular Events During Cell Cycle

During cell cycle, a series of events take place in a cell as it grows and divides. Interphase is the first phase of a cell cycle that occupies 60–80% of time of a cell cycle. In interphase, the cell grows, replicates its chromosomes, and prepares for cell division. The cell then enters the next phase, mitosis, and completes its division to generate two daughter cells. From here on, each daughter cell enters its own interphase and begins a new round of cell cycle. More specifically, the cell cycle is a four-stage process in which the cell increases in size (gap 1 or G1 phase), replicates its DNA (synthesis, or S phase), prepares to divide (gap 2 or G2 phase), and divides (mitosis or M phase). The stages G1, S, and G2 make up interphase, which accounts for the span between cell divisions. Mitosis can be further divided into prophase (indicated by centrosome maturation, chromosome condensation, and nuclear envelope breakdown), metaphase (evidenced by bipolar spindle formation and chromosome alignment at equator plane), anaphase (when sister chromatids segregation starts), telophase (when sister chromatids segregation completes), and cytokinesis. Based on the stimulatory and inhibitory messages a cell receives, it "decides" whether it should enter the cell cycle and divide.

2.2 The Regulation of the Cell Cycle

It is essential that the daughter cells produced be exact duplicates of the parent cell. To prevent a compromised cell from continuing to divide, there are tightly controlled mechanisms that operate at three main cell cycle checkpoints that occur near the end of G1, at the G2/M transition, and during metaphase of cell cycle. Among them, sequentially activation and inactivation of different CDKs/cyclin complexes are the major regulating machinery that controls the cell cycle progression of the four phases [3]. In addition, p53 and NF-κB pathways are also critical for the regulation of the cell cycle and especially the targets modulated by Par-4.

2.2.1 p53 Transcriptionally Activates p21 Expression and Arrests the Cell Cycle at G1 Stage

Tumor suppressor p53, whose various loss-of-function mutations have been identified in 50% of human cancers, is a transcription factor that is highly inducible by many stress signals such as DNA damage, oncogene activation, and nutrient deprivation. Cell-cycle arrest and the subsequent apoptosis are two most prominent outcomes of p53 activation [4]. Cell-cycle arrest by p53 activation is mainly mediated by the transcriptional activation of p21/WAF1. Briefly, p21 binds to cyclin E/Cdk2 and cyclin D/Cdk4 complexes to cause G1 arrest in the cell cycle and functions as a general Cdk repressor [5]. Interestingly, p21 also localizes at the cytoplasm and plays an oncogenic role for tumorigenesis [6], whereas Par-4 modulates its function and causes cell arrest at M phase.

2.2.2 NF-κB Activates the Transcription of BCL-2

The NF-κB (nuclear factor κB) transcription factor family is a pleiotropic regulator of many cellular pathways, providing a mechanism for the cell to respond to a wide variety of stimuli and environmental challenges. Therefore, it is not surprising that an important component of NF-κB's function includes regulation of the cell cycle. Upon stimulation, IKK is activated and phosphorylates IκBα and, thereby, triggers ubiquitination-dependent IκBα degradation via proteasome, resulting in rapid and transient nuclear translocation of canonical NF-κB members to activate downstream gene expression. Anti-apoptotic Bcl-2 is transcriptionally regulated by NF-κB [7] and acts mainly at the level of the G0/G1 phase transition of the cell cycle [8, 9], resulting in acceleration of cell proliferation. So far, accumulating evidence suggests that Par-4 inhibits NF-κB activity and represses the expression of Bcl-2, eventually resulting in a cell cycle arrest at G1 stage. We will review the role of Par-4 in the regulation of NF-κB in the following paragraph.

3 The Role of Par-4 Underlying Regulation of the Cell Cycle

Although Par-4 generally functions as a pro-apoptosis factor, amount of studies revealed its role in the regulation of cell cycle. Par-4 either represses the activity of NF-κB pathway and affects the level of BCL-2 or promotes p53 transcriptional activity and elevates the level of p21, consequently leading to cell cycle arrest at G1/S phase. Besides these, Par-4 is also involved in the regulation of S, M phase of cell cycle. In this part, we will review various functions of Par-4 according to its localization in the different phases of the cell cycle.

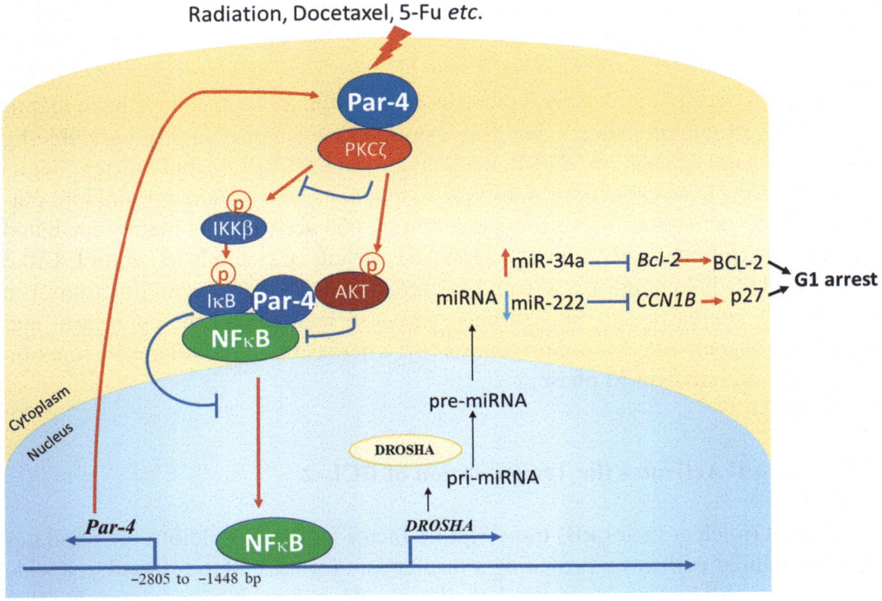

Fig. 1 Cytoplasmic Par-4 negatively regulates the NF-κB pathway through inhibiting NF-κB translocation into nucleus

3.1 The Role of Par-4 in G1 and S Phases

3.1.1 Cytosolic Par-4 Inhibits Nuclear Localization of NF-κB and Decreases Bcl-2 Expression, Leading to G1 Arrest

In normal cells and tissues, Par-4 localizes in the cytoplasm [10, 11]. Classically, cytoplasmic Par-4 regulates apoptosis through activating both the extrinsic [12] and intrinsic pathways of apoptosis [13]. Meanwhile, cytoplasmic localization of Par-4 also causes cell cycle arrest at G1 phase via directly or indirectly regulating NF-κB pathway and restricting BCL-2 expression (Fig. 1).

Par-4 Directly Interacts with NF-κB and Blocks Its Nuclear Localization

The translocation of NF-κB from cytoplasm to nucleus is the critical step of the activation of the NK-κB pathway [14]. Wang et al. [15] found that Par-4 binds to NF-κB in the cytoplasm but not in the nucleus and such binding facilitates apoptosis in the presence of 5-fluorouracil (5-FU). Mechanistically, the immunofluorescence of co-localization and co-immunoprecipitation results all suggested that Par-4 interacts with NF-κB subunit p65 and IκB in the cytosol and restricts NF-κB translocation and subsequent target gene regulation in the nucleus. In addition, one study revealed that pro-apoptotic activity of Par-4 is reduced when it is bound and phosphorylated by AKT1 in prostate cancer cells [16]. In agreement with that, these authors also discovered that AKT knockdown or inhibition enhanced the interaction of Par-4 and p65 in the cytoplasm and consequently blocks nuclear translocation of p65. Taken all together, Par-4 downregulates NF-κB transcriptional activity by directly interacting with p65 and inhibiting its nuclear localization, and this process can be negatively regulated by AKT.

Par-4 Represses ζPKC Activity to Indirectly Inhibit NF-κB Pathway

Using yeast two-hybrid assays, Dıaz-Meco et al. [17] found that Par-4 is an interacting protein of the regulatory domain of ζPKC (amino acids 1–250) and dramatically inhibits it enzymatic activity. Meanwhile, it is also reported that the atypical PKC including ζPKC binds to IKKs in vitro and in vivo and overexpression of ζPKC positively modulates IKKβ activity, but not that of IKKα in TNFα-stimulated cells [18]. Collectively, these studies demonstrate that Par-4 negatively regulates NF-κB pathway via inhibiting the activity of PKC to activate IKK.

Meanwhile, another study discovered that Par-4 deficiency in the lung leads to both enhanced NF-κB and increased AKT activity [19]. The authors demonstrated that Par-4 is highly expressed in normal lung but reduced in human lung cancer samples. The lack of Par-4 dramatically enhanced oncogenic Kras-induced lung carcinoma formation in vivo and was correlated with increased AKT phosphorylation at Ser473 in several cell systems including human lung cancer cells. Further evidence indicated that PKC directly phosphorylates AKT at Ser124 and subsequently affects phosphorylation of AKT at Ser473 and Thr308, two very critical events for the regulation of AKT activity and function [20]. Collectively, these results support a working model that Par-4 negatively regulates ζPKC kinase activity that activates AKT via direct phosphorylation of Ser124.

Par-4 Inhibits p65-Dependent DROSHA Expression and Regulates miRNA

Since Par-4 directly or indirectly inhibits NF-κB/p65 translocation, it is reasonable to speculate that Par-4 impacts the expression of the downstream genes of NF-κB. In an attempt to gain insight into the mechanism of Par-4-mediated susceptibility to apoptosis, Wang et al. [15] performed a microarray analysis to examine the difference of gene expression between control and Par-4-overexpressing HT29 cells. In agreement with Par-4's function during the cell cycle, 127 genes related to cell cycle were significantly and differentially expressed. Of note, a few genes associated with the microRNA pathway (e.g., *DROSHA*) were also affected by Par-4 overexpression. Further, these authors found that the expression level of *DROSHA* was downregulated and that NF-κB occupancy on the promoter of *DROSHA* was significantly decreased upon Par-4 overexpression in HT29 cells. DROSHA, a core nuclease that initiates miRNA processing in the nucleus, generates precursor molecules by cleaving primary miRNA transcripts that are subsequently exported into the cytoplasm for further processing [21]. Therefore, the downregulated DROSHA by Par-4 causes 13 microRNAs upregulated (e.g., miR-34a) and 9 microRNAs downregulated (e.g., miR-222). Consequently, as illustrated in Fig. 1, upregulated miR34a targets and inhibits BCL2 expression while downregulated miR-222 results in stabilization of *CCN1B* mRNA and elevation of its product p27, overall leading to G1 arrest.

Par-4 Is a Direct Target of NF-κB

Feedback regulation is a common mechanism in cells [22]. As discussed above, Par-4 directly and indirectly modulates NF-κB pathway and expression of its downstream targets. Reciprocally, it was reported that *Par-4* is also a direct target of NF-κB [23]. In normal endometrium, upregulated expression of Par-4 was detected in epithelial cells through the menstrual cycle and positively correlated with the phospho-p65 (pp65) level. On the contrary, most ECs (endometrial carcinomas) exhibited significant downregulation of Par-4 as compared to normal endometrium, with a positive correlation to the level of pp65. In EC cell lines, ectopic expression of p65 led to transactivation of Par-4 through specific binding to its promoter region. The reporter assay with a series of the 5′-truncated promoters of *Par-4* revealed that p65-responsive elements located from −2805 to −1448 nt. The ChIP assay revealed that p65 directly bound to the region that contains three κBBE (NF-κB consensus binding elements) sites. In addition, overexpression of Par-4 led to induction of not only apoptosis but also senescence, through changes in the expression of Bcl-2 and $p21^{WAF1}$, respectively. Taken all together, all the findings suggest that a signaling cascade involving sequential activation of NF-κB/p65 and Par-4 may participate in relatively early events of endometrial tumorigenesis, leading to modulation of cell kinetics including apoptosis and cell cycle progression.

Collectively, we may draw a whole picture that Par-4 inhibits the kinase activity of ζPKC and consequently represses the phosphorylation of IKKβ and AKT, respectively. On one hand, repressed IKK by Par-4 fails to phosphorylate IκB and causes its constitutive binding to NF-κB, eventually resulting in inhibition of the translocation of NF-κB into nucleus. On the other hand, unphosphorylated AKT due to the Par-4 and ζPKC interaction enhances the binding of Par-4 to NF-κB and further blocks its nuclear translocation (Fig. 1).

3.1.2 Nuclear Par-4 Functions as Co-transcriptional Factor of THPA for CCAR1 on the p53/p21 Axis

Par-4 contains two nuclear localization sequences (NLS1 and NLS2) at the N-terminal region that are high conserved across rat, mouse, and human species. At the same time, Par-4 exhibits a leucine zipper domain (LZ) with five leucine repeats

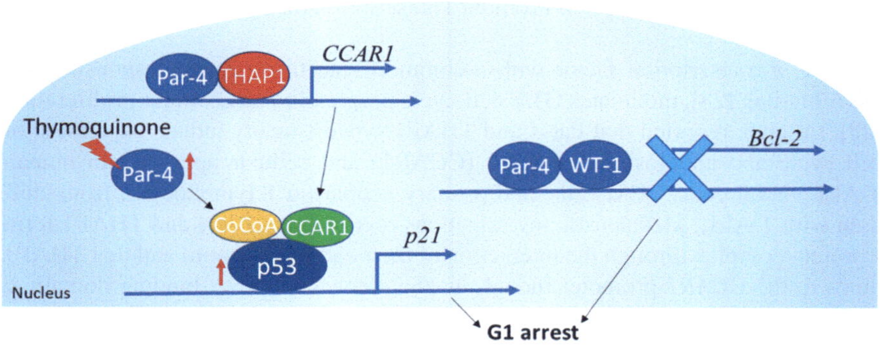

Fig. 2 Nuclear Par-4 regulates the transcriptional activity of p53, THAP1, and WT-1

at its carboxyl terminus spanning amino acids 290–332, of which the primary function is to allow protein–protein and protein–DNA interactions [24, 25]. El-Guendy et al. unveiled that NLS2 is essential for the translocation of Par-4 from the cytoplasm to the nucleus [26], suggesting that Par-4 must have a functional role in the nucleus (Fig. 2).

Par-4 Elevates p53 Level

Thymoquinone (TQ), a major bioactive ingredient of black cumin (*Nigella sativa* Linn.), suppresses growth without showing features of apoptosis, autophagy, necroptosis, and ferroptosis. It also induces cellular senescence associated with G1 phase cell-cycle arrest of glioma cells [27]. Western blot results showed that TQ treatment caused a marked decrease in the expression of cyclin E and cyclin-dependent kinase-2 (CDK-2) while persistent activation or overexpression of tumor suppressor proteins such as p53, p21 in a dose-dependent manner. Further study revealed that TQ treatment resulted in a significant dose- and time-dependent upregulation of Par-4 in U87MG cells. Most importantly, TQ-induced Par-4 expression triggers cellular senescence, as evidenced by increasing cellular size, β-galactosidase staining, G0/1 phase arrest, increased expression of senescence markers such as p53, p21, and Rb, and decreased expression of lamin B1, cyclin E, and CDK-2. Furthermore, siRNA-/shRNA-mediated knockdown of Par-4 reverses the TQ-induced effects. Collectively, we may propose that TQ-induced p53 upregulation is Par-4-dependent, which plays a critical role in cellular senescence in human malignant glioma cells.

Par-4 Interacts with THAP1 to Promote Transcription of CCAR1

THAP1, a transcription factor with a common zinc finger motif responsible for DNA binding [28], modulates G1/S cell-cycle progression and cellular proliferation [29]. Lu et al. reported that Par-4 and THAP1 synergistically induce expression of cell cycle and apoptosis regulator 1 (CCAR1) and cellular apoptosis in human T-ALL Jurkat cells, CEM cells, and primary neoplastic T lymphocytes from children with T-ALL. Mechanistic investigations revealed that Par-4 and THAP1 form a protein complex through the interaction of their carboxyl termini and that THAP1 binds to the *CCAR1* promoter though its zinc-dependent DNA-binding domain at amino terminus. Consequently, Par-4 and THAP1 collaborate to activate the promoter of *CCAR1* gene and promote its expression.

Another study reported that CCAR1 interacts with CoCoA (coiled-coil coactivator) and functions as a NR (nuclear receptor) coactivator [30]. These authors also found that purified CCAR1 and CoCoA protein bound specifically to p53 synthesized in vitro. Overexpression of either CCAR1 or CoCoA enhanced the expression of the reporter controlled by p53 binding sites in a p53-dependent manner in p53 null SAOS2 osteosarcoma cells. Moreover, ChIP assay results indicated that p53, CCAR1, and CoCoA were recruited to the region on the *CDKN1A/p21* promoter containing the p53 binding site in MCF-7 cells.

In this case, we may propose a working model that upon activation by TQ, Par-4 either elevates the p53 level or teams with THAP1 to promote the transcription of CCAR1 that subsequently functions as a coactivator of p53, eventually resulting in the upregulation of p21 and induction of cell arrest at G1 phase (Fig. 2).

3.1.3 Par-4 Interacts and Inhibits WT-1 on Bcl-2 Promoter

Besides the regulation of the transcriptional activity of p53, Par-4 also participates in the regulation of Bcl-2 at transcription level [31]. Cheema et al. discovered that both mRNA and protein levels of Bcl-2 were decreased in LNln3 (androgen-insensitive) but not in LNCaP (androgen-sensitive) cells upon *All-trans-retinoic Acid* (L-ATRA) treatment. Downregulated Bcl-2 correlates with G1 arrest and apoptosis in androgen-independent prostate cancer cell lines, accompanied by the translocation of Par-4 into nucleus and an induction of WT1 expression. Ectopic expression of Par-4 down-modulates while blocking WT1 results in increasing of Bcl-2 protein and transcript. Finally, these authors unveiled that Par-4 and WT1 proteins bind to a WT1-binding site on the Bcl-2 promoter (−1460) both in vitro and in vivo, leading to a decrease in transcription from the *Bcl-2* promoter. All in one, we can conclude that nuclear Par-4 regulates Bcl-2 expression through repressing the transcriptional activity of WT-1 on the Bcl-2 promoter and causes cell cycle arrest at G1 phase.

3.2 The Role of Par-4 in S Phase: Par-4 Inhibits TOP1 Activity

It was reported that Par-4 interacts with DNA topoisomerase 1 (TOP1) and inhibits its activity on DNA relaxation, resulting in S-phase arrest [32]. TOP1 is an enzyme that relaxes both negative and positive supercoils in the DNA through creating a single-strand nick, which allows rotation of the nicked strand around the intact strand and re-ligating the cleaved DNA [33, 34]. The DNA relaxation function of TOP1 is essential for the fundamental biological processes of DNA replication, transcription, and recombination, and accordingly, mutations completely lacking TOP1 activity are lethal.

Goswami et al. [32] performed a GST pull-down assay using GST-Par-4 and identified TOP1 as a direct interacting protein. They further found that Par-4 interacts with TOP1 via its leucine zipper domain and sequesters TOP1 from the DNA but does not induce apoptosis. Since the DNA relaxation activity of TOP1 is essential for DNA replication and S-phase progression [35], functionally, they discovered that Par-4 overexpression impedes DNA relaxation, leading to S-phase arrest of BPH-1 cells, a human benign prostatic hyperplasia cell line, and inhibition of its transformation. Conversely, Par-4 knockdown enhances DNA relaxation and gene transcription activities and promotes cellular transformation in a TOP1-dependent manner. Therefore, in the interphase of cell cycle, Par-4 functions to prevent TOP1-dependent DNA damage induced by chemotherapeutic agents and UV irradiation, resulting in an S-phase arrest in normal cells.

3.3 The Role of Par-4 in M Phase

3.3.1 Par-4 Regulates p-p21 in a p53-Independent Manner and Causes Mitosis Arrest

Classically, p21(CIP1/WAF1) is one of the downstream genes of p53, and the p21 protein binds to and inhibits the activity of cyclin-CDK2, -CDK1, and -CDK4/6 complexes and consequently functions as a repressor of cell cycle transition from G1 to S phase [36]. To support, it has been reported that overexpression of p21 results in G1- [37] or S-phase arrest [38]. However, p21 was recently showed to have oncogenic activity via its nontraditional targets in the cytoplasm [6]. In this case, Rasool et al. [39] revealed that 3-AWA (ER stress activator) diminishes the p21 levels in HCT116 (colorectal cancer) and PC3 (prostate cancer) cells by averting the senescent phenotype to commence G2/M arrest. Since 3-AWA is also a powerful inducer of Par-4, mechanistically, these authors showed that the downregulation of p21 level occurs through ER stress/JNK/caspase-3 axis via activation/induction of Par-4 and inhibition of AKT. Taken all together, Par-4 contributes to G2/M arrest via inhibiting the oncogenic role of p21 in cytoplasm.

3.3.2 The Distribution of Par-4 in Mitotic Cells Determines the Fate of Daughter Cells

Once cytokinesis, the last step of the mitosis, completes, the progenitor cells usually undergo a symmetric cell division and consequently divided into two daughter cells that are genetically identical. However, in some cases, cells also undergo asymmetric division. As a result, the fate of the daughter cells is different due to the different cytoplasmic components [40]. Bieberich et al. found that Par-4, ceramide, and the intermediate filament protein nestin are asymmetrically distributed during cell division of ES neural progenitor (NP) cells [41]. Par-4 and ceramide were correlated with TUNEL staining for apoptosis. However, the other daughter cell that did not express Par-4 was nestin positive and was not apoptotic. Therefore, it is asymmetric distribution of Par-4 that regulates ceramide-induced apoptosis during the proliferation and differentiation of stem cells.

4 PLK1 Function in Mitosis

The polo-like kinase (PLK) was first discovered in *Drosophila* in 1988, and its essential role in regulating cell division has been well demonstrated since then [42, 43]. Currently, five members of the PLK family (PLK1-PLK5) have been investigated in human, which function as serine-threonine kinases to regulate multiple intracellular processes, such as DNA replication, mitosis, and DNA damage response [44, 45]. PLK1 is the best-studied family member, involving in DNA damage response and cell cycle progression [46, 47]. PLK1 harbors two structural domains, the polo-box domain (PBD), related to substrates binding and subcellular distribution, and the kinase domain (KD), which could be regulated by upstream activating kinases [48–50]. Because of its crucial role in mitosis, the expression and activity of PLK1 are precisely and tightly controlled throughout the whole cell cycle. It has been well established that PLK1 is low in interphase, gradually increases during S and G2 phase, and reaches a peak during mitosis, which is conserved with the cell cycle progression [51–54]. Notably, PLK1 overexpresses in cancerous tissues and associates with the survival of cancer patients in numerous cancer types, including lung cancer, breast cancer, liver cancer, colon cancer, stomach cancer, and melanoma [55]. The disease-free survival and overall survival period remarkably shorten in patients with higher PLK1 expression level, in comparison to those with lower expression of PLK1, suggesting PLK1 as a potential and promising target for cancer therapy [55]. In this review, we will briefly discuss how PLK1 is regulated in both mRNA and protein levels during the cell cycle. The biological roles of PLK1 in cell cycle progression and DNA damage will be introduced as well.

4.1 Regulating PLK1 in Cell Cycle

4.1.1 Transcriptional Regulation

It has been well established that the promoter region of Plk1, located 5′ to the translation start site, is sufficient and crucial for mouse and human Plk1 expression [54, 56, 57]. There are three positive regulatory elements found between nucleotides −35 and −95, which could significantly induce Plk1 expression during G2/M [57]. Based on the consensus sequence motifs, several transcription factors have been identified to regulate Plk1 during the cell cycle, containing TCF3, AP1, AP2, SP-1, NF-Y/CBF, NF-κB [56], E2F1 [58], and E2F3 as well [59]. Moreover, FKH-TFs (forkhead transcription factors), regulators of the mitotic program, could also bind directly to the Plk1 promoter in vivo and lead to its activation [60]. Most importantly, Forkhead Box M1 (FoxM1) could dramatically stimulate Plk1 in a G2-specific manner [61], thus contributing to a cell cycle-dependent regulation of Plk1. In agreement, the deletion of FoxM1 or Plk1 in cells will result in similar pleiotropic mitotic defects, exhibiting unstable chromosome and inaccurate chromosome segregation [61, 62].

Compared to the transcriptional activation, the repression of Plk1 in the cell cycle is better studied. A bipartite repressor element, termed as CDE/CHR (cell cycle-dependent element/cell cycle genes homology region), has been identified as a crucial factor in the Plk1 repression [57]. The CDE/CHR element is a central regulator in cell cycle progression and presents at the promoter region of other G2/M--related genes as well, including cyclin B1, cyclin A, CDK1, Aurora A, or Cdc25C [63, 64]. The mutations in the CDE/CHR element could dramatically diminish the cell cycle-specific regulation of the Plk1 transcription [57]. Several transcription factors have been demonstrated to mediate the activity of CDE/CHR element, such as CDF-1 (CDE-CHR binding factor-1) [65], the DREAM (DP, RB-like, E2F4, and MuvB) complex [66], and the MMB (Myb-MuvB) complex [66]. Besides, RB and its related pocket proteins p107/p130 have been reported to suppress the expression of Plk1 [67]. E2F4 cooperates with p107/p130 and recruits SWI/SNF complex to the Plk1 promoter region, thus leading to the histone deacetylation and gene repression [67]. Furthermore, due to the role of PLK1 in the DNA damage response, PLK1 is tightly regulated by the DNA damage repair signaling pathway. DNA damage activates cell cycle checkpoints and allows DNA repair before mitotic entry through ATM, ATR, and their downstream CHK1 and CHK2 kinase [68–70], leading to the inhibition of Plk1 activity via both transcriptional regulation and protein modification [71, 72]. BRCA1 and its downstream CHK1 have been proven as a suppressor of Plk1 expression after ionizing radiation in breast cancer [73]. Most importantly, p53 could suppress Plk1 expression either by directly binding to the Plk1 promoter [74] or through its downstream p21 [75]. In response to DNA damage, p53 could bind to the Plk1 promoter region and then recruit histone deacetylases to inhibit Plk1 expression [74]. In parallel, as a mediator of the p53 repressive function, p21 could promote the switch from the MMB complex to the DREAM complex on the CHR element of the Plk1 promoter to block its transcription [75, 76]. Moreover, p53 also inhibits the transcription activity of FOXM1 and E2F1, thereby further diminishing the transcription of PLK1 [77, 78].

Along with the development of epigenetics, it has been revealed that the modification of DNA and histones tightly correlates with cell cycle regulation [79]. According to the previous publications, the methylation status in PLK1's promoter region varies in response to the environmental stimulus in cancer cells [80, 81]. However, the knowledge of how PLK1 is modulated by the epigenetic network at the transcription level is still limited and requires further investigation.

4.1.2 Post-translational Regulation

Based on the biochemical studies, the PBD of PLK1 binds to its KD, thus inhibits its catalytic activity [82, 83]. Deletion or mutations of PBD could dramatically increase the activity of KD during the G2 phase, suggesting the autoinhibitory machinery in the regulation of PLK1 activity [82, 83]. Later, its autoinhibitory state has been further understood and confirmed through analyzing the crystal structure of zPlk1. The binding of PBD and KD could either reduce the flexibility of the hinge region or sequestrate the activation loop, leading to the inhibition of kinase activity [84].

In contrast, the phosphorylation of T210 or S137 by Aurora A/Bora and other activating kinases could interrupt the intermolecular interaction and fully activate PLK1 during mitosis [50, 84–86]. During mitosis, mass spectrometry analysis has identified T210 as a main post-translational modification of PLK1 [50, 87, 88], which has been considered as the first step to initiate PLK1's activity [89]. Compared to wild-type PLK1, the substitution of aspartate for T210 results in constitutive activation, while the replacement of alanine or valine reduces its activity significantly [88, 89]. It has been well established that Aurora A cooperates with Bora to control the G2/M transition and mitotic entry, as well as activate PLK1 in the G2 phase [85, 86]. Bora, as a known cofactor of Aurora A, binds directly to PLK1 and increases the accessibility of the activation loop on PLK1, thus facilitates Aurora A to phosphorylate T210 before mitotic entry [85, 86]. Although Aurora A and Bora accumulate in the late G2 phase, they degrade and decrease drastically by proteasomes in mitosis, leaving the question of how PLK1 remains activated during mitosis [90, 91]. A study has demonstrated that once PLK1 is fully activated, minimal amounts of Aurora A/Bora complex could maintain PLK1's activity until the end of mitosis [90]. Interestingly, during mitosis, the proteolysis of Bora requires phosphorylation by PLK1, which could enhance its interaction with SCF-β-TrCP and promote degradation [91]. Once the PLK1 phosphorylation site is mutated, Bora would be stabilized in the mitosis, leading to prolonged metaphase and delayed anaphase in cell division [91]. Apart from T210, phosphorylation at S137 is another important post-translational modification found in PLK1, regarded as the prerequisite for the appropriate mitotic progression. Unlike T210, the phosphorylation of S137 occurs in late mitosis [92]. Expression of S137D in cells results in improper activation of APC/C (anaphase-promoting complex) and eventually leads to mitotic catastrophe [92]. However, further investigation is indispensable to validate and understand the role of S137 in PLK1's function. Moreover, Pak1 (p21-Activated kinase1) has been reported to phosphorylate PLK1 at S49, which is essential for the establishment of a functional bipolar spindle and maintenance of proper spindle tension [93]. Besides, another study based on mass spectrometry has identified additional phosphorylation sites on the catalytic domain of PLK1 showing a robust induction during mitosis, while the regulator and function of which are still unclear [87]. Distinct from phosphorylation, the study of how phosphatases control PLK1 spatiotemporal activity is limited. MYPT1/PP1C and MYPT1/PP1β could bind to PLK1 to diminish its phosphorylation at T210, thus tightly controlling PLK's activity during mitosis as previously demonstrated [94, 95]. Besides, cyclin A/CDK1 could negatively modulate the activity of PLK1 in prometaphase by priming its binding with MYPT1, which destabilizes the KT-MT attachment and ensures faithful chromosome segregation [96]. Moreover, PP2A binding with different regulatory subunits could dephosphorylate PLK1 in distinct stages and exhibit divergent effects. PP2A/B56 regulates PLK1 to maintain the balance of kinetochore-microtubule attachments during mitosis [97], whereas PP2A/B55α associates with PLK1 to regulate checkpoint recovery after DNA damage [98].

Similar to phosphorylation, ubiquitination is a reversible protein modification to regulate the spatially and temporally activity of PLK1 [99]. During the G2/M transition, PLK1 would be ubiquitinated by the checkpoint protein Chfr and finally subject to degradation under the mitotic stress, therefore prevent Cdc25C from activation and delay the mitotic entry [100]. On the contrary, once cells proceed to normal G2/M transition, USP16 (ubiquitin-specific peptidase 16) would deubiquitinate PLK1 to ensure its recruitment and activity on kinetochores. Consequently, proper kinetochore-microtubule (KT-MT) attachment and accurate chromosome alignment could be established and guaranteed in the metaphase [101]. Interestingly, cullin3 (CUL3)-based E3 ligase antagonizes to USP16 and contributes to PLK1's dynamic cellular localization instead of degradation [102]. Once the chromosome is properly aligned, CUL3/KLHL22 ubiquitinates K492 at PBD of PLK1 and disassociates PLK1 from kinetochores, which satisfies the spindle assembly checkpoint (SAC) and promotes mitotic progression [102]. Most importantly, PLK1's expression and activity maintain high in the early mitosis; however, PLK1 has to go through proteolysis in anaphase to ensure successful cytokinesis and proper mitotic exit [103]. It is well established that APC/C, an E3 ligase of diverse mitotic proteins, also contributes to the proteolysis of PLK1 in anaphase [103, 104]. Cdc20 binds with APC/C to initiate mitotic exit by targeting mitotic proteins containing a destruction box (D box), such as securin [105, 106]. Hereafter, Cdc20 is degraded by itself and substituted by Cdh1, through which APC/C gains a broader specificity toward mitotic protein [104, 106]. PLK1 contains a conserved RxxL D-box-like motif that could be recognized and ubiquitinated by APC/C^{Cdh1}. Subsequently, PLK1 goes through proteasome degradation to achieve a proper mitotic exit eventually [103, 107]. Moreover, CHIP (C-terminal Hsc70-interacting protein) is also reported as an E3 ligase to modulate PLK1's homeostasis during mitosis in prostate cancer [108].

Recently, innovative post-translational modifications of PLK1, other than phosphorylation and ubiquitination, have also been revealed. After phosphorylation and activation by CDK1/cyclin B1, SUMO-conjugating enzyme UBC9 interacts with PLK1 to conjugates SUMOs (small ubiquitin-like modifiers) [109]. This SUMOylation of PLK1 enhances its nuclear translocation and prevents it from proteasome degradation [109]. Besides, SUMO-1 could regulate PLK1 in microtubules and the spindle pore, while SUMO-2/3 is related to PLK1's localization on kinetochores, as reported recently [110]. Moreover, monomethylation at K209 by methyltransferase G9a silences PLK1 through antagonizing T210 phosphorylation under DNA damage stress [111]. PLK1 with methyl-deficient K209A mutant results in defects in DNA damage repair and DNA replication, whereas cells with methyl-mimic K209M mutant PLK1 exhibit prolonged metaphase-to-anaphase transition due to the improper separation of the chromosome [111]. Furthermore, SET7/9 could suppress PLK1 kinase activity through methylation of PLK1 at K191 in early mitosis, promoting a dynamic attachment of KT-MT for error correction [112].

4.2 PLK1's Biological Function

4.2.1 PLK1 in Cell Cycle

PLK1 is a well-known cell cycle-regulating kinase and plays a vital role in regulating cell division (Fig. 3). Collective data have demonstrated that PLK1 is critical for the early development of mouse embryos, the complete depletion of which causes extensive mitotic aberrations and embryonic lethal [113, 114]. Preclinical studies have revealed that PLK1 participates in multiple cell cycle-related processes, involving mitotic entry, centrosome maturation, spindle assembly, chromosome aggregation, and cytokinesis [115].

PLK1 with CDK1/Cyclin B1

It is well established that the activity of the CDK1/cyclin B1 complex is crucial for the G2/M transition and is precisely mediated by a protein network to ensure the appropriate commitment to a mitotic state [116]. Intriguingly, CDK1-mediated phosphorylation can not only directly regulate the stability and activity of its substrate but may also provide a docking site of PBD to prime the further phosphorylation by PLK1 [117, 118]. At the onset of mitosis, Wee1, the CDK1 inhibitory kinase, could be phosphorylated by CDK1 and PLK1 at S123 and S53 separately to generate a phosphor-degron [119]. As a consequence, Wee 1 is recognized and ubiquitinated by β-TrCP to degradation, leading to the rapid activation of CDK1 [119]. Similarly, the Myt1-mediated inhibition toward CDK1 would diminish once dual-phosphorylated by CDK1 and PLK1 [120, 121]. In parallel, PLK1-dependent phosphorylation of Cdc25c enhances its translocation to the nucleus, thus

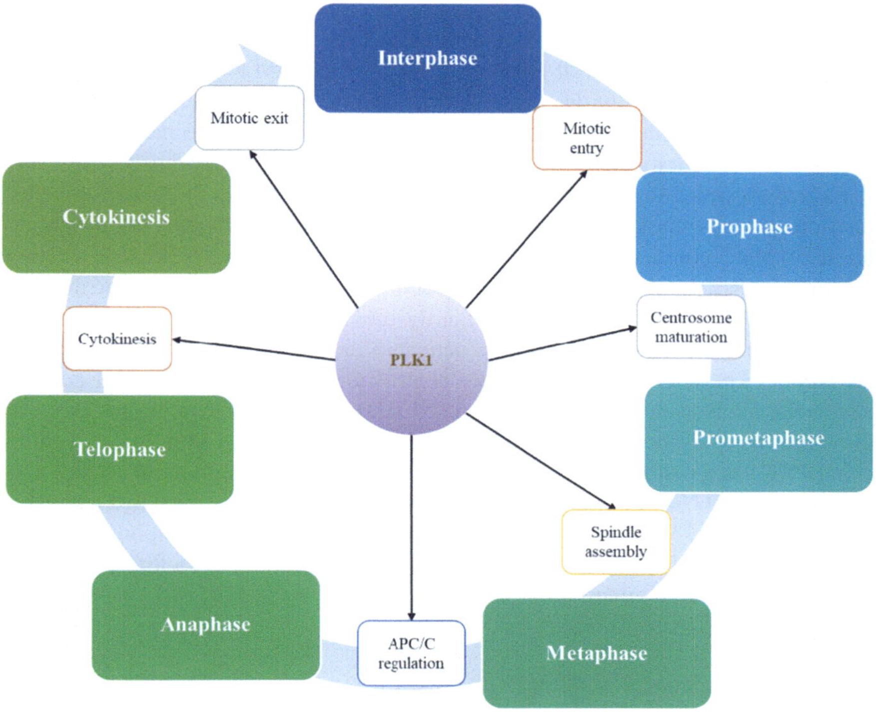

Fig. 3 PLK1 plays an important role in cell cycle

dephosphorylating and activating CDK1 during the prophase [122]. Besides, PLK1 phosphorylates cyclin B1 during prophase to stimulate its rapid nuclear entry [123, 124]. Moreover, FOXM1, functions as a master transcriptional factor of multiple mitotic genes including cyclin B1 and PLK1 itself, is also demonstrated as a substrate of PLK1, which would dramatically enhance its transcriptional activity and promote mitotic progression [125].

PLK1 at Centrosomes

In the G2 phase, PLK1 is recruited by Gravin to localize at the centrosomes, a prerequisite for the formation of mature centrosomes [126, 127]. Henceforward, the pericentriolar matrix (PCM) of centrosomes undergoes a dramatic expansion to organize spindle microtubules in advance of the formation of a bipolar spindle [128]. Ninein-like protein (NLP) plays a crucial role in microtubule organization and interacts with γ-tubulin ring complexes (γ-TuRCs) to stimulate microtubule nucleation [129]. Once phosphorylated by PLK1, NLP fails to associate with either centrosomes or γ-tubulin, thus grants the proper establishment of the mitotic spindle

[129]. In parallel, PLK1 phosphorylates pericentrin (PCNT) to initiate the centrosomes maturation, enhance recruitment of centrosomal proteins, and ensure fidelity of centriole separation as well [130, 131]. Recently, Sas-4 is identified as a novel PLK1 substrate in *Drosophila*, related to the expansion of PCM [132]. At the onset of mitosis, the phosphorylation of PLK1 is essential for Sas-4 to recruit γ-tubulin and Cnn and enable its localization to expand outward from the centrosome [132].

Except for centrosome maturation, PLK1 also participates in the centrosome disjunction and separation during late G2/prophase [47]. PLK1-dependent phosphorylation of Mst2 could enhance the activity of its downstream Nek2A to remove C-Nap1 and Rootletin from centrioles, consequently activating centrosome disjunction [133, 134]. In addition, Kinesin family member 11 (KIF11), a key factor to drive apart duplicated centrosomes, is activated indirectly by PLK1 through Nek9 and Nek6/7 kinase cascade [135]. Once centrosomes are separated, PLK1 contributes to the nuclear envelope breakdown (NEBD) via phosphorylating $p150^{Glued}$, to allow the interplay between chromosome and spindle [136]. Furthermore, centrosomal PLK1 is a mediator of switching dynein/dynactin through regulating dynactin members and NuMA (nuclear mitotic apparatus protein 1) to establish a dictated mitotic spindle [137]. Last but not the least, the sequential phosphorylation of Aurora A and PLK1 stabilizes 53BP1 (p53-binding protein 1) to facilitate the maintenance of centrosome integrity [138].

PLK1 at Kinetochores

Except for centrosomes, PLK1 locates at kinetochores and plays a crucial role in the regulation of KT-MT attachment [139]. In the interphase and early mitosis, PLK1 binds with PBIB1 via PBD and subsequently could be recruited to kinetochores [140]. Intriguingly, PLK1 phosphorylates PBIB1 at T78 to induce its degradation at kinetochores, thereby enabling PLK1 to associate with other components [140]. To maintain localization and function in kinetochores, PLK1 interacts with several scaffold proteins after PBIB1 degradation, including INCENP (inner centromere protein), BUB1 (benzimidazole 1), BUBR1, and RSF1 [141–144]. BUBR1 could be phosphorylated by PLK1 during prometaphase, thus promoting the enrichment of PP2A/B56α phosphatase by BUBR1 at kinetochores [143, 145]. PP2A/B56α complex mediates the dephosphorylation of Aurora B to counter its activity on the outer kinetochores, leading to the stability of MT-KT attachment [145]. Interestingly, PLK1 could also phosphorylate BRCA2 and facilitate it to form a complex with BUBR1 and PP2A, to establish a stable KT-MT interaction and ensure the proper chromosome alignment [146, 147]. Besides, PLK1 could promote the establishment of proper KT-MT attachment by phosphorylating CLIP-170 [148]. The knockdown of CLIP-170 results in the defects of KT-MT attachment in human cells and leads to mitotic arrest [149]. CK2 (casein kinase 2)-induced phosphorylation of CLIP-170 is essential for its dynein-/dynactin-dependent kinetochore localization, the process of which is facilitated and enhanced by PLK1 [148]. Moreover, Tex14 (testis expressed protein 14) is recruited to kinetochores by PLK1 to regulate the KT-MT attachment during early mitosis [150]. Hereafter, the PLK1-mediated phosphorylation of Tex14 also results in its ubiquitination by APC/C and subsequent degradation, to guarantee the proper transition from metaphase to anaphase [150]. Also, PLK1 phosphorylates MDC1 to promote prometaphase-metaphase transition and maintain genomic stability, independent of MDC1's role in DNA damage repair [151]. Furthermore, PLK1 positively regulates Sgt1 and enhances its association with MIS12 complex at kinetochores, which promotes the recruitment of NDC80 and promises efficient microtubule-binding sites [152]. Above all, both unstable KT-MT attachment and centromere disintegration are demonstrated as the mechanisms of chromosome alignment defects induced by PLK1 inhibition, suggesting PLK1's essential function in both centrosome and kinetochores during mitosis [153].

PLK1 in Spindle Formation and Chromosome Segregation

For accurate chromosome segregation, the formation of KT-MT attachment is tightly controlled and under the surveillance of the spindle assembly checkpoint [154]. Monopolar spindle 1 (MPS1) is the principal organizer of the SAC signaling cascade, which could bind to the unattached kinetochores and recruit SAC-related proteins, thus preventing the activation of APC/C and exit of mitosis [154]. It has been well established that PLK1 cooperates with MPS1 to initiate, maintain, and regulate the activity of SAC [155–158]. PLK1 directly phosphorylates MPS1 to enhance its catalytic activity and initiate SAC signaling [158]. Meanwhile, PLK1-phosphorylated BUB1 and KNL1 facilitate the recruitment of other SAC components to kinetochores [156–158]. Besides, PLK1 is necessary for the localization and activity of Aurora B at kinetochores during SAC [155] but also contributes to its removal after KT-MT formation as previously described [145].

Once SAC is satisfied, mitotic checkpoint complex [159], the inhibitory machinery of APC/C, is disassembled due to PLK1's negative regulation of $p31^{comet}$ [160]. Significantly, PLK1 initiates the E3 ligase activity of APC/C through dual phosphorylation along with CDK1/cyclin B1, leading to the degradation of cyclins and other mitotic regulators [161]. In parallel, PLK1 controls the destruction of APC/C inhibitor EMI1 (early mitotic inhibitor 1) as well, inducing rapid activation of APC/C and entry to anaphase [162]. In addition, sister chromatids experience segregation and have to be distributed equally to different poles in anaphase, the process of which is also mediated by PLK1 [163]. Cohesin functions to hold two sister chromatids together since the S phase, the deactivation of which is the fundamental basis for chromatids separation [163]. In the prophase and prometaphase, PLK1 phosphorylates the subunits of cohesin, subsequently disassociates cohesin from chromosome arms, and enhances its cleavability, contributing to the proper separation of chromosomes [163, 164].

PLK1 in Cytokinesis

After chromosome segregation, cells must undergo cleavage under the guidance of the midzone to achieve a successful cell division [165]. During anaphase, PLK1 could translocate from centrosomes and kinetochores to the mitotic midzone with the facilitation of PRC1 (protein regulator of cytokinesis 1) [166]. CDK1 involves in the PRC1 regulation and acts as an inhibitor toward the interaction between PRC1 and PLK1 before anaphase, as previously reported [166]. However, another study proposes that PLK1 is the dominant regulator of PRC1's activity in both pre-anaphase and post-anaphase, instead of CDK1 [165]. The phosphorylation of PLK1 could inhibit PRC1's activity in organizing the midzone complex, preventing the premature of the midzone for cells in metaphase [165]. Another well-studied target of PLK1 during cytokinesis is RhoA signaling. In late mitosis, RhoGEF ECT2 is recruited and activated by HsCyk-4 to the central spindle, thus facilitating the targeting and activation of RhoA GTPase, which is crucial for the maintenance of cortical contractility and the specification of cleavage plane [167]. Significantly, PLK1 promotes the central spindle recruitment of ECT2 through phosphorylation, triggering the initiation of cytokinesis and the integration of cleavage furrow [167, 168]. As a consequence, the inhibition of PLK1 abolishes the localization of RhoA GTPase at the equatorial cortex, eventually fails cytokinesis [169–171]. Besides, PLK1 and RhoA function together to maximize the activity of ROCK2 kinase, thus promoting the contraction of actomyosin during cleavage [172]. In the meantime, PLK1 also associates with other proteins located at the central spindle, including MKLP2 (mitotic kinesin-like protein2), NUDC (nuclear distribution gene C), SVIL (supervillin), and CEP55 (centrosomal protein of 55 kDa), to ensure appropriate cytokinesis [173–176].

4.2.2 PLK1 in DNA Damage

Due to its crucial role in cell cycle regulation, it is not surprising that PLK1 is inactivated and subjected to degradation after DNA damage in G2 and mitosis [71, 177]. Once DNA damage repair is completed, PLK1 is also responsible for shutting down the DNA damage checkpoint and promoting mitotic progression [47].

As previously mentioned, Aurora A and Bora phosphorylate PLK1 at T210 to enhance its activity and promote mitotic entry [85]. Under DNA double-strand break in the G2 phase, Bora is phosphorylated at T501 by ATM/ATR directly and then ubiquitinated by E3-ligase SCF-β-TrCP and degraded [178]. As a consequence, the degradation of Bora finally leads to the inhibition of PLK1, which would be further enhanced by phosphatase PP2A/B55α [98, 178]. In parallel, G9A monomethylates PLK1 at Y209 to antagonize the T210 phosphorylation and contributes to the PLK1 inhibition as well. [111]. Meanwhile, in response to DNA damage in the G2 phase, Cdc14B translocates to the nucleoplasm and activates E3 ligase APC/C^{Cdh1} to degrade PLK1 and induce arrest [177]. Recently, a study has shown that ATM spreads all over the chromatin and prevents mitotic progression under DNA damage, but its inhibition toward PLK1 could counteract by chromatin-bound phosphatase Wip1 [179]. Wip1 antagonizes ATM's activity to reactive PLK1 and allows the cell cycle to restart even though ATM is still present at DNA lesions [179]. Besides, Claspin is stabilized and activated dependent on ATR/CHK1 activity, triggering the DNA damage checkpoint [180, 181]. Once PLK1's activity recuperates, PLK1 phosphorylates Claspin, thus promoting its degradation via E3-ligase SCF-β-TrCP, to recover from the DNA damage checkpoint and move forward to mitosis [181]. Moreover, PLK1 and CK2 sequentially phosphorylate Rad51 to enhance its recruitment to the damage site and promote DNA damage repair through HR (homologous recombination) [182]. Also, PLK1 phosphorylates Mre11, a component of MRN (Mre11/Rad50/Nbs1), to prevent the loading of the MRN complex to damaged DNA and subsequently inhibit DNA damage repair [183]. Furthermore, under DNA replication stress, PLK1 suppresses DNA damage checkpoint through its interaction with RAD9 [184]. Recently, it has also been indicated that PLK1 cooperates with CDK1/cyclin B1 in targeting CtIP to induce the error-prone microhomology-mediated end joining and recover from the G2/M checkpoint [185].

In addition to the DNA damage signaling cascade, PLK1 also involves in the regulation of p53, a key mediator of DNA damage checkpoint as indicated in Fig. 4 [186]. Upon DNA damage, p53 is activated and stabilized by ATM/ATR and CHK1/CHK2, either inducing the cell cycle arrest or leading to the cell apoptosis [187–189]. As previously discussed, p53 and its downstream p21 could inhibit the transcription of PLK1 [74–78, 80, 93], while reactivated PLK1 diminishes p53's activity to enter the mitosis after DNA damage repair. PLK1 co-localizes with and directly phosphorylates p53 in the nucleus, which abrogates its transcription activity, facilitates its nuclear export, and induces its degradation [190, 191]. Besides, Topors (topoisomerase I-binding protein), functions as ubiquitin and SUMO-1 E3 ligase to stabilize p53, is a substrate of PLK1, which would be inhibited and led to degradation by PLK1's phosphorylation [192, 193]. Moreover, PLK1 phosphorylates

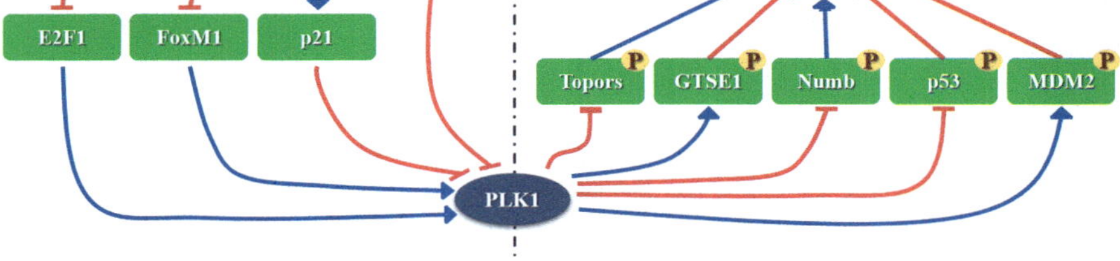

Fig. 4 Crosstalk between PLK1 and p53

GTSE1 (G2 and S-phase-expressed 1) to induce its nuclear translocation and activity, shuttling p53 out of the nucleus [194]. Furthermore, Numb is identified as a novel substrate of PLK1, the phosphorylation of which uncouple Numb from p53 and causes the proteasome degradation of p53 [195].

4.2.3 PLK1 Is Correlated with NF-κB Signaling

It has been reported that PLK1 could participate in multiple cancer-related pathways, especially involves in the regulation of the NF-κB signaling. A study published previously has demonstrated that NF-κB is a transcription factor of PLK1 in esophageal squamous cell carcinoma [196]. Through binding to the promoter region of PLK1, NF-κB could dramatically induce the expression of PLK1, which could protect cells from detachment or oxidative stress [196, 197]. In esophageal squamous cell carcinoma, RelA could modulate the transcriptional expression of PLK1 and β-catenin to protect cells from anoikis induced by the cell detachment [196]. In the prostate cancer model, oxidative stress induced by castration could also activate NF-κB [197]. Under this circumstance, the expression level of PLK1 is enhanced by NF-κB, which could promote the activity of PI3K and AR signaling pathways in a PLK1-dependent manner. Eventually, the high expression of PLK1 induced by NF-κB could lead to the occurrence of castration-resistant prostate cancer in the patients [197].

On the contrary, PLK1 has been well demonstrated as a negative regulator of the NF-κB signaling pathway [198]. Through phosphorylating the scaffold protein TANK/I-TRAF, PLK1 could prevent TNF receptor-induced NF-κB transactivation and suppress the DNA-binding activity of NF-κB as well [198]. Moreover, PLK1 could also modulate the activity of the IKK complex through its interaction with NEMO [198]. Upon the phosphorylation by PLK1, the ubiquitination of NEMO induced by TNF-α will be suppressed in a TANK-dependent manner [198]. In parallel, IKKβ, another component of the IKK complex, has been identified as a substrate of PLK1 as well, which could inhibit the TNF-induced IKK activation [199]. Furthermore, CYLD could remove ubiquitin chains from the IKK signaling components, leading to the inhibition of the NF-κB pathway. Importantly, CYLD is a substrate of PLK1, the activity of which could be regulated by PLK1 during mitosis [200]. To summarize, PLK1 could negatively regulate the activity of the NF-κB signaling via its phosphorylation to the IKK complex, scaffold protein TANK, and deubiquitinase CYLD.

5 Conclusion and Future Perspectives

PLK1 is the prerequisite in the regulation of multiple cell processes, the functions of which are vital and irreplaceable for cell proliferation and cancer progression. Recently, except for its classical role in cell cycle and DNA damage, plenty of

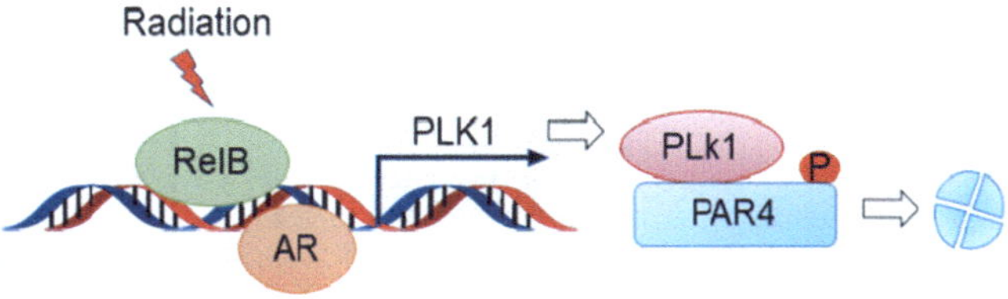

Fig. 5 A work model to be tested. We hypothesize that PLK1 is gradually elevated during long-term radiotherapy in a manner dependent on RelB. Elevated PLK1 binds to and phosphorylates Par-4 to drive its protein degradation and inhibition of cell death, eventually resulting in radioresistance of PCa

reports also reveal PLK1's role in cancer cell metabolism, metastasis, and progression, targeting at SREBP1, MTHFR, TSC1, G6PD, PTEN, CRAF, FBW7, etc. [201–207]. Consequently, PLK1 appears to be a promising and feasible target for cancer therapy, and the inhibitors aiming at PLK1 are under investigation and clinical trials [115, 208]. However, there is still a lot to learn about PLK1's function and regulation in the cell cycle, as well as cancer progression. For instance, epigenetic, immunology, and microenvironment affect gene expression drastically and control tumor development dominantly. PLK1's role in these cellular processes remains unclear and needs further investigation, which is essential for understanding PLK1's role in tumor progression and the application of PLK1 inhibitors. With the development of innovative tools and technologies, such as PLK1 mice models and FRET (fluorescence resonance energy transfer) biosensors [209], it is possible to further investigate and understand PLK1's interaction with other molecules within cells.

Based on all the literature reviewed above, we are proposing a very intriguing working model that might integrate NF-κB, androgen receptor (AR), PLK1, and Par-4 during radiation therapy of prostate cancer (PCa) (Fig. 5).

5.1 *Preliminary Data Available to Support the Working Model*

We first asked whether RelB regulates the transcription of PLK1. As indicated, we found that enforced expression of RelB dramatically enhanced the PLK1-luciferase activity (Fig. 6a) and that both RelB and AR bind to the PLK1 promoter (Fig. 6b). To support, analysis of PCa data set of TCGA reveals that RelB positively correlates to PLK1 expression (Fig. 6c). Interestingly, we found that depletion of RelB resulted in reduction of PLK1 expression in PC3 cells, but not in C4-2B cells (Fig. 6d), and that re-introduction of AR in PC3 cells antagonized RelB depletion-induced PLK1 inhibition (Fig. 6f), suggesting that AR is involved in RelB-regulated PLK1 expression. In agreement, AR inhibition with enzalutamide caused decrease of PLK1, AR, and RelB (Fig. 6e). In sum, we concluded that both RelB and AR act upstream of PLK1 in PCa (Fig. 5). We then asked whether PLK1 directly targets Par-4. As indicated, PLK1 interacts with (Fig. 7a) and phosphorylates (Fig. 7b) Par-4.

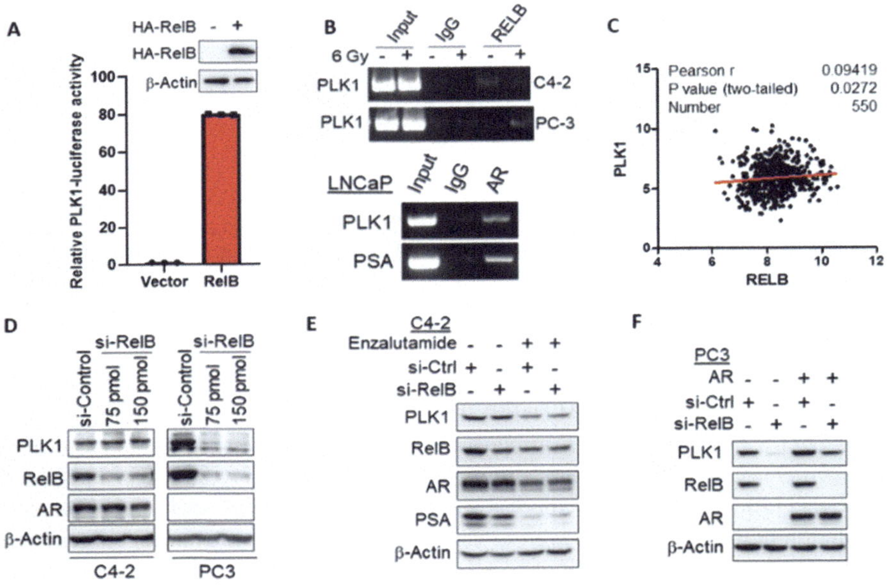

Fig. 6 RelB regulates PLK1. (**a**) Enforced expression of RelB enhanced PLK1-luciferase activity. PLK1-luciferase reporter was co-transfected with RelB for 48 h and harvested for analysis by a Promega Dual-luc activity kit. (**b**) RelB and AR bind to the promoter of PLK1. Lysates from LNCaP, C4-2B, and PC3 cells (irradiated or not) were immunoprecipitated by antibodies against RelB or AR, followed by PCR with specific pairs of primers of the PLk1 promoter. (**c**) RelB shows a positive correlation with PLK1 from TCGA PCa database. (**d**) Knocking down of RelB caused reduction of PLK1 in PC3 cells. C4-2B and PC-3 cells were transfected with siRNA against RelB for 48 h and harvested for immunoblotting (IB). (**e**) AR blockage resulted in reduction of both PLK1 and RelB. C4-2B cells were transfected with siRelB and treated with enzalutamide for 48 h. (**f**) Restoring of AR led to PLK1 stabilization in RelB-depleted PC3 cells. PC-3 cells (WT or AR expressing) were transfected with siRelB for 48 h

Overexpression of PLK1 WT and constitutively active T210D mutant, but not kinase-dead K82M mutant, resulted in Par-4 degradation (Fig. 7c). Further, MG132 reverses PLK1-induced degradation of Par-4, indicating that PLK1 phosphorylation of Par-4 results in proteasome-dependent degradation (Fig. 7d).

5.2 *Future Research Needed to Further Test the Working Model*

To investigate the mechanism underlying RelB-dependent regulation of PLK1 expression during acquisition of radioresistance, one has to compare the levels of RelB and PLK1 in isogenic PCa cells with different radioresistance, such as PC3 vs radioresistant PC3 cells and DU145 vs radioresistant DU145 cells. Of note,

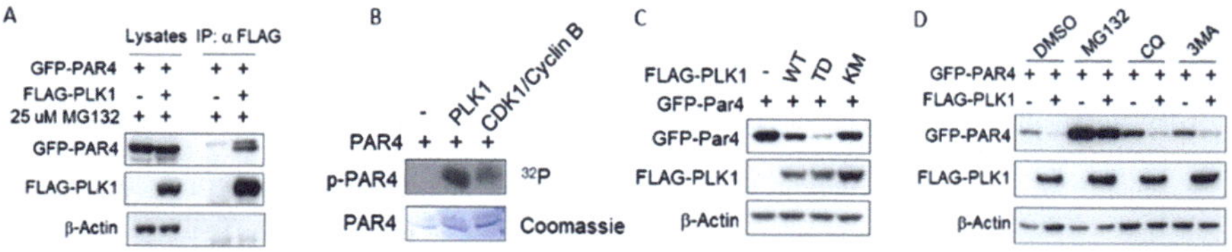

Fig. 7 Par-4 (labeled here as PAR4) is a PLK1 substrate. (**a**) Par-4 interacts with PLK1. GFP-Par-4 and Flag-PLK1 were co-transfected into HEK293T cells and harvested for anti-Flag IP. (**b**) PLK1 and Cdk1/CyclinB1 phosphorylate Par-4. Purified Par-4 was incubated with PLK1 or Cdk1/CyclinB1 complex in the presence of [γ- ^{32}P]ATP. (**c**) PLK1 promotes Par4 degradation. 293 T cells were co-transfected with GFP-Par-4 and Flag-PLK1 (WT, constitutively active T210D, kinase-dead K82M). (**d**) Proteasome inhibitor blocked PLK1-induced Par4 degradation. 293 T cells were co-transfected with GFP-Par-4 and Flag-PLK1 and treated with proteasome inhibitor MG132, lysosome inhibitor chloroquine, or autophagy inhibitor 3MA

radioresistant PC3 and DU145 cells were generated with increasing irradiation dosage over several months by the lab of Dr. Daret St. Clair at University of Kentucky and available to our lab now. Second, one has to determine whether PLK1 is decreased upon RelB depletion in radioresistant PC3 and DU145 cells. Third, whether overexpression of RelB and PLK1 drives radioresistance in otherwise radiosensitive PC3 and DU145 cells has to be determined. Fourth, one also needs to generate radioresistant LNCaP or C4-2B cells and ask whether AR is involved in RelB-dependent PLK1 transcription during acquisition of radioresistance. Finally, it is of high significance to ask whether the RelB/AR-dependent PLK1 transcription is a conserved mechanism for development of resistance to enzalutamide and docetaxel. In this regard, various isogenic PCa cells (LNCaP vs its enzalutamide-resistant derivative MR49F, C4-2B vs enzalutamide-resistant C4-2R, 22Rv1 vs 22Rv1-EnzR, PC3 vs docetaxel-resistant PC3-TR, DU145 vs docetaxel-resistant DU145-TR) should be used to rigorously test the working model.

The second future direction will be dissecting the functional significance of PLK1 phosphorylation of Par-4. Toward that end, efforts are needed to map the PLK1 phosphorylation sites by a combination of mass spectrometry and in vitro kinase assays with various S/T to A mutants. To firmly establish the kinase/substrate partnership between PLK1 and Par-4, one needs to generate phosphor-specific antibody, which can be further used to validate the phosphorylation events in vivo. To study the functional significance of the phosphorylation event, one could generate various Par-4 constructs (WT, S/T to A or S/T to D). While SA mutant is Plk1 unphosphorylatable, SD mutant is mimicking PLK1-induced phosphorylation. Cells stably expressing Par-4 (WT, SA, SD) can be subjected to radiation or enzalutamide, followed by phenotype analyses, including cell proliferation, anchorage-independent growth, apoptosis, and tumor formation. Further, PCa cells expressing Par-4 (WT, SA, SD) can be inoculated into nude mice to form tumors, followed by treatment with radiation or enzalutamide. Finally, one can test if a combination of Plk1 inhibitor and Par-4 activator enhances the efficacy of radiotherapy in PCa patient-derived xenograft tumors.

References

1. Evan GI, Brown L, Whyte M, Harrington E (1995) Apoptosis and the cell cycle. Curr Opin Cell Biol 7(6):825–834
2. Pucci B, Kasten M, Giordano A (2000) Cell cycle and apoptosis. Neoplasia 2(4):291–299
3. Barnum KJ, O'Connell MJ (2014) Cell cycle regulation by checkpoints. Methods Mol Biol 1170:29–40
4. Chen J (2016) The cell-cycle arrest and apoptotic functions of p53 in tumor initiation and progression. Cold Spring Harb Perspect Med 6(3):a026104
5. Abbas T, Dutta A (2009) p21 in cancer: intricate networks and multiple activities. Nat Rev Cancer 9(6):400–414
6. Gorospe M, Cirielli C, Wang X, Seth P, Capogrossi MC, Holbrook NJ (1997) p21(Waf1/Cip1) protects against p53-mediated apoptosis of human melanoma cells. Oncogene 14(8):929–935
7. Catz SD, Johnson JL (2001) Transcriptional regulation of bcl-2 by nuclear factor kappa B and its significance in prostate cancer. Oncogene 20(50):7342–7351
8. Zinkel S, Gross A, Yang E (2006) BCL2 family in DNA damage and cell cycle control. Cell Death Differ 13(8):1351–1359
9. Bonnefoy-Berard N, Aouacheria A, Verschelde C, Quemeneur L, Marcais A, Marvel J (2004) Control of proliferation by Bcl-2 family members. Biochim Biophys Acta 1644(2–3):159–168
10. Boghaert ER, Sells SF, Walid AJ, Malone P, Williams NM, Weinstein MH et al (1997) Immunohistochemical analysis of the proapoptotic protein Par-4 in normal rat tissues. Cell Growth Differ 8(8):881–890
11. Guo Q, Fu W, Xie J, Luo H, Sells SF, Geddes JW et al (1998) Par-4 is a mediator of neuronal degeneration associated with the pathogenesis of Alzheimer disease. Nat Med 4(8):957–962
12. Chakraborty M, Qiu SG, Vasudevan KM, Rangnekar VM (2001) Par-4 drives trafficking and activation of Fas and Fasl to induce prostate cancer cell apoptosis and tumor regression. Cancer Res 61(19):7255–7263
13. Boehrer S, Chow KU, Beske F, Kukoc-Zivojnov N, Puccetti E, Ruthardt M et al (2002) In lymphatic cells par-4 sensitizes to apoptosis by down-regulating bcl-2 and promoting disruption of mitochondrial membrane potential and caspase activation. Cancer Res 62(6):1768–1775
14. Zhang Q, Lenardo MJ, Baltimore D (2017) 30 years of NF-kappaB: a blossoming of relevance to human pathobiology. Cell 168(1–2):37–57
15. Wang BD, Kline CL, Pastor DM, Olson TL, Frank B, Luu T et al (2010) Prostate apoptosis response protein 4 sensitizes human colon cancer cells to chemotherapeutic 5-FU through mediation of an NF kappaB and microRNA network. Mol Cancer 9:98
16. Goswami A, Burikhanov R, de Thonel A, Fujita N, Goswami M, Zhao Y et al (2005) Binding and phosphorylation of par-4 by akt is essential for cancer cell survival. Mol Cell 20(1):33–44
17. Diaz-Meco MT, Municio MM, Frutos S, Sanchez P, Lozano J, Sanz L et al (1996) The product of par-4, a gene induced during apoptosis, interacts selectively with the atypical isoforms of protein kinase C. Cell 86(5):777–786
18. Lallena MJ, Diaz-Meco MT, Bren G, Paya CV, Moscat J (1999) Activation of IkappaB kinase beta by protein kinase C isoforms. Mol Cell Biol 19(3):2180–2188
19. Joshi J, Fernandez-Marcos PJ, Galvez A, Amanchy R, Linares JF, Duran A et al (2008) Par-4 inhibits Akt and suppresses Ras-induced lung tumorigenesis. EMBO J 27(16):2181–2193
20. Cantley LC (2002) The phosphoinositide 3-kinase pathway. Science 296(5573):1655–1657
21. Lee Y, Ahn C, Han J, Choi H, Kim J, Yim J et al (2003) The nuclear RNase III Drosha initiates microRNA processing. Nature 425(6956):415–419
22. Harris SL, Levine AJ (2005) The p53 pathway: positive and negative feedback loops. Oncogene 24(17):2899–2908
23. Saegusa M, Hashimura M, Kuwata T, Okayasu I (2010) Transcriptional regulation of pro-apoptotic Par-4 by NF-kappaB/p65 and its function in controlling cell kinetics during early events in endometrial tumourigenesis. J Pathol 221(1):26–36

24. Hebbar N, Wang C, Rangnekar VM (2012) Mechanisms of apoptosis by the tumor suppressor Par-4. J Cell Physiol 227(12):3715–3721
25. El-Guendy N, Zhao Y, Gurumurthy S, Burikhanov R, Rangnekar VM (2003) Identification of a unique core domain of par-4 sufficient for selective apoptosis induction in cancer cells. Mol Cell Biol 23(16):5516–5525
26. El-Guendy N, Rangnekar VM (2003) Apoptosis by Par-4 in cancer and neurodegenerative diseases. Exp Cell Res 283(1):51–66
27. Subburayan K, Thayyullathil F, Pallichankandy S, Rahman A, Galadari S (2018) Par-4-dependent p53 up-regulation plays a critical role in thymoquinone-induced cellular senescence in human malignant glioma cells. Cancer Lett 426:80–97
28. Campagne S, Saurel O, Gervais V, Milon A (2010) Structural determinants of specific DNA-recognition by the THAP zinc finger. Nucleic Acids Res 38(10):3466–3476
29. Cayrol C, Lacroix C, Mathe C, Ecochard V, Ceribelli M, Loreau E et al (2007) The THAP-zinc finger protein THAP1 regulates endothelial cell proliferation through modulation of pRB/E2F cell-cycle target genes. Blood 109(2):584–594
30. Kim JH, Yang CK, Heo K, Roeder RG, An W, Stallcup MR (2008) CCAR1, a key regulator of mediator complex recruitment to nuclear receptor transcription complexes. Mol Cell 31(4):510–519
31. Cheema SK, Mishra SK, Rangnekar VM, Tari AM, Kumar R, Lopez-Berestein G (2003) Par-4 transcriptionally regulates Bcl-2 through a WT1-binding site on the bcl-2 promoter. J Biol Chem 278(22):19995–20005
32. Goswami A, Qiu S, Dexheimer TS, Ranganathan P, Burikhanov R, Pommier Y et al (2008) Par-4 binds to topoisomerase 1 and attenuates its DNA relaxation activity. Cancer Res 68(15):6190–6198
33. Champoux JJ (2001) DNA topoisomerases: structure, function, and mechanism. Annu Rev Biochem 70:369–413
34. Wang JC (2002) Cellular roles of DNA topoisomerases: a molecular perspective. Nat Rev Mol Cell Biol 3(6):430–440
35. Miao ZH, Player A, Shankavaram U, Wang YH, Zimonjic DB, Lorenzi PL et al (2007) Nonclassic functions of human topoisomerase I: genome-wide and pharmacologic analyses. Cancer Res 67(18):8752–8761
36. Gartel AL, Radhakrishnan SK (2005) Lost in transcription: p21 repression, mechanisms, and consequences. Cancer Res 65(10):3980–3985
37. el-Deiry WS, Harper JW, O'Connor PM, Velculescu VE, Canman CE, Jackman J et al (1994) WAF1/CIP1 is induced in p53-mediated G1 arrest and apoptosis. Cancer Res 54(5):1169–1174
38. Radhakrishnan SK, Feliciano CS, Najmabadi F, Haegebarth A, Kandel ES, Tyner AL et al (2004) Constitutive expression of E2F-1 leads to p21-dependent cell cycle arrest in S phase of the cell cycle. Oncogene 23(23):4173–4176
39. Rasool RU, Nayak D, Chakraborty S, Faheem MM, Rah B, Mahajan P et al (2017) AKT is indispensable for coordinating Par-4/JNK cross talk in p21 downmodulation during ER stress. Oncogenesis 6:e341
40. Knoblich JA (2010) Asymmetric cell division: recent developments and their implications for tumour biology. Nat Rev Mol Cell Biol 11(12):849–860
41. Bieberich E, MacKinnon S, Silva J, Noggle S, Condie BG (2003) Regulation of cell death in mitotic neural progenitor cells by asymmetric distribution of prostate apoptosis response 4 (PAR-4) and simultaneous elevation of endogenous ceramide. J Cell Biol 162(3):469–479
42. Sunkel CE, Glover DM (1988) Polo, a mitotic mutant of Drosophila displaying abnormal spindle poles. J Cell Sci 89(Pt 1):25–38
43. Llamazares S, Moreira A, Tavares A, Girdham C, Spruce BA, Gonzalez C et al (1991) Polo encodes a protein kinase homolog required for mitosis in Drosophila. Genes Dev 5(12A):2153–2165
44. van de Weerdt BC, Medema RH (2006) Polo-like kinases: a team in control of the division. Cell Cycle 5(8):853–864

45. Andrysik Z, Bernstein WZ, Deng L, Myer DL, Li YQ, Tischfield JA et al (2010) The novel mouse polo-like kinase 5 responds to DNA damage and localizes in the nucleolus. Nucleic Acids Res 38(9):2931–2943
46. Petronczki M, Lenart P, Peters JM (2008) Polo on the rise-from mitotic entry to cytokinesis with Plk1. Dev Cell 14(5):646–659
47. Bruinsma W, Raaijmakers JA, Medema RH (2012) Switching polo-like kinase-1 on and off in time and space. Trends Biochem Sci 37(12):534–542
48. van de Weerdt BC, Littler DR, Klompmaker R, Huseinovic A, Fish A, Perrakis A et al (2008) Polo-box domains confer target specificity to the polo-like kinase family. Biochim Biophys Acta 1783(6):1015–1022
49. Lee KS, Grenfell TZ, Yarm FR, Erikson RL (1998) Mutation of the polo-box disrupts localization and mitotic functions of the mammalian polo kinase Plk. Proc Natl Acad Sci U S A 95(16):9301–9306
50. Jang YJ, Ma S, Terada Y, Erikson RL (2002) Phosphorylation of threonine 210 and the role of serine 137 in the regulation of mammalian polo-like kinase. J Biol Chem 277(46):44115–44120
51. Fenton B, Glover DM (1993) A conserved mitotic kinase active at late anaphase—telophase in syncytial Drosophila embryos. Nature 363(6430):637–640
52. Lee KS, Yuan YL, Kuriyama R, Erikson RL (1995) Plk is an M-phase-specific protein kinase and interacts with a kinesin-like protein, CHO1/MKLP-1. Mol Cell Biol 15(12):7143–7151
53. Lake RJ, Jelinek WR (1993) Cell cycle- and terminal differentiation-associated regulation of the mouse mRNA encoding a conserved mitotic protein kinase. Mol Cell Biol 13(12):7793–7801
54. Martin BT, Strebhardt K (2006) Polo-like kinase 1: target and regulator of transcriptional control. Cell Cycle 5(24):2881–2885
55. Liu Z, Sun Q, Wang X (2017) PLK1, a potential target for cancer therapy. Transl Oncol 10(1):22–32
56. Bräuninger A, Strebhardt K, Rübsamen-Waigmann H (1995) Identification and functional characterization of the human and murine polo-like kinase (Plk) promoter. Oncogene 11(9):1793–1800
57. Uchiumi T, Longo DL, Ferris DK (1997) Cell cycle regulation of the human polo-like kinase (PLK) promoter. J Biol Chem 272(14):9166–9174
58. Tategu M, Nakagawa H, Sasaki K, Yamauchi R, Sekimachi S, Suita Y et al (2008) Transcriptional regulation of human polo-like kinases and early mitotic inhibitor. Journal of genetics and genomics =. Yi Chuan Xue Bao 35(4):215–224
59. Olsson AY, Feber A, Edwards S, Te Poele R, Giddings I, Merson S et al (2007) Role of E2F3 expression in modulating cellular proliferation rate in human bladder and prostate cancer cells. Oncogene 26(7):1028–1037
60. Alvarez B, Martinez AC, Burgering BM, Carrera AC (2001) Forkhead transcription factors contribute to execution of the mitotic programme in mammals. Nature 413(6857):744–747
61. Laoukili J, Kooistra MR, Bras A, Kauw J, Kerkhoven RM, Morrison A et al (2005) FoxM1 is required for execution of the mitotic programme and chromosome stability. Nat Cell Biol 7(2):126–136
62. Liu X, Erikson RL (2002) Activation of Cdc2/cyclin B and inhibition of centrosome amplification in cells depleted of Plk1 by siRNA. Proc Natl Acad Sci U S A 99(13):8672–8676
63. Zwicker J, Lucibello FC, Wolfraim LA, Gross C, Truss M, Engeland K et al (1995) Cell cycle regulation of the cyclin A, cdc25C and cdc2 genes is based on a common mechanism of transcriptional repression. EMBO J 14(18):4514–4522
64. Lucibello FC, Truss M, Zwicker J, Ehlert F, Beato M, Muller R (1995) Periodic cdc25C transcription is mediated by a novel cell cycle-regulated repressor element (CDE). EMBO J 14(1):132–142
65. Zwicker J, Lucibello FC, Jerome V, Brusselbach S, Muller R (1998) CDF-1 mediated repression of cell cycle genes targets a specific subset of transactivators. Nucleic Acids Res 26(4):4926–4932

66. Muller GA, Quaas M, Schumann M, Krause E, Padi M, Fischer M et al (2012) The CHR promoter element controls cell cycle-dependent gene transcription and binds the DREAM and MMB complexes. Nucleic Acids Res 40(4):1561–1578
67. Gunawardena RW, Siddiqui H, Solomon DA, Mayhew CN, Held J, Angus SP et al (2004) Hierarchical requirement of SWI/SNF in retinoblastoma tumor suppressor-mediated repression of Plk1. J Biol Chem 279(28):29278–29285
68. Bakkenist CJ, Kastan MB (2003) DNA damage activates ATM through intermolecular autophosphorylation and dimer dissociation. Nature 421(6922):499–506
69. Matsuoka S, Huang M, Elledge SJ (1998) Linkage of ATM to cell cycle regulation by the Chk2 protein kinase. Science 282(5395):1893–1897
70. Yarden RI, Pardo-Reoyo S, Sgagias M, Cowan KH, Brody LC (2002) BRCA1 regulates the G2/M checkpoint by activating Chk1 kinase upon DNA damage. Nat Genet 30(3):285–289
71. Smits VA, Klompmaker R, Arnaud L, Rijksen G, Nigg EA, Medema RH (2000) Polo-like kinase-1 is a target of the DNA damage checkpoint. Nat Cell Biol 2(9):672–676
72. van Vugt MA, Smits VA, Klompmaker R, Medema RH (2001) Inhibition of polo-like kinase-1 by DNA damage occurs in an ATM- or ATR-dependent fashion. J Biol Chem 276(45):41656–41660
73. Ree AH, Bratland A, Nome RV, Stokke T, Fodstad O (2003) Repression of mRNA for the PLK cell cycle gene after DNA damage requires BRCA1. Oncogene 22(55):8952–8955
74. McKenzie L, King S, Marcar L, Nicol S, Dias SS, Schumm K et al (2010) p53-dependent repression of polo-like kinase-1 (PLK1). Cell Cycle 9(20):4200–4212
75. Fischer M, Quaas M, Steiner L, Engeland K (2016) The p53-p21-DREAM-CDE/CHR pathway regulates G2/M cell cycle genes. Nucleic Acids Res 44(1):164–174
76. Quaas M, Muller GA, Engeland K (2012) p53 can repress transcription of cell cycle genes through a p21(WAF1/CIP1)-dependent switch from MMB to DREAM protein complex binding at CHR promoter elements. Cell Cycle 11(24):4661–4672
77. Zhou Z, Cao JX, Li SY, An GS, Ni JH, Jia HT (2013) p53 suppresses E2F1-dependent PLK1 expression upon DNA damage by forming p53-E2F1-DNA complex. Exp Cell Res 319(20):3104–3115
78. Pandit B, Halasi M, Gartel AL (2009) p53 negatively regulates expression of FoxM1. Cell Cycle 8(20):3425–3427
79. Li X (2020) Epigenetics and cell cycle regulation in cystogenesis. Cell Signal 68:109509
80. Ward A, Hudson JW (2014) p53-dependent and cell specific epigenetic regulation of the polo-like kinases under oxidative stress. PLoS One 9(1):e87918
81. Ward A, Sivakumar G, Kanjeekal S, Hamm C, Labute BC, Shum D et al (2015) The deregulated promoter methylation of the polo-like kinases as a potential biomarker in hematological malignancies. Leuk Lymphoma 56(7):2123–2133
82. Mundt KE, Golsteyn RM, Lane HA, Nigg EA (1997) On the regulation and function of human polo-like kinase 1 (PLK1): effects of overexpression on cell cycle progression. Biochem Biophys Res Commun 239(2):377–385
83. Jang YJ, Lin CY, Ma S, Erikson RL (2002) Functional studies on the role of the C-terminal domain of mammalian polo-like kinase. Proc Natl Acad Sci U S A 99(4):1984–1989
84. Xu J, Shen C, Wang T, Quan J (2013) Structural basis for the inhibition of polo-like kinase 1. Nat Struct Mol Biol 20(9):1047–1053
85. Seki A, Coppinger JA, Jang CY, Yates JR, Fang G (2008) Bora and the kinase Aurora a cooperatively activate the kinase Plk1 and control mitotic entry. Science 320(5883):1655–1658
86. Macurek L, Lindqvist A, Lim D, Lampson MA, Klompmaker R, Freire R et al (2008) Polo-like kinase-1 is activated by aurora A to promote checkpoint recovery. Nature 455(7209):119–123
87. Daub H, Olsen JV, Bairlein M, Gnad F, Oppermann FS, Korner R et al (2008) Kinase-selective enrichment enables quantitative phosphoproteomics of the kinome across the cell cycle. Mol Cell 31(3):438–448
88. Lee KS, Erikson RL (1997) Plk is a functional homolog of Saccharomyces cerevisiae Cdc5, and elevated Plk activity induces multiple septation structures. Mol Cell Biol 17(6):3408–3417

89. Paschal CR, Maciejowski J, Jallepalli PV (2012) A stringent requirement for Plk1 T210 phosphorylation during K-fiber assembly and chromosome congression. Chromosoma 121(6):565–572
90. Bruinsma W, Macurek L, Freire R, Lindqvist A, Medema RH (2014) Bora and Aurora-a continue to activate Plk1 in mitosis. J Cell Sci 127(Pt 4):801–811
91. Seki A, Coppinger JA, Du H, Jang CY, Yates JR 3rd, Fang G (2008) Plk1- and beta-TrCP-dependent degradation of Bora controls mitotic progression. J Cell Biol 181(1):65–78
92. van de Weerdt BC, van Vugt MA, Lindon C, Kauw JJ, Rozendaal MJ, Klompmaker R et al (2005) Uncoupling anaphase-promoting complex/cyclosome activity from spindle assembly checkpoint control by deregulating polo-like kinase 1. Mol Cell Biol 25(5):2031–2044
93. Maroto B, Ye MB, von Lohneysen K, Schnelzer A, Knaus UG (2008) P21-activated kinase is required for mitotic progression and regulates Plk1. Oncogene 27(36):4900–4908
94. Yamashiro S, Yamakita Y, Totsukawa G, Goto H, Kaibuchi K, Ito M et al (2008) Myosin phosphatase-targeting subunit 1 regulates mitosis by antagonizing polo-like kinase 1. Dev Cell 14(5):787–797
95. Kachaner D, Filipe J, Laplantine E, Bauch A, Bennett KL, Superti-Furga G et al (2012) Plk1-dependent phosphorylation of optineurin provides a negative feedback mechanism for mitotic progression. Mol Cell 45(4):553–566
96. Dumitru AMG, Rusin SF, Clark AEM, Kettenbach AN, Compton DA (2017) Cyclin a/Cdk1 modulates Plk1 activity in prometaphase to regulate kinetochore-microtubule attachment stability. elife 6
97. Foley EA, Maldonado M, Kapoor TM (2011) Formation of stable attachments between kinetochores and microtubules depends on the B56-PP2A phosphatase. Nat Cell Biol 13(10):1265–1271
98. Wang L, Guo Q, Fisher LA, Liu D, Peng A (2015) Regulation of polo-like kinase 1 by DNA damage and PP2A/B55alpha. Cell Cycle 14(1):157–166
99. Liu J, Zhang C (2017) The equilibrium of ubiquitination and deubiquitination at PLK1 regulates sister chromatid separation. Cell Mol Life Sci 74(12):2127–2134
100. Kang D, Chen J, Wong J, Fang G (2002) The checkpoint protein Chfr is a ligase that ubiquitinates Plk1 and inhibits Cdc2 at the G2 to M transition. J Cell Biol 156(2):249–259
101. Zhuo X, Guo X, Zhang X, Jing G, Wang Y, Chen Q et al (2015) Usp16 regulates kinetochore localization of Plk1 to promote proper chromosome alignment in mitosis. J Cell Biol 210(5):727–735
102. Beck J, Maerki S, Posch M, Metzger T, Persaud A, Scheel H et al (2013) Ubiquitylation-dependent localization of PLK1 in mitosis. Nat Cell Biol 15(4):430–439
103. Lindon C, Pines J (2004) Ordered proteolysis in anaphase inactivates Plk1 to contribute to proper mitotic exit in human cells. J Cell Biol 164(2):233–241
104. Wasch R, Engelbert D (2005) Anaphase-promoting complex-dependent proteolysis of cell cycle regulators and genomic instability of cancer cells. Oncogene 24(1):1–10
105. Morgan DO (1999) Regulation of the APC and the exit from mitosis. Nat Cell Biol 1(2):E47–E53
106. Eckerdt F, Strebhardt K (2006) Polo-like kinase 1: target and regulator of anaphase-promoting complex/cyclosome-dependent proteolysis. Cancer Res 66(14):6895–6898
107. Glotzer M, Murray AW, Kirschner MW (1991) Cyclin is degraded by the ubiquitin pathway. Nature 349(6305):132–138
108. Chen L, Li J, Farah E, Sarkar S, Ahmad N, Gupta S et al (2016) Cotargeting HSP90 and its client proteins for treatment of prostate Cancer. Mol Cancer Ther 15(9):2107–2118
109. Wen D, Wu J, Wang L, Fu Z (2017) SUMOylation promotes nuclear import and stabilization of polo-like kinase 1 to support its mitotic function. Cell Rep 21(8):2147–2159
110. Feitosa WB, Hwang K, Morris PL (2018) Temporal and SUMO-specific SUMOylation contribute to the dynamics of polo-like kinase 1 (PLK1) and spindle integrity during mouse oocyte meiosis. Dev Biol 434(2):278–291

111. Li W, Wang HY, Zhao X, Duan H, Cheng B, Liu Y et al (2019) A methylation-phosphorylation switch determines Plk1 kinase activity and function in DNA damage repair. Sci Adv 5(3):eaau7566
112. Yu R, Wu H, Ismail H, Du S, Cao J, Wang J et al (2020) Methylation of PLK1 by SET7/9 ensures accurate kinetochore-microtubule dynamics. J Mol Cell Biol 12(6):462–476
113. Wachowicz P, Fernandez-Miranda G, Marugan C, Escobar B, de Carcer G (2016) Genetic depletion of polo-like kinase 1 leads to embryonic lethality due to mitotic aberrancies. BioEssays 38(Suppl 1):S96–S106
114. Lu LY, Wood JL, Minter-Dykhouse K, Ye L, Saunders TL, Yu X et al (2008) Polo-like kinase 1 is essential for early embryonic development and tumor suppression. Mol Cell Biol 28(22):6870–6876
115. Liu X (2015) Targeting polo-like kinases: a promising therapeutic approach for cancer treatment. Transl Oncol 8(3):185–195
116. Lindqvist A, Rodriguez-Bravo V, Medema RH (2009) The decision to enter mitosis: feedback and redundancy in the mitotic entry network. J Cell Biol 185(2):193–202
117. Elia AE, Cantley LC, Yaffe MB (2003) Proteomic screen finds pSer/pThr-binding domain localizing Plk1 to mitotic substrates. Science 299(5610):1228–1231
118. Elia AE, Rellos P, Haire LF, Chao JW, Ivins FJ, Hoepker K et al (2003) The molecular basis for phosphodependent substrate targeting and regulation of Plks by the polo-box domain. Cell 115(1):83–95
119. Watanabe N, Arai H, Nishihara Y, Taniguchi M, Watanabe N, Hunter T et al (2004) M-phase kinases induce phospho-dependent ubiquitination of somatic Wee1 by SCFbeta-TrCP. Proc Natl Acad Sci U S A 101(13):4419–4424
120. Booher RN, Holman PS, Fattaey A (1997) Human Myt1 is a cell cycle-regulated kinase that inhibits Cdc2 but not Cdk2 activity. J Biol Chem 272(35):22300–22306
121. Nakajima H, Toyoshima-Morimoto F, Taniguchi E, Nishida E (2003) Identification of a consensus motif for Plk (polo-like kinase) phosphorylation reveals Myt1 as a Plk1 substrate. J Biol Chem 278(28):25277–25280
122. Toyoshima-Morimoto F, Taniguchi E, Nishida E (2002) Plk1 promotes nuclear translocation of human Cdc25C during prophase. EMBO Rep 3(4):341–348
123. Toyoshima-Morimoto F, Taniguchi E, Shinya N, Iwamatsu A, Nishida E (2001) Polo-like kinase 1 phosphorylates cyclin B1 and targets it to the nucleus during prophase. Nature 410(6825):215–220
124. Yuan J, Eckerdt F, Bereiter-Hahn J, Kurunci-Csacsko E, Kaufmann M, Strebhardt K (2002) Cooperative phosphorylation including the activity of polo-like kinase 1 regulates the subcellular localization of cyclin B1. Oncogene 21(54):8282–8292
125. Fu Z, Malureanu L, Huang J, Wang W, Li H, van Deursen JM et al (2008) Plk1-dependent phosphorylation of FoxM1 regulates a transcriptional programme required for mitotic progression. Nat Cell Biol 10(9):1076–1082
126. Lane HA, Nigg EA (1996) Antibody microinjection reveals an essential role for human polo-like kinase 1 (Plk1) in the functional maturation of mitotic centrosomes. J Cell Biol 135(6 Pt 2):1701–1713
127. Canton DA, Keene CD, Swinney K, Langeberg LK, Nguyen V, Pelletier L et al (2012) Gravin is a transitory effector of polo-like kinase 1 during cell division. Mol Cell 48(4):547–559
128. Job D, Valiron O, Oakley B (2003) Microtubule nucleation. Curr Opin Cell Biol 15(1):111–117
129. Casenghi M, Meraldi P, Weinhart U, Duncan PI, Korner R, Nigg EA (2003) Polo-like kinase 1 regulates Nlp, a centrosome protein involved in microtubule nucleation. Dev Cell 5(1):113–125
130. Kim J, Lee K, Rhee K (2015) PLK1 regulation of PCNT cleavage ensures fidelity of centriole separation during mitotic exit. Nat Commun 6:10076
131. Lee K, Rhee K (2011) PLK1 phosphorylation of pericentrin initiates centrosome maturation at the onset of mitosis. J Cell Biol 195(7):1093–1101

132. Ramani A, Mariappan A, Gottardo M, Mandad S, Urlaub H, Avidor-Reiss T et al (2018) Plk1/polo phosphorylates Sas-4 at the onset of mitosis for an efficient recruitment of pericentriolar material to centrosomes. Cell Rep 25(13):3618–30 e6
133. Mardin BR, Agircan FG, Lange C, Schiebel E (2011) Plk1 controls the Nek2A-PP1gamma antagonism in centrosome disjunction. Curr Biol 21(13):1145–1151
134. Bahe S, Stierhof YD, Wilkinson CJ, Leiss F, Nigg EA (2005) Rootletin forms centriole-associated filaments and functions in centrosome cohesion. J Cell Biol 171(1):27–33
135. Bertran MT, Sdelci S, Regue L, Avruch J, Caelles C, Roig J (2011) Nek9 is a Plk1-activated kinase that controls early centrosome separation through Nek6/7 and Eg5. EMBO J 30(13):2634–2647
136. Li H, Liu XS, Yang X, Song B, Wang Y, Liu X (2010) Polo-like kinase 1 phosphorylation of p150Glued facilitates nuclear envelope breakdown during prophase. Proc Natl Acad Sci U S A 107(33):14633–14638
137. Kiyomitsu T, Cheeseman IM (2012) Chromosome- and spindle-pole-derived signals generate an intrinsic code for spindle position and orientation. Nat Cell Biol 14(3):311–317
138. Yim H, Shin SB, Woo SU, Lee PC, Erikson RL (2017) Plk1-mediated stabilization of 53BP1 through USP7 regulates centrosome positioning to maintain bipolarity. Oncogene 36(7):966–978
139. Lenart P, Petronczki M, Steegmaier M, Di Fiore B, Lipp JJ, Hoffmann M et al (2007) The small-molecule inhibitor BI 2536 reveals novel insights into mitotic roles of polo-like kinase 1. Curr Biol 17(4):304–315
140. Kang YH, Park JE, Yu LR, Soung NK, Yun SM, Bang JK et al (2006) Self-regulated Plk1 recruitment to kinetochores by the Plk1-PBIP1 interaction is critical for proper chromosome segregation. Mol Cell 24(3):409–422
141. Goto H, Kiyono T, Tomono Y, Kawajiri A, Urano T, Furukawa K et al (2006) Complex formation of Plk1 and INCENP required for metaphase-anaphase transition. Nat Cell Biol 8(2):180–187
142. Qi W, Tang Z, Yu H (2006) Phosphorylation- and polo-box-dependent binding of Plk1 to Bub1 is required for the kinetochore localization of Plk1. Mol Biol Cell 17(8):3705–3716
143. Elowe S, Hummer S, Uldschmid A, Li X, Nigg EA (2007) Tension-sensitive Plk1 phosphorylation on BubR1 regulates the stability of kinetochore microtubule interactions. Genes Dev 21(17):2205–2219
144. Lee HS, Park YY, Cho MY, Chae S, Yoo YS, Kwon MH et al (2015) The chromatin remodeller RSF1 is essential for PLK1 deposition and function at mitotic kinetochores. Nat Commun 6:7904
145. Suijkerbuijk SJ, Vleugel M, Teixeira A, Kops GJ (2012) Integration of kinase and phosphatase activities by BUBR1 ensures formation of stable kinetochore-microtubule attachments. Dev Cell 23(4):745–755
146. Ehlen A, Martin C, Miron S, Julien M, Theillet FX, Ropars V et al (2020) Proper chromosome alignment depends on BRCA2 phosphorylation by PLK1. Nat Commun 11(1):1819
147. Lee M, Daniels MJ, Venkitaraman AR (2004) Phosphorylation of BRCA2 by the polo-like kinase Plk1 is regulated by DNA damage and mitotic progression. Oncogene 23(4):865–872
148. Li H, Liu XS, Yang X, Wang Y, Wang Y, Turner JR et al (2010) Phosphorylation of CLIP-170 by Plk1 and CK2 promotes timely formation of kinetochore-microtubule attachments. EMBO J 29(17):2953–2965
149. Tanenbaum ME, Galjart N, van Vugt MA, Medema RH (2006) CLIP-170 facilitates the formation of kinetochore-microtubule attachments. EMBO J 25(1):45–57
150. Mondal G, Ohashi A, Yang L, Rowley M, Couch FJ (2012) Tex14, a Plk1-regulated protein, is required for kinetochore-microtubule attachment and regulation of the spindle assembly checkpoint. Mol Cell 45(5):680–695
151. Li Z, Shao C, Kong Y, Carlock C, Ahmad N, Liu X (2017) DNA damage response-independent role for MDC1 in maintaining genomic stability. Mol Cell Biol 37(9)

152. Liu XS, Song B, Tang J, Liu W, Kuang S, Liu X (2012) Plk1 phosphorylates Sgt1 at the kinetochores to promote timely kinetochore-microtubule attachment. Mol Cell Biol 32(19):4053–4067
153. Addis Jones O, Tiwari A, Olukoga T, Herbert A, Chan KL (2019) PLK1 facilitates chromosome biorientation by suppressing centromere disintegration driven by BLM-mediated unwinding and spindle pulling. Nat Commun 10(1):2861
154. Hayward D, Alfonso-Perez T, Gruneberg U (2019) Orchestration of the spindle assembly checkpoint by CDK1-cyclin B1. FEBS Lett 593(20):2889–2907
155. O'Connor A, Maffini S, Rainey MD, Kaczmarczyk A, Gaboriau D, Musacchio A et al (2015) Requirement for PLK1 kinase activity in the maintenance of a robust spindle assembly checkpoint. Biol Open 5(1):11–19
156. Espeut J, Lara-Gonzalez P, Sassine M, Shiau AK, Desai A, Abrieu A (2015) Natural loss of Mps1 kinase in nematodes uncovers a role for polo-like kinase 1 in spindle checkpoint initiation. Cell Rep 12(1):58–65
157. Ikeda M, Tanaka K (2017) Plk1 bound to Bub1 contributes to spindle assembly checkpoint activity during mitosis. Sci Rep 7(1):8794
158. von Schubert C, Cubizolles F, Bracher JM, Sliedrecht T, Kops G, Nigg EA (2015) Plk1 and Mps1 cooperatively regulate the spindle assembly checkpoint in human cells. Cell Rep 12(1):66–78
159. Kuntz K, Salovich J, Mook R, Emmitte K, Chamberlain S, Rheault T et al (2007) Identification of GSK461364, a novel small molecule polo-like kinase 1 inhibitor for the treatment of cancer. Cancer Res 67(9 Supplement):4171
160. Kaisari S, Shomer P, Ziv T, Sitry-Shevah D, Miniowitz-Shemtov S, Teichner A et al (2019) Role of polo-like kinase 1 in the regulation of the action of p31(comet) in the disassembly of mitotic checkpoint complexes. Proc Natl Acad Sci U S A 116(24):11725–11730
161. Golan A, Yudkovsky Y, Hershko A (2002) The cyclin-ubiquitin ligase activity of cyclosome/APC is jointly activated by protein kinases Cdk1-cyclin B and Plk. J Biol Chem 277(18):15552–15557
162. Hansen DV, Loktev AV, Ban KH, Jackson PK (2004) Plk1 regulates activation of the anaphase promoting complex by phosphorylating and triggering SCFbetaTrCP-dependent destruction of the APC Inhibitor Emi1. Mol Biol Cell 15(12):5623–5634
163. Hauf S, Roitinger E, Koch B, Dittrich CM, Mechtler K, Peters JM (2005) Dissociation of cohesin from chromosome arms and loss of arm cohesion during early mitosis depends on phosphorylation of SA2. PLoS Biol 3(3):e69
164. Sumara I, Vorlaufer E, Stukenberg PT, Kelm O, Redemann N, Nigg EA et al (2002) The dissociation of cohesin from chromosomes in prophase is regulated by polo-like kinase. Mol Cell 9(3):515–525
165. Hu CK, Ozlu N, Coughlin M, Steen JJ, Mitchison TJ (2012) Plk1 negatively regulates PRC1 to prevent premature midzone formation before cytokinesis. Mol Biol Cell 23(14):2702–2711
166. Neef R, Gruneberg U, Kopajtich R, Li X, Nigg EA, Sillje H et al (2007) Choice of Plk1 docking partners during mitosis and cytokinesis is controlled by the activation state of Cdk1. Nat Cell Biol 9(4):436–444
167. Wolfe BA, Takaki T, Petronczki M, Glotzer M (2009) Polo-like kinase 1 directs assembly of the HsCyk-4 RhoGAP/Ect2 RhoGEF complex to initiate cleavage furrow formation. PLoS Biol 7(5):e1000110
168. Petronczki M, Glotzer M, Kraut N, Peters JM (2007) Polo-like kinase 1 triggers the initiation of cytokinesis in human cells by promoting recruitment of the RhoGEF Ect2 to the central spindle. Dev Cell 12(5):713–725
169. Santamaria A, Neef R, Eberspacher U, Eis K, Husemann M, Mumberg D et al (2007) Use of the novel Plk1 inhibitor ZK-thiazolidinone to elucidate functions of Plk1 in early and late stages of mitosis. Mol Biol Cell 18(10):4024–4036
170. Brennan IM, Peters U, Kapoor TM, Straight AF (2007) Polo-like kinase controls vertebrate spindle elongation and cytokinesis. PLoS One 2(5):e409

171. Burkard ME, Randall CL, Larochelle S, Zhang C, Shokat KM, Fisher RP et al (2007) Chemical genetics reveals the requirement for polo-like kinase 1 activity in positioning RhoA and triggering cytokinesis in human cells. Proc Natl Acad Sci U S A 104(11):4383–4388
172. Lowery DM, Clauser KR, Hjerrild M, Lim D, Alexander J, Kishi K et al (2007) Proteomic screen defines the polo-box domain interactome and identifies Rock2 as a Plk1 substrate. EMBO J 26(9):2262–2273
173. Neef R, Preisinger C, Sutcliffe J, Kopajtich R, Nigg EA, Mayer TU et al (2003) Phosphorylation of mitotic kinesin-like protein 2 by polo-like kinase 1 is required for cytokinesis. J Cell Biol 162(5):863–875
174. Zhou T, Aumais JP, Liu X, Yu-Lee LY, Erikson RL (2003) A role for Plk1 phosphorylation of NudC in cytokinesis. Dev Cell 5(1):127–138
175. Bastos RN, Barr FA (2010) Plk1 negatively regulates Cep55 recruitment to the midbody to ensure orderly abscission. J Cell Biol 191(4):751–760
176. Hasegawa H, Hyodo T, Asano E, Ito S, Maeda M, Kuribayashi H et al (2013) The role of PLK1-phosphorylated SVIL in myosin II activation and cytokinetic furrowing. J Cell Sci 126(Pt 16):3627–3637
177. Bassermann F, Frescas D, Guardavaccaro D, Busino L, Peschiaroli A, Pagano M (2008) The Cdc14B-Cdh1-Plk1 axis controls the G2 DNA-damage-response checkpoint. Cell 134(2):256–267
178. Qin B, Gao B, Yu J, Yuan J, Lou Z (2013) Ataxia telangiectasia-mutated- and Rad3-related protein regulates the DNA damage-induced G2/M checkpoint through the Aurora A cofactor Bora protein. J Biol Chem 288(22):16139–16144
179. Jaiswal H, Benada J, Mullers E, Akopyan K, Burdova K, Koolmeister T et al (2017) ATM/Wip1 activities at chromatin control Plk1 re-activation to determine G2 checkpoint duration. EMBO J 36(14):2161–2176
180. Kumagai A, Dunphy WG (2000) Claspin, a novel protein required for the activation of Chk1 during a DNA replication checkpoint response in Xenopus egg extracts. Mol Cell 6(4):839–849
181. Mamely I, van Vugt MA, Smits VA, Semple JI, Lemmens B, Perrakis A et al (2006) Polo-like kinase-1 controls proteasome-dependent degradation of Claspin during checkpoint recovery. Curr Biol 16(19):1950–1955
182. Yata K, Lloyd J, Maslen S, Bleuyard JY, Skehel M, Smerdon SJ et al (2012) Plk1 and CK2 act in concert to regulate Rad51 during DNA double strand break repair. Mol Cell 45(3):371–383
183. Li Z, Li J, Kong Y, Yan S, Ahmad N, Liu X (2017) Plk1 phosphorylation of Mre11 antagonizes the DNA damage response. Cancer Res 77(12):3169–3180
184. Wakida T, Ikura M, Kuriya K, Ito S, Shiroiwa Y, Habu T et al (2017) The CDK-PLK1 axis targets the DNA damage checkpoint sensor protein RAD9 to promote cell proliferation and tolerance to genotoxic stress. elife 6
185. Wang H, Qiu Z, Liu B, Wu Y, Ren J, Liu Y et al (2018) PLK1 targets CtIP to promote microhomology-mediated end joining. Nucleic Acids Res 46(20):10724–10739
186. Riley T, Sontag E, Chen P, Levine A (2008) Transcriptional control of human p53-regulated genes. Nat Rev Mol Cell Biol 9(5):402–412
187. Shieh SY, Ahn J, Tamai K, Taya Y, Prives C (2000) The human homologs of checkpoint kinases Chk1 and Cds1 (Chk2) phosphorylate p53 at multiple DNA damage-inducible sites. Genes Dev 14(3):289–300
188. Cheng Q, Chen J (2010) Mechanism of p53 stabilization by ATM after DNA damage. Cell Cycle 9(3):472–478
189. Chehab NH, Malikzay A, Appel M, Halazonetis TD (2000) Chk2/hCds1 functions as a DNA damage checkpoint in G(1) by stabilizing p53. Genes Dev 14(3):278–288
190. Ando K, Ozaki T, Yamamoto H, Furuya K, Hosoda M, Hayashi S et al (2004) Polo-like kinase 1 (Plk1) inhibits p53 function by physical interaction and phosphorylation. J Biol Chem 279(24):25549–25561

191. Chen J, Dai G, Wang YQ, Wang S, Pan FY, Xue B et al (2006) Polo-like kinase 1 regulates mitotic arrest after UV irradiation through dephosphorylation of p53 and inducing p53 degradation. FEBS Lett 580(15):3624–3630
192. Yang X, Li H, Zhou Z, Wang WH, Deng A, Andrisani O et al (2009) Plk1-mediated phosphorylation of Topors regulates p53 stability. J Biol Chem 284(28):18588–18592
193. Yang X, Li H, Deng A, Liu X (2010) Plk1 phosphorylation of Topors is involved in its degradation. Mol Biol Rep 37(6):3023–3028
194. Liu XS, Li H, Song B, Liu X (2010) Polo-like kinase 1 phosphorylation of G2 and S-phase-expressed 1 protein is essential for p53 inactivation during G2 checkpoint recovery. EMBO Rep 11(8):626–632
195. Shao C, Chien SJ, Farah E, Li Z, Ahmad N, Liu X (2018) Plk1 phosphorylation of numb leads to impaired DNA damage response. Oncogene 37(6):810–820
196. Lin DC, Zhang Y, Pan QJ, Yang H, Shi ZZ, Xie ZH et al (2011) PLK1 Is transcriptionally activated by NF-kappaB during cell detachment and enhances anoikis resistance through inhibiting beta-catenin degradation in esophageal squamous cell carcinoma. Clin Cancer Res 17(13):4285–4295
197. Zhang Z, Hou X, Shao C, Li J, Cheng JX, Kuang S et al (2014) Plk1 inhibition enhances the efficacy of androgen signaling blockade in castration-resistant prostate cancer. Cancer Res 74(22):6635–6647
198. Zhang W, Wang J, Zhang Y, Yuan Y, Guan W, Jin C et al (2010) The scaffold protein TANK/I-TRAF inhibits NF-kappaB activation by recruiting polo-like kinase 1. Mol Biol Cell 21(14):2500–2513
199. Higashimoto T, Chan N, Lee YK, Zandi E (2008) Regulation of I(kappa)B kinase complex by phosphorylation of (gamma)-binding domain of I(kappa)B kinase (beta) by polo-like kinase 1. J Biol Chem 283(51):35354–35367
200. Stegmeier F, Sowa ME, Nalepa G, Gygi SP, Harper JW, Elledge SJ (2007) The tumor suppressor CYLD regulates entry into mitosis. Proc Natl Acad Sci U S A 104(21):8869–8874
201. Ma X, Wang L, De H, Li Y, Yang D, Li T et al (2017) Polo-like kinase 1 coordinates biosynthesis during cell cycle progression by directly activating pentose phosphate pathway. Nat Commun 8(1):1506
202. Li Z, Kong Y, Song L, Luo Q, Liu J, Shao C et al (2018) Plk1-mediated phosphorylation of TSC1 enhances the efficacy of rapamycin. Cancer Res 78(11):2864–2875
203. Li X, Nai S, Ding Y, Geng Q, Zhu B, Yu K et al (2017) Polo-like kinase 1 (PLK1)-dependent phosphorylation of methylenetetrahydrofolate reductase (MTHFR) regulates replication *via* histone methylation. Cell Cycle 16(20):1933–1942
204. Bengoechea-Alonso MT, Ericsson J (2016) The phosphorylation-dependent regulation of nuclear SREBP1 during mitosis links lipid metabolism and cell growth. Cell Cycle 15(20):2753–2765
205. Xiao D, Yue M, Su H, Ren P, Jiang J, Li F et al (2016) Polo-like Kinase-1 regulates Myc stabilization and activates a feedforward circuit promoting tumor cell survival. Mol Cell 64(3):493–506
206. Wu J, Ivanov AI, Fisher PB, Fu Z (2016) Polo-like kinase 1 induces epithelial-to-mesenchymal transition and promotes epithelial cell motility by activating CRAF/ERK signaling. elife 5
207. Li Z, Li J, Bi P, Lu Y, Burcham G, Elzey BD et al (2014) Plk1 phosphorylation of PTEN causes a tumor-promoting metabolic state. Mol Cell Biol 34(19):3642–3661
208. Strebhardt K (2010) Multifaceted polo-like kinases: drug targets and antitargets for cancer therapy. Nat Rev Drug Discov 9(8):643–660
209. Colicino EG, Hehnly H (2018) Regulating a key mitotic regulator, polo-like kinase 1 (PLK1). Cytoskeleton (Hoboken) 75(11):481–494

Conformational Studies of the Par-4 C-Terminal Domain

David S. Libich, Samjhana Pandey, and Steven M. Pascal

Abstract Par-4 is an approximately 40 kilodalton tumor suppressor protein that is critical for apoptotic processes that affect a wide variety of cancer cells. Full-length Par-4 is known to self-associate via its C-terminal domain, although it has also been shown to be largely disordered in vitro. However, various isolated regions of Par-4 have been shown to contain structure. The C-terminal quarter of Par-4 consists of a coiled coil that includes a leucine zipper as its C-terminal half. The isolated leucine zipper forms a stable helical dimer in vitro only under conditions of either low pH or high salt concentration. The reason for this requirement has been traced to a negative-negative charge interaction across the dimer interface. The cl-Par-4 fragment, which comprises the C-terminal half of Par-4, has also been studied. Again, a high degree of helicity is evident in the presence of low pH or high salt, though the self-association state differs between an apparent coiled coil dimer and a non-coiled tetramer, respectively, under these conditions. These results also suggest that the SAC domain, which is the minimal region required for triggering apoptosis selectively in cancer cells, may be disordered in FL-Par-4 but ordered in cl-Par-4. Possible implications for apoptotic processes are discussed.

Keywords Par-4 · PAWR · cl-Par-4 · SAC · Coiled coil · Apoptosis · Cancer · IDP · Disorder · Circular dichroism · Light scattering · NMR

D. S. Libich
Department of Biochemistry and Structural Biology, University of Texas Health Science Center at San Antonio, San Antonio, TX, USA

S. Pandey · S. M. Pascal (✉)
Department of Chemistry and Biochemistry, Old Dominion University, Norfolk, VA, USA
e-mail: spascal@odu.edu

V. M. Rangnekar (ed.), *Tumor Suppressor Par-4*,
https://doi.org/10.1007/978-3-030-73572-2_3

1 Introduction

1.1 Par-4 Discovery and Function

The Par-4 tumor suppressor, also known as PAWR, was first discovered in a differential screen of immediate-early genes upregulated in prostate cancers cells undergoing apoptosis [1]. Par-4 has since been linked to a wide variety of human cancers (e.g. [1–5]), as well as to neurodegenerative disease [6–10]. Essentially, downregulation of Par-4 activity allows tumor cells to proliferate. Conversely, Par-4 upregulation is linked to excessive cell death in neurodegenerative disease. The ability to control Par-4 function, either via activation or suppression, has therefore been of medical interest for some time. Structural knowledge would clearly assist these efforts.

1.2 Rat vs. Human cl-Par-4 and Domain Structure

To date, all structural work with Par-4 has utilized either the rat (332 aa) or human (340 aa) forms. These two sequences are 74% identical. This review will focus on the Par-4 C-terminus and its domains. For this purpose, the C-terminus will be defined as the C-terminal fragment released after cleavage by caspase-3, as part of the caspase cascade activated in cells undergoing apoptosis [11]. This 209 amino acid C-terminal fragment is referred to as cl-Par-4, where "cl" denotes cleaved.

The N-terminal fragment released by caspase-3 cleavage is called PAF, for Par-4 amino-terminal fragment. The rat and human PAF regions have only 58% identity, with ten amino acids requiring gaps for alignment. In contrast, cl-Par-4 is far better conserved, with 84% identity between rat and human forms and no gaps (see Fig. 1). The cl-Par-4 fragment can be further divided into three domains: (1) the SAC domain (selective for apoptosis in cancer cells), which is the minimal region required to induce apoptosis in tumor cells without killing healthy cells [12]; (2) the coiled coil (CC) domain that is responsible for Par-4 self-association [13]; and (3) the linker between the SAC and CC domains. As can be seen in Table 1, both the SAC and CC domains are highly conserved between rat and human, while the linker is not. A leucine zipper (LZ) is also present within the C-terminal half of the CC domain, and its identity between rat and human forms is higher than is the identity for the remainder of the CC domain. This general pattern of highest conservation within the SAC and LZ domains is conserved in other organisms as well. In fact, across a broader range of organisms, the LZ is more highly conserved than is the SAC domain (e.g. see multi-sequence alignment in [14]).

```
                                                 SAC
Rat    124   SAPEKGRSSGPSARKGKGQIEKRKLREKRRSTGVVNIPAAECLDEYEDDEAGQKERKRED  183
               PEKG+SSGPSARKGKGQIEKRKLREKRRSTGVVNIPAAECLDEYEDDEAGQKERKRED
Human  132   GVPEKGKSSGPSARKGKGQIEKRKLREKRRSTGVVNIPAAECLDEYEDDEAGQKERKRED  191

Rat    184   AITQQNTIQNEAASLPDPGTSYLPQDPSRTVPGRYKSTISAPEEEILNRYPRTDRSGFSR  243
             AITQQNTIQNEA +L DPG+SYL Q+P RTV GRYKST S  EE++ +RY RTDRSGF R
Human  192   AITQQNTIQNEAVNLLDPGSSYLLQEPPRTVSGRYKSTTSVSEEDVSSRYSRTDRSGFPR  251

                                  CC                             LZ
Rat    244   HNRDTSAPANFASSSTLEKRIEDLEKEVLRERQENLRLTRLMQDKEEMIGKLKEEIDLLN  303
             +NRD +      SSSTLEK+IEDLEKEV+RERQENLRL RLMQDKEEMIGKLKEEIDLLN
Human  252   YNRDANVSGTLVSSSTLEKKIEDLEKEVVRERQENLRLVRLMQDKEEMIGKLKEEIDLLN  311

Rat    304   RDLDDMEDENEQLKQENKTLLKVVGQLTR  332
             RDLDD+EDENEQLKQENKTLLKVVGQLTR
Human  312   RDLDDIEDENEQLKQENKTLLKVVGQLTR  340
```

Fig. 1 Sequence alignment of rat and human cl-Par-4. Amino acid numbering is based on full-length Par-4. The positions of the SAC, CC, and LZ domains are shaded. Within the LZ domain, the heptad repeats of Leu residues in the "d" positions are colored green, while the one Met residue at the "d" position is brown, and the N residues at the "a" positions are colored magenta. Repelling D and E residues (see Fig. 2) are colored red

Table 1 Sequence conservation between rat and human cl-Par-4 and its domains

Domain	cl-Par-4	SAC	Linker	CC	LZ
# of aa	209	59	71	79	39
% identity	84	100	48	92	97

1.3 Brief Summary of Structural Work

Knowledge of Par-4 domain structure arose gradually. Early analyses of the Par-4 amino acid sequence revealed no recognizable motifs, other than the possibility of a LZ domain or a death domain at the Par-4 C-terminus [1]. A LZ is simply a special subset of the CC category, in which a heptad repeat of (mostly) leucines is present. As can be seen from the helical wheel diagram of Fig. 2, this requirement is met by the C-terminal 39 residues of rat Par-4, which contains 5 leucine residues at the "d" position. Interestingly, the only non-conserved residue in the LZ is at the "d" position, where M_{309} in rat is replaced by I_{317} in human Par-4. The potential for death domain formation is more difficult to assess by amino acid analysis alone, since sequence identity between death domains in unrelated proteins is often only marginally higher than random [15].

Experimental analyses were therefore undertaken to study the rat Par-4 "LZ" region [16–18]. These initial studies revealed a conformational environment dependence. Subsequent studies suggested that the LZ region forms a coiled coil in the context of the full-length Par-4 protein but provided little if any evidence for

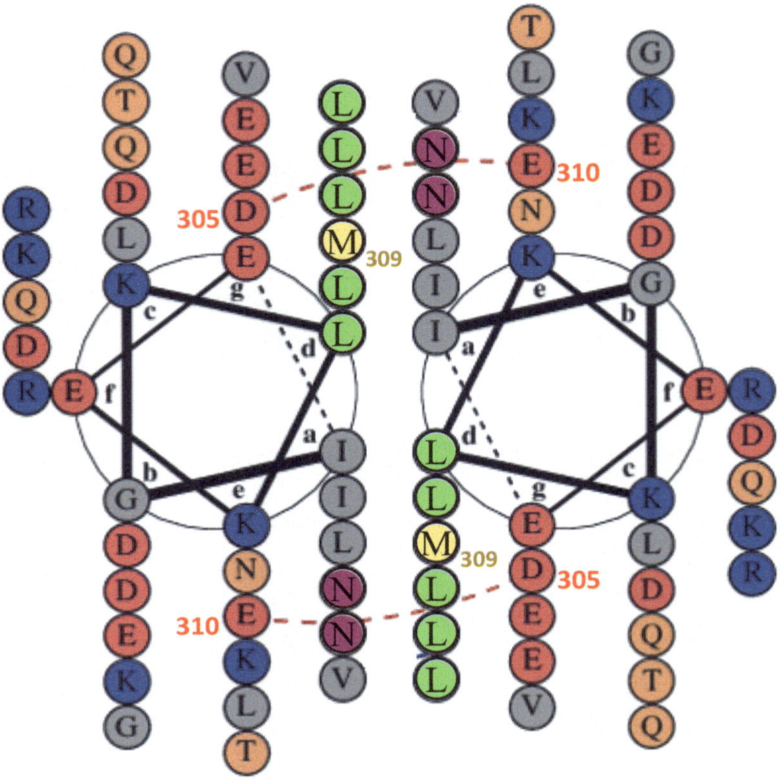

Fig. 2 Helical wheel diagram of the rat Par-4 LZ region. In the "d" positions, the Leu heptad repeat residues are colored green, and the Met residue (M_{309}) is yellow. In the "a" position, hydrophobic residues are colored gray while the polar Asn residues are colored magenta. At all other positions, acidic residues are red, basic residues are blue, polar uncharged residues are orange, and hydrophobic residues are gray. The dashed red line indicates a charge-charge repulsion between D_{305} and E_{310}. The human sequence of the LZ is identical, except that M_{309} is replaced by Ile and the numbering is increased by eight

structure outside of the coiled coil [14]. Some years later, the CC domain was shown experimentally by X-ray crystallography to form a dimeric coiled coil [19]. This crystal structure is discussed in chapter "Structural Analysis of Par-4 and Crystallographic Analysis of the Regulatory Domain" of this volume.

The ability of caspase-3 to cleave Par-4 into the PAF and cl-Par-4 fragments is a relatively recent discovery [11]. The PAF fragment is believed to serve mainly as a decoy that protects full-length Par-4 from ubiquitin-based degradation [20]. Conversely, the cl-Par-4 fragment remains active in apoptotic pathways [21, 22] and in fact may be the principal active form. The cl-Par-4 fragment enters the nucleus where it helps to tip the survival/death balance to death. Recent biophysical analyses of the intact cl-Par-4 fragment revealed an environmental dependency related to that seen for the LZ [23, 24]. However, interesting new information was obtained, including evidence of α–helix formation outside of the CC region, most likely in the SAC domain. This raises the possibility that cleavage by caspase-3 may be necessary and perhaps even sufficient for formation of the active/folded conformation of the SAC domain, which lies just C-terminal to the caspase-3 cleavage site. Furthermore, conditions were identified in which cl-Par-4 can form higher-order structures. This suggests that environment-dependent quaternary structure may also play a role in regulating cl-Par-4 translocation and activity.

2 Overview of Techniques Utilized

A wide range of biophysical techniques have been used to characterize the C-terminal domain of Par-4. The principles of those techniques will be briefly discussed.

Circular Dichroism (CD) – CD spectroscopy measures the differential absorption of left vs. right circularly polarized light. In order to display dichroism, the sample first must absorb light at the wavelength sampled, and second it must contain a chiral center. In the case of proteins, the most useful wavelength range is 200–250 nm (within the far UV range). The absorbing electronic transitions are the π to π* and n to π* transitions of electrons from molecular orbitals (MOs) involving the peptide bond. These MOs are influenced by the nearby C_α chiral centers. The observed dichroism is strongly dependent upon the relative positioning of the backbone atoms around these chiral centers. For this reason, CD spectroscopy is highly sensitive to backbone dihedral angles and is an effective means to assess protein secondary structure. Figure 3a shows basic CD spectral shapes for pure α-helical, β-sheet, and disordered or random coil structures. These three basic elements can easily be distinguished, when only one type is present. For proteins with a mixture of secondary structure, deconvolution programs are used to estimate the linear combination of these shapes, along with additional secondary structure variations such as beta turns and 3_{10} helix, that could contribute to the overall observed CD spectrum. This process yields an estimate of secondary structure composition.

Dynamic Light Scattering (DLS) – DLS reports on the size distribution of particles in solution and thus is useful for investigating the associative behavior of proteins and other biomolecules. In DLS, light scattering is monitored as a function of time, as a means to assess the rate of translational motion of the sample molecules. For example, if the protein is aggregated, then the resultant large particles move slowly, and the scattering pattern therefore changes slowly. From this correlation, a Stokes radius (R_S) is generated: R_S is the radius of a spherical particle that theoretically would translate at the observed rate. While other factors such as particle shape may also affect particle translation motion, DLS nonetheless can provide a useful estimate of particle size, and thus self-association state.

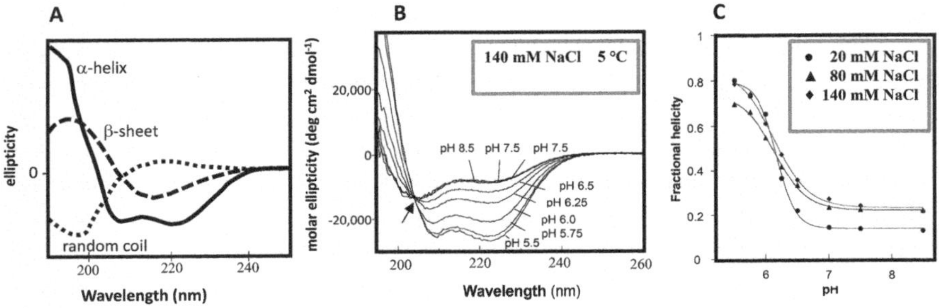

Fig. 3 Effect of pH and salt on the secondary structure of rat Par-4 LZ. (**a**) Idealized CD spectral traces expected for isolated α-helix (solid trace), β-sheet (dashed line), and random coil (dotted line). (**b**) CD spectra of Par-4 LZ through a pH range of 5.5 to 8.5. As pH is reduced, spectrum becomes more characteristic of α-helix, with coiled coil character. (**c**) Fractional helicity as a function of pH, as determined by CD spectrum deconvolution, at three different salt concentrations. pH reduction increases helicity. At neutral pH, increasing salt increases helicity. This trend reverses at acidic pH. Sample conditions are 22 μM Par-4 LZ with 140 mM NaCl and 12 mM NaP_i at 5 °C for part (**b**) and the same but with 27 μM Par-4 LZ for part (**c**). (Modified from the original figures in Ref. [16])

Size-Exclusion Chromatography with Multiple Angle Light Scattering (SEC-MALS) – SEC-MALS is a hyphenated technique in which the eluent from an SEC column is fed continuously into a MALS device. The SEC column serves to separate sample particles based on their relative sizes, with larger particles emerging first due to their exclusion from small pathways within the beads of the column. When the eluted fractions reach the DLS device, UV absorbance (A_{280}), refractive index, and light scattering at multiple (three or more) distinct angles are recorded. Through a combined analysis of these data, an estimate of the molecular weight of the particles is made. From these data, the oligomeric state of the protein (monomer, dimer, tetramer, etc.) may be determined. The radius of gyration (R_g), the weighted average distance of the atoms from the geometrical center of the particle, is also extracted. Therefore, R_g is complimentary to the R_S obtained from DLS. SEC-MALS analysis can also provide a shape factor, which can distinguish, for instance, spherical from elongated particle shapes. SEC can also be used without MALS (albeit less accurately for non-spherical proteins) to estimate particle size via column retention time.

Nuclear Magnetic Resonance (NMR) – NMR is capable of a wide range of analyses. In these studies, chemical shift analysis was used to assess the degree of disorder and secondary structure. Disorder is indicated by a narrow range of backbone HN chemical shifts. Secondary structure is determined by computationally comparing measured backbone C_α, C_β, and C-carbonyl chemical shifts to a database of chemical shifts from proteins of known secondary structure. This is the CSI or chemical shift indexing procedure. Analysis of ^{15}N relaxation experiments provides an estimate of the degree of mobility of each residue, since relaxation is related to molecular dynamics. Peak intensities can also be used to assess molecular dynamics (more intense peaks are associated with rapidly moving groups) and to detect the possibility of ms-μs time scale conformational interconversion, which can broaden and therefore reduce the intensity of peaks. Measurement of backbone three-bond $^{3}J_{HNH\alpha}$ coupling constants can also be used to assess secondary structure.

Molecular Modeling (MM) – MM also has a wide range of capabilities. Here it was used to create dimer and tetramer models that are consistent with the data. While these models cannot be taken as high-resolution experimental structures, the features of the models are useful for discussion of structural and functional concepts.

H-MBP-3C Expression/Purification System – Par-4 has been difficult to express and purify in large amounts and high concentration, due to its disordered and aggregating nature. For this reason, a customized expression vector, H-MBP-3C, was developed to create a one-step purification protocol that produces soluble protein [25]. In brief, the resulting construct codes for a fusion protein of maltose-binding protein (MBP) and Par-4, with a cleavable linker between the two. "H" represents a six-histidine tag that is N-terminal to the MBP. MBP helps to solubilize the Par-4 protein in cell extracts, while the His-tag allows the fusion protein to be immobilized on a nickel affinity column. Subsequent addition of a His-tagged protease to the column separates the MBP tag from the Par-4 protein. Since the protease and the MBP tag each contain a His-tag, they are retained on the column, while the Par-4 protein elutes (~95% purity). An additional ion exchange column step can be used to improve the protein purity to ~99% if desired. For some constructs, fusion with MBP was not required for solubility, and in these cases, a vector coding for only a cleavable N-terminal His-tag was used for purification.

3 Rat LZ (Amino Acids 286–332)

3.1 Helicity as a Function of pH, Salt, and Temperature

The first biophysical and structural studies of Par-4 [16] were undertaken with a fragment of the rat protein representing the LZ (292–332) preceded by a polar stretch of six amino acids (286–291) that were included to improve solubility. Initial studies at neutral pH (6.5 or 7.0) and with low salt or PBS showed weak dichroism (see Fig. 3b, pH 7.0 trace), with a global minimum at 200 nm, which is characteristic of "random coil" or disorder. Deconvolution algorithms indicated only approximately 20% helicity. This was surprising, since a helical wheel diagram (see Fig. 2) suggested that a heptad repeat of leucine residues at the "d" position should create an amphipathic LZ, with a hydrophobic core formed by the "a" and "d" residues, and a highly polar/charged surface formed by most of the remaining residues.

In seeking an explanation for this lack of order, it was noted that other acidic coiled coils (note the high number of red residues in Fig. 2) have been shown to lack structure at neutral pH and to increase in helical content with acidic pH (e.g.

[26–29]). This effect is due to pH-induced protonation of acidic groups that would otherwise repel each other via their mutual negative charges. While none of the acidic residues in the helical wheel diagram lie at the core “a” and “d” positions, it has been established that side chain groups of “e” and “g” position residues can also interact across the dimer interface [26, 28, 30–32]. Therefore, it was hypothesized that the negatively charged E_{310} residue in the “e” position was repelling one of the E residues in the “g” position of the dimer partner molecule. To test this hypothesis, CD spectroscopy was obtained as a function of pH.

It can be seen (Fig. 3b) that lowering the pH increased the dichroicity of the sample, with the most rapid change taking place near the pH 6.0 range, indicating a conformational pK_a near 6.0. Concentrating on the lowest pH tested, pH 5.5, two minima are seen, at 208 and 222 nm. This pattern is characteristic of α-helix formation (see Fig. 3a). Deconvolution reveals that the Par-4 LZ is approximately 80% helical at pH 5.5 (Fig. 3c). The shapes of the fractional helicity curves in Fig. 3c are consistent with the presence of a cooperative transition from a single dominant conformation at neutral pH to a second dominant conformation at acidic pH, with a conformational pK_a near 6.1. Furthermore, the fact that the 222 nm minimum is more intense than the 208 nm minimum is a clear indication of coiled coil formation [26, 33, 34], in which α-helices wrap around each other slightly.

Also of interest was the isodichroic point at 203 nm: regardless of the pH, the dichroism at this wavelength remains unchanged. This feature is also highly indicative of a two-state conformational equilibrium [27, 30, 31]. Essentially, each of the two predominant states apparently has the same molar ellipticity at 203 nm, so that no matter which state predominates, the reading at 203 nm does not change. It is highly unlikely that a third significantly populated state would also have the same molar ellipticity at 203 nm. This finding was consistent with what is known about coiled coil formation: typically, helix formation and coiling of the helices to form the dimer (or higher order structure) are highly coupled; the helices are not stable until interactions with the dimer partner are established [27, 35–39]. Therefore, it appears that upon reduction of pH, a disordered structure cooperatively transitions to a highly helical LZ.

Additional CD spectra were obtained as a function of temperature and Par-4 peptide concentration to determine melting temperatures. The results were consistent with the above interpretation: melting temperatures increased (indicating increased LZ stability) with reduction in pH, increase in salt, and increase in Par-4 concentration. The stabilizing effect of salt is related to the stabilizing effect of acidic pH: positively charged ions in solution help screen the destabilizing “e”–“g” negative-negative repulsions across the dimer interface [27, 29, 40]. Increase in stability with increased Par-4 concentration is a clear indication of the coupling between dimerization and helix formation.

3.2 Site-Directed Mutagenesis (E310K, E310Q, D305K, and D305N)

To test the hypothesis that "e"–"g" repulsions involving the E_{310} residue were responsible for the observed pH dependence of the LZ conformation, a series of point mutants were made [17]. These included E310K and E310Q, in which the negatively charged E_{310} side chain was replaced by either a positively charged side chain (K) or a similarly sized polar but neutral side chain (Q). Also, LZ modeling suggested that the E_{310} residue should interact across the dimer interface most intimately with the D_{305} residue. Therefore, the analogous D305K and D305N mutants were made, since removing either one of the two allegedly repelling negative charges (E_{310} or D_{305}) should produce similar stabilizing effects.

The results can be seen in the CD spectra of Fig. 4. Each of the neutral polar substitutions reduced the degree of pH dependence: significant helical character (~35%) was evident even at neutral pH. The effects of the K mutations were more pronounced: changing either E_{310} or D_{305} to lysine completely eliminated the pH dependence, leading to stable LZ formation throughout the tested pH range. This result confirms that the D_{305}–E_{310} repulsion is responsible for the pH-dependent two-state conformational transition with pH that is seen in the wild-type (WT) LZ.

Additional CD spectroscopy as a function of temperature produced experimental melting temperatures for the WT and mutants (Fig. 5a). Melting here refers to the cooperative conversion of the helical LZ into a disordered state. At neutral pH, the LZ of the K mutants are the most stable, while the WT LZ is the least stable. As pH is lowered, the temperature stabilities converge somewhat. From this melting temperature data, enthalpies of melting were obtained (Fig. 5b), with analogous results:

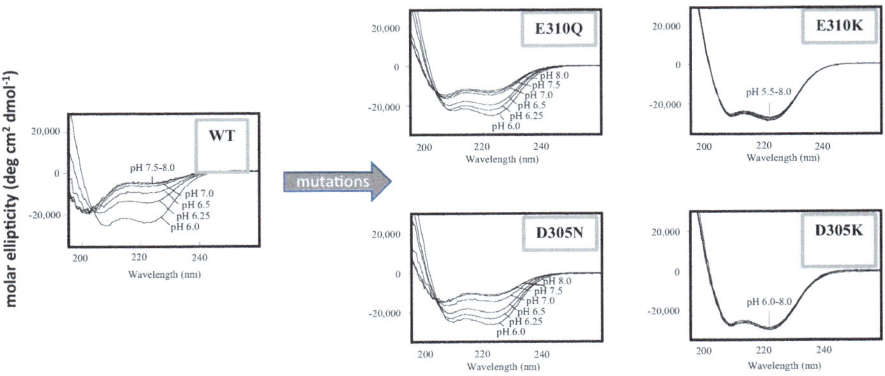

Fig. 4 CD spectroscopy of wild-type and mutant rat Par-4 LZ region as a function of pH. The marked pH dependence of the wild-type spectrum (left panel) is reduced by introduction of neutral polar mutants (center panels) and eliminated by the introduction of Lysine mutants (right panels). Sample conditions are approximately 40 mM Par-4 LZ peptide with 20 mM NaCl and 12 mM NaP_i at 5 °C. (Modified from the original figures in Ref. [17])

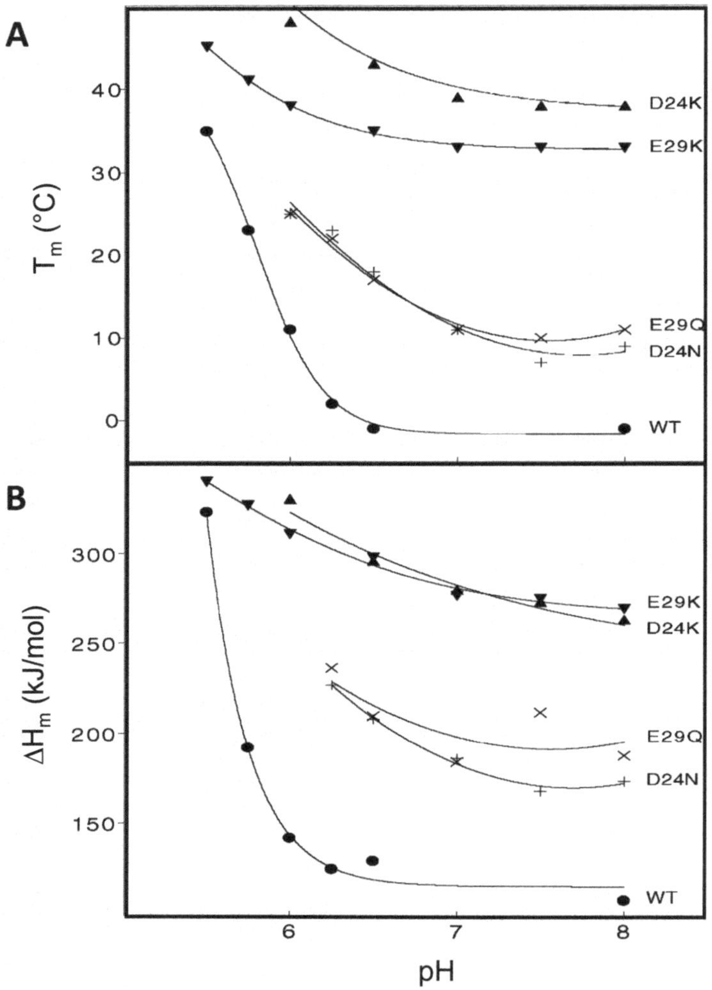

Fig. 5 Thermodynamic analysis of wild-type and mutant rat Par-4 LZ region as a function of pH. (**a**) Melting temperature as a function of pH, determined via analysis of observed changes in CD spectra. Melting temperature goes up as pH goes down. (**b**) Enthalpy of melting as a function of pH. ΔH_m goes up as pH goes down. These effects are more intense for the WT construct. (Modified from the original figures in Ref. [17])

more energy is required to melt the K mutants, while little is needed to melt the WT LZ, though these values also converge at acidic pH, which is consistent with the above observation that acidic pH stabilizes the WT LZ.

3.3 Hydrodynamic Studies

The above studies established that the Par-4 LZ populates a two-state conformational equilibrium as a function of pH but did not determine the specific nature of the two states. Analysis of CD spectra determined approximate helical content and detected the presence of coiled coil formation as a function of pH, salt concentration, or residue substitution. However, these studies did not probe the oligomeric state of Par-4 LZ: several different types of coiled coils are possible, including dimeric, trimeric, and tetrameric. To further investigate these possibilities, the WT rat LZ was studied via light scattering and chromatographic techniques.

The obtained SEC and DLS results correlate well with helicity: conditions and mutants which favor helicity were associated with a larger hydrodynamic size. This is consistent with the concept of intermolecular coiled coil formation at low pH, high salt, and by the mutant LZ at neutral pH. The SEC elution profiles (data not shown) were non-Gaussian and occurred earlier than predicted by their calculated monomeric MW, indicating non-ideal behavior due to rod-like extended conformations, or intramolecular associations, or some combination thereof. DLS (data not shown) exhibits similar behavior, with broad distributions that are indicative of non-ideal shapes (extended) and interconverting conformations. At pH 7.0, the D305K and E310K mutants exhibit larger measured R_s than the WT LZ. The R_s of the WT LZ increases with decreasing pH.

Although SEC and DLS analysis established oligomerization, these techniques cannot accurately establish the self-association state. For this, SEC-MALS was used. At neutral pH and ambient temperature, SEC-MALS analysis of the WT LZ produced a particle molecular weight of 5.8 kDa (Fig. 6a, red trace), a monomer. The D305K mutant produced a complex peak, consistent with a predominant dimer conformation (11.5 kDa) in exchange with a minor monomer (5.8 kDa) conformation (Fig. 6a, blue trace). When combined with results from the above CD analysis, these data indicate that the WT LZ is predominantly monomeric and disordered at neutral pH, while the stabilized mutants (D305K and E310K) are predominantly dimeric and helical under these conditions. Lowering the pH alters the monomer-dimer equilibrium of the WT LZ, to favor the dimer conformation. These experiments definitively establish that the previously reported two state equilibrium in the Par-4 LZ is in fact a monomer-dimer equilibrium. This specification provides insights into interpretation of the subsequent NMR-based investigations discussed below.

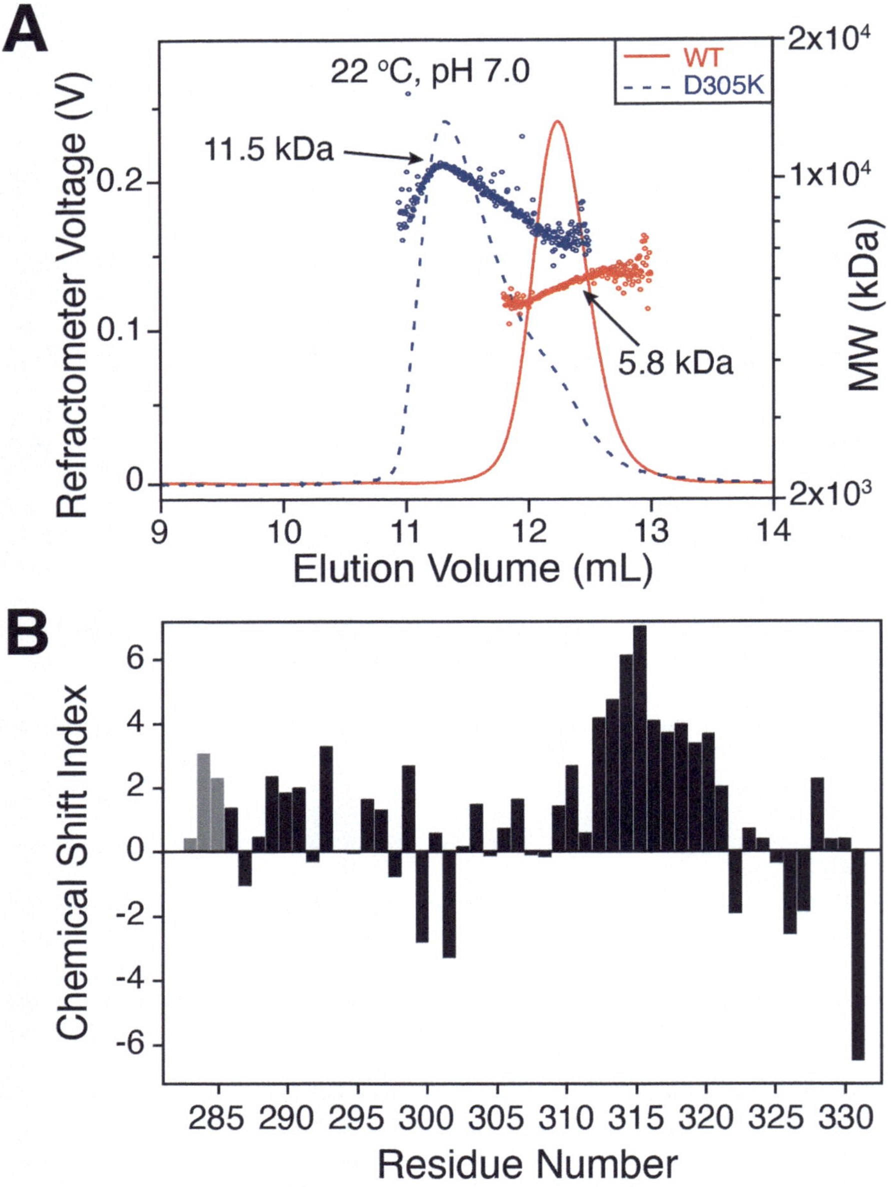

Fig. 6 Rat Par-4 LZ particle size and secondary structure analysis. (**a**) SEC-MALS elution (traces) and molecular weight profiles (dots) for WT LZ (red) and D305K mutant (blue). The data were normalized to scale peak heights for presentation. (**b**) Par-4 WT LZ chemical shift index at pH 7 and low salt. The LZ domain spans residues 286–332 (black bars). Protease site residues remaining after cleavage of the purification tag are denoted by gray bars. (Modified from the original figures in Ref. [18])

3.4 NMR Studies

Chemical Shift Index Analysis of the Disordered State – Resonance assignments for the rat LZ domain were obtained using standard approaches on a sample at pH 6.5 and 5 °C, conditions under which the LZ is primarily monomeric and disordered. Use of these conditions alleviated sample instability issues and minimized conformation exchange processes, each of which could cause peak broadening and reduce spectral quality. Using the carbon backbone shifts, CSI analysis (Fig. 6b) was performed via the Wishart/Sykes method [41]. Consistently positive CSI values for the residues from Glu312 through to Lys321 reveal a moderate to strong propensity for the formation of α-helix in this region. These results show that even under conditions in which the WT LZ is primarily monomeric, it possesses partial secondary structure ordering and an overall ensemble bias to helical conformations, likely a necessary condition for coiled-coil formation [18].

NMR Spectral Analysis – A series of 2D ^{1}H,^{15}N-SOFAST-HMQC spectra were recorded [18], for WT and mutant samples at two temperatures (25 °C and 5 °C) and three distinct pH values (7.0, 6.0, and 4.4). At pH 7.0, the WT LZ spectra at both temperatures (Fig. 7a, b, black contours) feature uniformly narrow, sharp peaks with limited ^{1}H chemical shift dispersion, characteristic features of disordered proteins. Reduction of the pH to 4.4 (Fig. 7a, green contours) or use of the D305K mutant (Fig. 7a, red contours) cause an increase in the ^{1}H chemical shift dispersion (blue dashed box) that is accompanied by broadening of some resonances, including those in the box. A similar but less pronounced effect is seen at pH 6.0 (Fig. 7b, blue contours). This results in two groups of peaks for the samples with broadening: a set of sharp peaks and a set of broadened peaks. The sharp peaks are consistent with the disordered monomer conformation. The broadening of the second set of peaks could be due to slower molecular tumbling (e.g. of a dimer relative to a monomer) or to conformational exchange (e.g. between a monomer and dimer or between different dimer forms). These two sources of broadening could both be present. It should also be noted that the appearance of broadened peaks correlates with reduction in intensity of the sharp peaks.

To further investigate the source of the broadening, an E310K mutant LZ sample was prepared using the REDPRO sample protocol [42]. In REDPRO, the protein is expressed from *E. coli* grown in a D_2O-based medium. This results in the deuteration of most sites, including the alpha hydrogens. Upon lyophilizing the protein and re-dissolving in H_2O, the exchangeable amide protons return, but their shape characteristics are improved by proximity to alpha deuterons rather than the more highly magnetic alpha protons. When combined with the TROSY-HSQC variant, this procedure can greatly sharpen peaks that had been broadened by slow tumbling but would not have a significant effect on broadening due to chemical exchange.

As can be seen in Fig. 7b (red contours), the two sets of peaks are now more clearly distinguished, with very narrow peaks mixed with quite broad peaks. This result clarifies that chemical exchange is a significant factor in the observed broadening, confirming that the coiled coil dimer interconverts on a ms-μs time scale

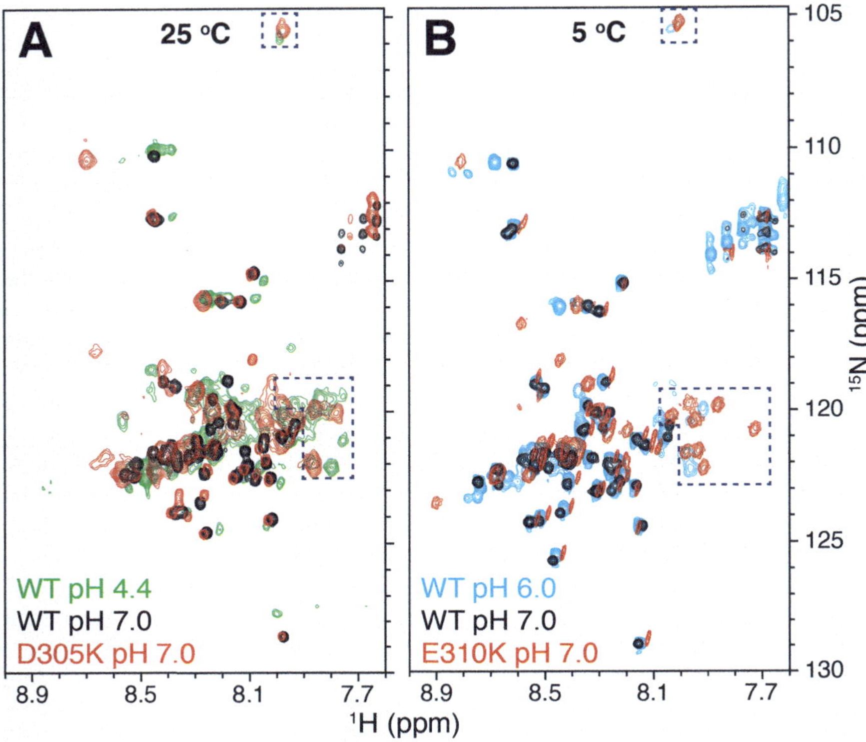

Fig. 7 $^{1}H,^{15}N$-SOFAST-HMQC spectra of WT LZ and point mutants. (**a**) WT LZ at pH 4.4 (green contours), pH 7.0 (black contours), and D305K mutant at pH 7.0 (red contours), each recorded at 25 °C. (**b**) WT LZ at pH 6.0 (blue contours), pH 7.0 (black contours), and REDPRO-labeled E310K mutant at pH 7.0 (red contours), each recorded at 5 °C. All spectra were recorded with 24 transients and processed with identical parameters to facilitate direct comparison. The REDPRO sample utilized a $^{1}H,^{15}N$-TROSY-HSQC pulse sequence. Blue dashed boxes denote peaks arising from the ordered, α-helical conformation of the coiled coil. (Modified from the original figures in Ref. [18])

either with a subset of the monomer population, with alternative dimeric forms, or with other (hidden) minor conformations.

4 Rat FL, ΔLZ, and SAC Domain Constructs

To investigate how the LZ structure is affected by its context in the larger Par-4 molecule, CD spectroscopy, NMR, and light scattering studies were expanded to include FL-Par-4, the isolated SAC domain, and a deletion mutant missing only the LZ (ΔLZ). These studies, together with sequence analysis and proteolysis studies, established the intrinsically disordered character of Par-4 [14].

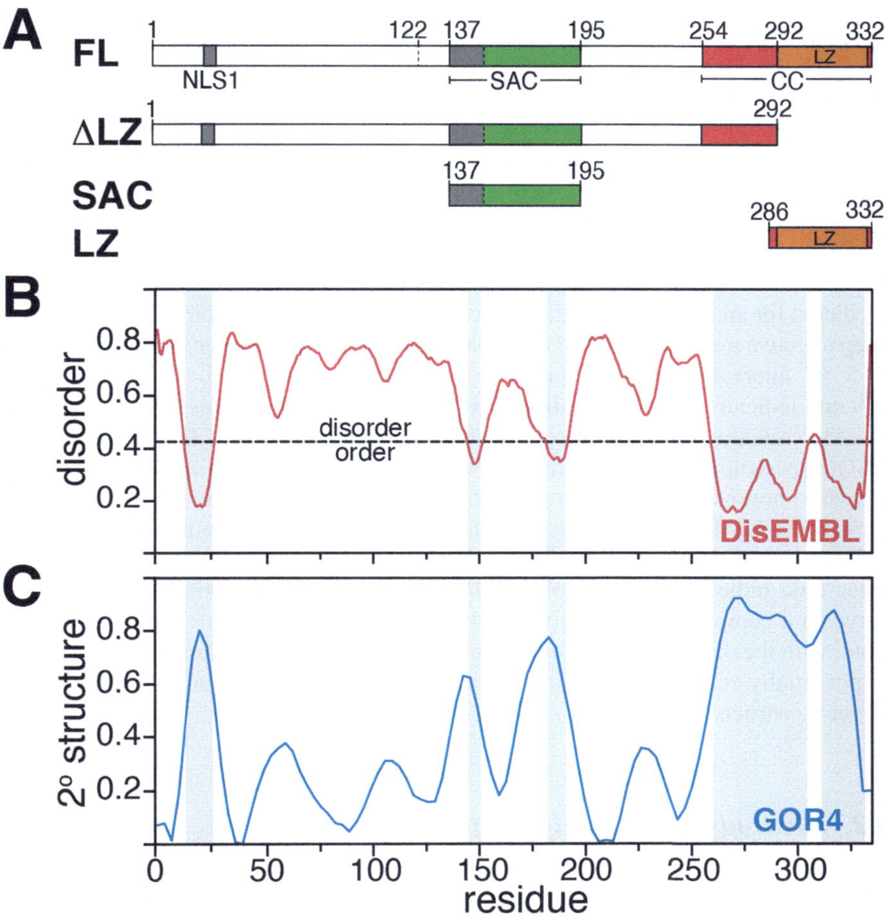

Fig. 8 Par-4 constructs and order prediction. (**a**) Constructs of rat Par-4 used in the studies described herein. Regions of functional importance are highlighted including NLS1 (20–25), NLS2 (137–153), the SAC region (137–195), the coiled coil region (254–332), and the Leucine zipper (292–332). (**b**) Prediction of disordered regions of rat Par-4 (DisEMBL). The dashed line represents a threshold value between predicted order and disorder. (**c**) Secondary structure prediction using the GOR4 algorithm. For clarity, only the prediction for α-helix is shown. The gray bars in panels B and C show that the regions of predicted order align with regions of predicted α-helix. (Modified from the original figure in Ref. [14])

The domain structure of full-length rat Par-4 is shown in Fig. 8a. The two domains with established function are the SAC domain and the CC. Bioinformatic analysis of Par-4 primary structure with the disorder prediction algorithm DisEMBL suggest a mostly intrinsically disordered protein (Fig. 8b). The primary ordered region is the C-terminal CC domain. In addition, three short stretches of predicted order occur at each nuclear localization signal (NLS1 near the N-terminus and NLS2 within the SAC domain) and near the C-terminus of the SAC domain. Further, the GOR4 algorithm identifies these regions as highly probable to form an α-helix (Fig. 8c).

4.1 Structural Analysis of the SAC Domain

The SAC domain (residues 137–195) is defined functionally as the minimal portion of Par-4 capable of eliciting apoptosis, rather than as a structural domain. As mentioned above, sequence analysis does predict the potential for structure, though far less strongly than for the CC domain. The R_S of the isolated SAC domain measured by DLS was 21 Å, larger than expected for a monomeric globular protein of comparable size, yet smaller than predicted for a fully extended or disordered chain. The addition of 1 M urea caused the R_S to increase to 28 Å, a value in line with that predicted for an extended and disordered conformation. CD spectra of SAC show a deep negative transition at ~200 nm indicative of intrinsic disorder. Spectra recorded at 5 °C intervals reveal the presence of an isodichroic point at approximately 209 nm, indicative of conformational exchange between two states (Fig. 9a), though disorder remains dominant throughout the temperature range. Finally, a $^{1}H,^{15}N$-HSQC spectrum of the SAC domain was obtained, and the observed peaks were shown to not overlay well with peaks from either the FL Par-4 or ΔLZ Par-4 samples. These NMR results suggest that the SAC domain conformation when isolated differs significantly from its conformation when in the larger constructs. Together, these data indicate that the isolated SAC domain is in conformational exchange between a compact, at least partially structured state, and an extended, disordered state, with the disordered state dominating the ensemble. The minor structured state is potentially related to the conformation that the SAC domain may take in the larger constructs.

4.2 Hydrodynamic Analysis of FL and ΔLZ Par-4

Given the calculated molecular weights of monomeric FL Par-4 (36.2 kDa) and ΔLZ Par-4 (31.2 kDa), these two molecules, if spherical, would be predicted to have Stokes radii of 26.5 Å and 25.1 Å, respectively. Experimentally, DLS data leads to extracted R_s values of 189 Å and 32.5 Å for the FL and ΔLZ proteins, respectively. This result clearly shows that FL Par-4 forms aggregates in solution, containing on the order of hundreds of molecules. The ΔLZ DLS data is closer to being consistent with monomer formation but indicates a slightly elevated R_s value relative to that expected of a globular monomer. In fact, the 32.5 Å R_s for ΔLZ is consistent with a globular protein with a molecular weight of approximately 64 kDa. This indicates that the ΔLZ DLS data is consistent with either globular dimer formation, or a partially disordered monomer, which would be less compact than an ordered monomer.

Next, disorder was artificially induced by addition of denaturing agents, for the purpose of comparing the resulting data with the above data in non-denaturing conditions. The addition of 1 M urea resulted in the increase of the observed R_s for ΔLZ to 44 Å, consistent with that expected for a highly disordered monomer. This confirms that some order, causing some compactness of the ΔLZ fragment, does

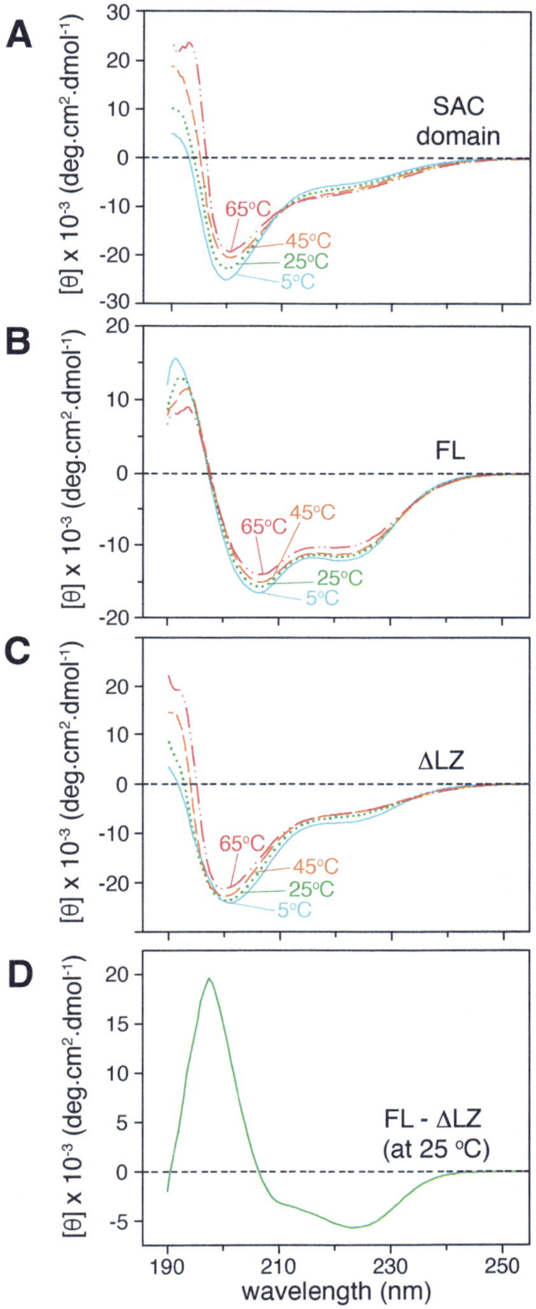

Fig. 9 CD spectra of various Par-4 constructs. (**a**) The isolated Par-4 SAC domain, (**b**) FL rat Par-4, and (**c**) rat ΔLZ. (**d**) Difference between the 25 °C traces in **b** and **c**, showing a ($\theta_{222}/\theta_{208}$) ratio > 1, indicative of a coiled coil. All samples are in 20 mM Tris pH 7.0, 20 mM NaCl. All spectra except (**d**) were recorded at 5 °C intervals from 5–75 °C, with only four equally spaced traces drawn for clarity. (Modified from the original figures in Ref. [14])

exist in the absence of urea. In contrast, addition of 1 M urea to FL protein resulted in a marked decrease of the R_s to 78 Å. Thus, the large aggregates are broken up by 1 M urea, but the resulting species still have a Stokes radius larger than the predicted disordered monomer value (50 Å), but close to the size expected for a highly disordered dimer. At 6 M urea, the measured R_s of FL Par-4 drops to 58 Å, very close to the predicted value of a disordered monomer. From this data, it is not possible to interpret the degree of order in aggregated FL protein in non-denaturing conditions.

4.3 CD Spectroscopy of FL and ΔLZ Par-4

CD spectra of FL Par-4 taken at a range of temperatures are shown in Fig. 9b. The minima near 208 nm and 222 nm are consistent with α-helix formation, but the intensity ratio of these two minima is opposite of what would be expected for coiled coil formation. However, it should be understood that this spectrum is expected to depict a combination of α-helical and disordered regions, which would enhance the 208 nm minimum relative to the 222 nm minimum. There appears to be an isodichroic point just below 200 nm, which would be consistent with a two-state equilibrium, with changes in temperature causing a shift in the equilibrium. However, the ellipticity is at or very near to zero at the isodichroic point. This sort of isodichroic point can result from a mere change in intensity due to changes in, for instance, sample concentration, or effective sample concentration, due to scattering. Also, the overall shape of the CD spectrum does not change with temperature. Taken together, the most supportable interpretation of these spectra is that the FL Par-4 protein remains folded and highly helical throughout the temperature range tested, but higher temperature may drive additional aggregation which causes a slight intensity reduction in the CD spectra.

In contrast, CD spectra of the ΔLZ construct (Fig. 9c) produce a minimum near 200 nm and only a very minor minimum near 222 nm. This indicates a highly disordered structure, with a small amount of α-helix formation. The absence of an isodichroic point suggests that no simple two-state equilibrium is present. Like the FL protein, the ΔLZ fragment shows only modest changes with temperature, indicating thermal stability.

Figure 9d shows the difference between the CD spectra of the FL (Fig. 9b) and the ΔLZ (Fig. 9c). This difference spectrum has the model shape of a coiled coil, with minima at 208 nm and 222 nm, and the 222 minimum is more intense. This clearly shows that the residues forming the LZ, in the context of the FL protein, form a coiled coil and quite possibly assist the remainder of the CC region (see Fig. 8a) to form a coiled coil as well. Note that these spectra were taken at pH 7.0, an environment in which the isolated LZ is highly disordered. Thus, it appears that the remainder of the protein helps the LZ to fold under neutral conditions and vice versa: the LZ helps the remainder of the protein to fold or at least the remainder of the CC. As mentioned above, folding is also associated with aggregation under the conditions tested.

4.4 NMR Studies of FL and ΔLZ Par-4

The ^{1}H,^{15}N-HSQC spectra of FL Par-4 and the ΔLZ fragment display characteristic features of intrinsic disorder including sharp peaks with narrow line widths and poor chemical shift dispersion in both ^{1}H and ^{15}N dimensions (Fig. 10). The number of peaks, particularly for the FL, are far less than one per amino acid, indicating significant exchange-induced broadening, possibly due to a hidden conformation. The observed peaks are therefore most likely from disordered regions of the protein which undergo localized fast motion, while ordered regions are hidden due to the overall slow molecular tumbling of the aggregates. Overlaying the spectra reveals that the ΔLZ peaks are not a subset of the FL peaks, indicating distinct local chemical environments for the nuclei of each construct. This supports the DLS and CD-based observations regarding the global influence of the LZ region: absence of the LZ (in ΔLZ and SAC constructs) enables these constructs to sample different regions of conformational space, involving mostly disorder.

5 The Human cl-Par-4 Fragment

In 2012, it was reported that as part of the caspase cascade during apoptosis, human Par-4 is cleaved into two pieces by caspase-3 [11]. The resulting C-terminal fragment (cl-Par-4) contains 209 amino acids (see Fig. 1). The cl-Par-4 fragment continues to be active in apoptosis: it migrates to the nucleus and alters gene expression in favor of cell death [11, 21, 22]. It is interesting to note that the cl-Par-4 fragment contains the entire CC domain and the entire SAC domain. In fact, the cleavage site is only 13 amino acids N-terminal to the start of the SAC domain. This opens the possibility that the SAC domain, for which there is currently little evidence of structure in full-length Par-4, may be induced to fold by the caspase-3-mediated cleavage of Par-4.

Initial structural studies of the human cl-Par-4 fragment indicated a pronounced tendency to aggregate in vitro at neutral pH with low ionic strength. Reduction of pH and increase in ionic strength each greatly reduced this aggregation, and conditions were identified under which cl-Par-4 forms well-ordered dimers [23] or a tetramer [24]. Biophysical analyses under various conditions suggest that the SAC domain may be helical in the context of the cl-Par-4 fragment.

5.1 The Effects of pH (Coiled Coil Dimer)

The helicity and self-association state of human cl-Par-4 were assessed as a function of pH, at low ionic strength (20 mM NaCl) and ambient temperature [23]. The resulting CD spectra are presented in Fig. 11. The spectrum at pH 4 is clearly more

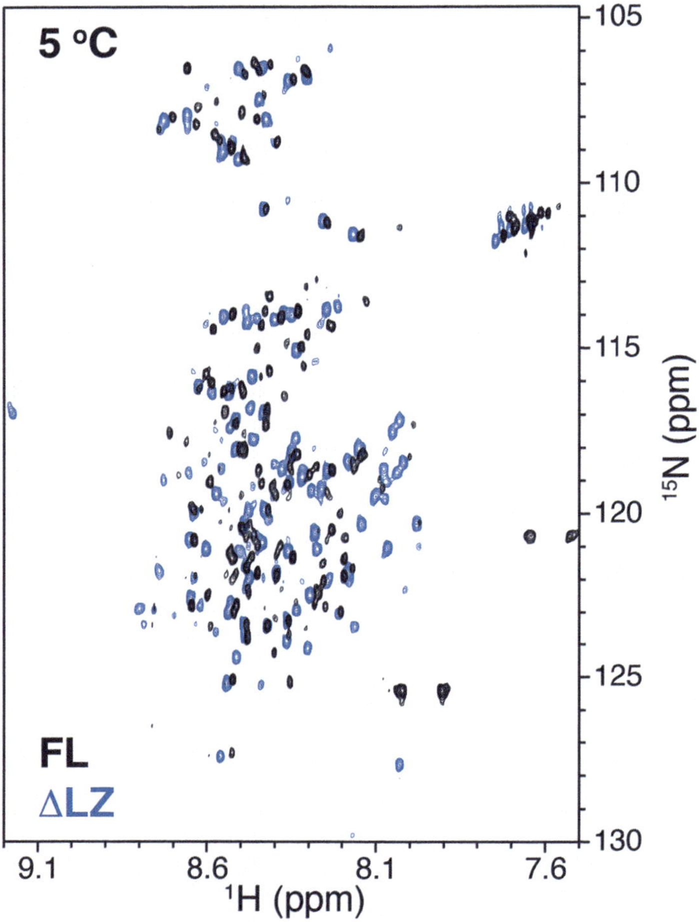

Fig. 10 ^{1}H,^{15}N-HSQC spectra of larger Par-4 constructs. Spectrum of FL rat Par-4 (black contours) and ΔLZ (blue contours) were each recorded at 5 °C. (Modified from the original figures in Ref. [14])

intense than those at neutral or basic pH. The shape and intensity at pH 4 are consistent with a high degree of helicity (~80%), as shown by the deconvolution results in Table 2. Spectra at each pH, however, show some indication of helicity, with characteristic minima at 208 and 222 nm. Spectral deconvolution (Table 2) suggests 62% helicity at pH 10 and 50% helicity at pH 7, with beta content increasing to make up the difference. Little to no beta content is indicated at pH 4. For comparison, the CD spectrum of a highly disordered conformation induced by SDS is also shown in Fig. 11.

Interestingly, the intensity ratio of the two minima ($\theta_{222}/\theta_{208}$) was highest (1.6) at pH 7, followed by 1.2 and 1.1 at pH 4 and 10, respectively (Table 2). Each of these

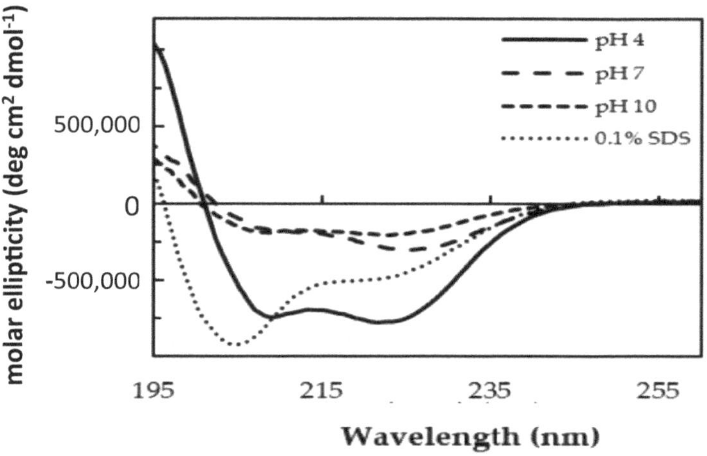

Fig. 11 CD spectra of cl-Par-4 at acidic, neutral, and basic pH. All spectra were recorded at ambient temperature and at 20 mM NaCl. For comparison, the CD spectrum under denaturing conditions (0.1% SDS) is also shown. (Modified from the original figures in Ref. [23])

Table 2 Data analysis of cl-Par-4 vs. pH

Attributes	pH 4	pH 7	pH 10
% α-helix	81	50	62
% β-content	~0	30	15
$\theta_{222}/\theta_{208}$	1.2	1.6	1.1
R_s (nm)	43	483	339

CD deconvolution using Selcon3 was used to estimate secondary structure content. The dichroism ratios ($\theta_{222}/\ \theta_{208}$) suggest coiled coil formation. Stokes radii (R_s) vary greatly

ratios, being greater than 1, would seem to suggest coiled coil formation. However, this ratio is simply related to coiled coil formation only in the case of isolated helices: significant beta or other secondary structure content can skew the results. In this case, only the ratio at pH 4, which is apparently 80% helical, can be simply interpreted to show that *coiled coil* is present.

Investigation of particle size at these pH values was performed using DLS. The data showed large particles at pH 7 and 10 (mean R_s of 483 and 339 nm, respectively) and significantly smaller particles at pH 4 (mean R_s of 43 nm) (Table 2). Moreover, Stokes radii at pH 7 and 10 were found to drift substantially over a 7-day period, while little variation in R_S with time was observed at pH 4. These data clearly show that cl-Par-4 forms large aggregates at neutral or basic pH. This suggests that the relatively low intensity of dichroism recorded at neutral and basic pH could be due at least in part to scattering losses and/or a lower effective protein cross section due to non-uniform distribution of molecules [43]. This then adds uncertainty to the analysis of secondary structure by deconvolution. However, it is safe to say that at pH 4, which produces small particles, cl-Par-4 forms a highly helical *coiled coil*.

Unfortunately, achievable concentrations at pH 4 were not sufficient to use SEC-MALS to more precisely determine the self-association state. However, the estimate of a dimer at pH 4 can be arrived at by comparison with DLS and SEC-MALS results for cl-Par-4 under high salt conditions, as discussed in the following section.

5.2 *The Effects of Sodium (Non-coiled Coil Tetramer)*

The helicity and self-association state of cl-Par-4 were also assessed as a function of ionic strength (NaCl), at neutral pH and ambient temperature [24]. Eight different NaCl concentrations in the range of 0.02–3.0 M were used. CD spectra indicated helicity throughout the salt concentration range, with observed characteristic minima at 208 and 222 nm (Fig. 12). The magnitude of dichroism was found to increase with ionic strength up to 0.4 M, above which magnitude did not change significantly. Spectral deconvolution suggests approximately 80% helix at each NaCl concentration above 0.25 M. Below this range, helical prediction decreases, with compensating increase in beta content. Interestingly, at least 20% beta content is indicated throughout the NaCl range, which differs from the case seen at low pH, where little to no beta content was indicated. This difference may or may not correspond to an actual difference in beta content, as the uncertainty of deconvolution techniques in detecting beta content might be expected to be in the 10–20% range [44].

The intensity ratios of the two minima ($\theta_{222}/\theta_{208}$) revealed a decreasing trend from 1.71 to 0.92 with increase in ionic strength from 0.02 to 3 M (Table 3). The

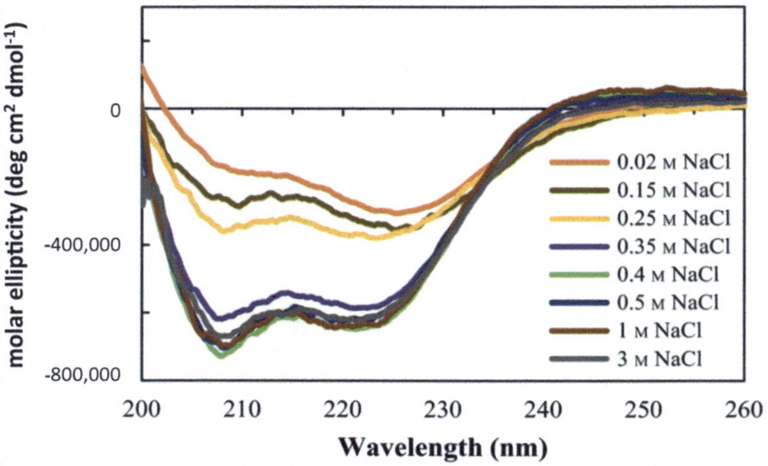

Fig. 12 CD spectra of cl-Par-4 vs. NaCl concentration. All spectra were recorded at ambient temperature and pH 7.0. (Modified from the original figures in Ref. [24])

Table 3 Data analysis of cl-Par-4 vs. NaCl concentration

	[NaCl] (M)			
Attributes	0.02	0.25	0.50	1.0
% α-helix	46 (19*)	80	80	80
% β-content	30 (60*)	21	20	20
$\theta_{222}/\theta_{208}$	1.7	1.0	0.88	0.88
R_s (nm)	420	455	115	86

CD deconvolution using Selcon3 was used to estimate secondary structure content. The asterisks (*) indicate CD deconvolution results using BeStSel. The dichroism ratios ($\theta_{222}/\theta_{208}$) suggest the presence of non-coiled coil helices at high salt. Stokes radii (R_s) vary greatly

ratios greater than 1 observed at 0.25 M and below would seem to suggest coiled coil formation. However, again, the possible presence of significant non-helical content and large particles at lower salt concentrations could skew these results. At higher salt concentrations (0.35 M and above), where dichroism intensity and helical content are both high, the observed ($\theta_{222}/\theta_{208}$) ratios below 1 can more reliably be interpreted to indicate that a significant amount of *non-coiled coil* helix is present. This does not preclude the presence of some amount of coiled coil. To further assess the possibility of significant beta content at low salt, an additional spectral deconvolution tool (BeStSel) was used (Table 3). BeStSel [45, 46] deconvolution for the 0.02 M NaCl spectrum suggests approximately 60% beta content (sheet + turn) against 30% beta content (sheet + turn) obtained with the Selcon3 [47] algorithm.

DLS was used to analyze the effect of salt on hydrodynamic size. Stokes radii measured at low salt (0.02 and 0.25 M NaCl) showed large aggregate formation with R_s larger than 400 nm (Table 3), consistent with the large particles observed at neutral pH in the previous section. Particle size significantly decreases with high salt (0.5 and 1.0 M NaCl). Interestingly, the detected R_s at 1.0 M NaCl (86 nm) is almost precisely double than discussed in the previous section for low salt and pH 4 (R_S of 43 nm). This suggests a higher-order structure under the influence of high salt than under the influence of low pH.

Achievable cl-Par-4 concentration at various salt concentrations were sufficient to allow further analysis of the self-association state via SEC-MALS. At 1 M salt, the obtained average molar mass of 94 kDa is consistent with tetramer formation. Lower salt concentrations produced molar masses consistent with the presence of aggregates in the 20mer range. Larger aggregates may have difficulty traversing the SEC column. Again, the presence of these aggregates could complicate secondary structure deconvolution at low salt. However, the results at high salt clearly support the presence of a highly helical tetramer, with *low coiled coil* content. Furthermore, since the tetramer produces an R_S value significantly larger (approximately double) than observed at pH 4, this suggests that the pH 4 coiled coil is a dimer.

6 Discussion

6.1 Folding of the Isolated LZ

Conformational instability of the Par-4 LZ appears to arise from charge-charge repulsions between acidic residues occupying the "g" and "e" positions (D_{305} and E_{310}, respectively, in rat Par-4; see Fig. 2). A homology model of the Par-4 LZ (Fig. 13a), built using the GCN4 structure (PDB 2ZTA) as a template, shows that acidic residues D_{305} and E_{310} are part of an extensive negatively charged surface and so would repel across the dimer interface. Such a charge mismatch could potentially be accommodated by a coiled coil due to the considerable stability inherent in hydrophobic interactions at the core of the dimer (the "a" and "d" positions). However, in the case of the Par-4 LZ, two of the "a" positions are occupied by polar asparagine residues (N_{313} and N_{320} in rat, colored magenta in Fig. 2 and in Fig. 11a), reducing the hydrophobic attraction. The net effect of the placement of these acidic

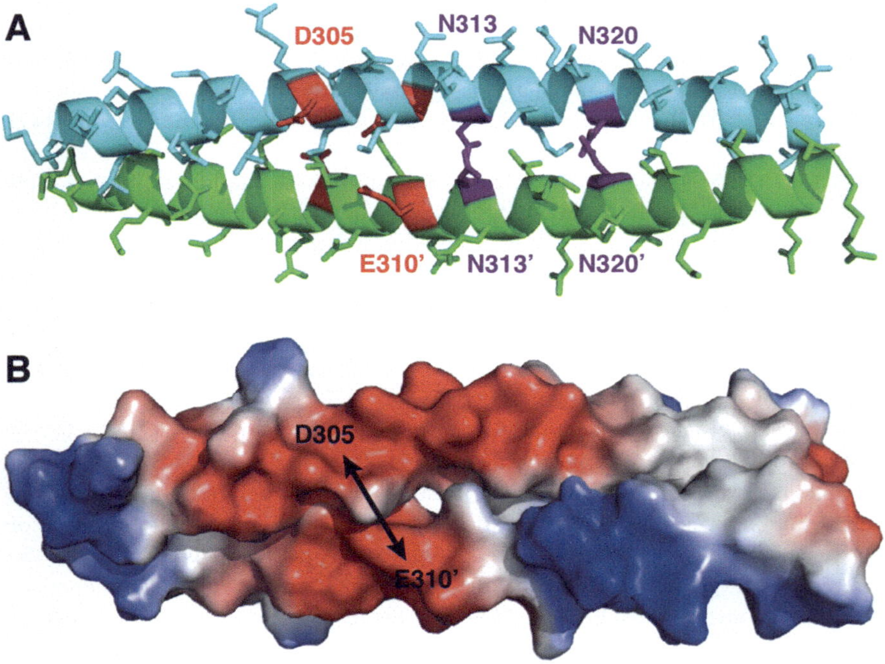

Fig. 13 Model of the rat Par-4 LZ parallel dimer produced using the GCN4 structure (PDB 2ZTA) as a template. (**a**) Cartoon representation highlighting the repelling negatively charged residues D_{305} and E_{310} (red) and the destabilizing Asn residues N_{313} and N_{320} (purple). (**b**) Electrostatic surface representation in the same orientation as **a**. Repulsive interaction between D_{305} and E_{310} of the opposite chain is labeled with a black arrow. (Modified from the original figure in Ref. [18])

and polar residues is a destabilization of the LZ dimer and formation of a largely disordered, or partially ordered, monomer (POM).

Stabilization of the isolated LZ to form the coiled coil dimer (CCD) was achieved experimentally via a number of strategies: (1) mutation of either D_{305} or E_{310} to lysine to create an attraction rather than a repulsion; (2) screening of the D_{305}-E_{310} repulsion by sodium (or other) counter-ions; and (3) acidic pH, which partially protonates the acidic residues, lessening their repulsion. It should be noted that the pK_a of the acidic side chains of Asp and Glu are nominally near 4 or below. However, in a highly acidic or negatively charged microenvironment such as the red surface shown in Fig. 11b, solvent protons are attracted, which would raise the effective pK_a of the acidic side chains to the 5–6 range. Thus, lowering the pH to this range is sufficient to significantly reduce the D_{305}-E_{310} charge-charge repulsion.

6.2 *Other Par-4 Constructs*

The above considerations are for the isolated LZ. In the context of FL Par-4, additional stabilizing factors are potentially present. Thus, at pH 7 and in low salt, the FL Par-4 is not monomeric, but is highly aggregated, with some α-helix present (see Fig. 9b). At pH 7 and in low salt, the cl-Par-4 fragment behaves similarly to the FL Par-4, with aggregation evident. However, conditions which stabilize the isolated LZ, namely, high salt or acidic pH, were shown to disaggregate the cl-Par-4 fragment into well-ordered dimers or tetramers. It is not yet known whether FL Par-4 would similarly disaggregate under these conditions. These results suggest that depending upon the environment, the particle size of the naturally occurring cl-Par-4 fragment can vary considerably. This effect is apparently linked to the charge characteristics in the LZ region, as those conditions that favor LZ formation also favor formation of small, well-folded cl-Par-4 particles. It may be that misfolding at the dimer interface leads to aggregation, while a more stable dimer interface favors smaller particles (dimers or tetramers).

This linked aggregation/misfolding event may play a role in post-translational regulation of Par-4. Cells undergoing apoptosis have been shown to be subject to significant changes in pH and ionic environment [48–52]. Each of these actions could induce large and possibly disordered cl-Par-4 particles, or even aggregated FL Par-4, to fold into smaller ordered particles that are able to function more efficiently in apoptotic pathways. It seems likely that formation of smaller particles may also uncover the NLS2 and the remainder of the SAC domain, at least in cl-Par-4, allowing both more efficient nuclear translocation and interaction with GRP78 at the cell surface. Translocation to the nucleus or to the cell membrane may also be facilitated simply by size considerations.

The SAC domain is the minimal region of Par-4 that can cause apoptosis in certain cells. When isolated in vitro, the SAC domain was shown to be highly disordered and monomeric, with some helical propensities. In other words, it is a POM. Altering the salt and pH conditions did not significantly alter the SAC conformation (unpublished results). However, in the context of the cl-Par-4 domain, the SAC domain appears to be helical, at least in either high salt or low pH conditions. The evidence for this lies in the fact that cl-Par-4 approaches 80% helicity under these conditions (Figs. 11 and 12, Tables 2 and 3). The CC domain comprises only 38% of the cl-Par-4 fragment (see Fig. 1). Thus, an additional stretch of helix, about the same size as the CC, must also be formed. The SAC domain is just about the right size and has the most helical propensity (see Fig. 8c) of any remaining part of the cl-Par-4 fragment. Interestingly, the CD spectra of the FL protein at pH 7 (Fig. 9b) and low salt is consistent with 24% helicity, which suggests that in the context of FL Par-4 under these conditions, the SAC domain is not helical. It has not yet been determined how pH and salt affect the helicity and self-association state of FL Par-4 and hence how the SAC domain fold is affected by environment in the context of the FL protein. It could be that folding of the SAC domain requires caspase-induced cleavage of Par-4 to form the cl-Par-4 fragment. This cleavage event may not only allow disaggregation of Par-4 (in the form of cl-Par-4), resulting in exposure of the NLS2 and the SAC domain, but it may also trigger folding of the newly exposed SAC domain.

It was mentioned in the introduction that the PAF fragment, the N-terminal fragment released after caspase-induced cleavage, is thought to function mainly as a decoy that protects cl-Par-4 from degradation. But it may also function in the context of FL-Par-4. If the above analysis is correct, the PAF region may help cover the SAC domain and its NLS2 signal, preventing the SAC domain from folding and preventing the formation of small, more active, and more easily translocated particles. Thus, the PAF region may help the FL Par-4 remain in a less active, aggregated conformation, awaiting the signals of caspase-induced cleavage and environmental changes to fold and expose the SAC domain.

6.3 Conclusions and Future Perspectives

The first reports of the formation of cl-Par-4 via the action of caspases appeared in 2012 [11]. As described above, it has since been established that cl-Par-4 can form highly helical coiled coil dimers, highly helical mostly non-coiled tetramers, and larger aggregates, in vitro. The conditions favoring cl-Par-4 dimers and tetramers mirror the conditions favoring folding of the isolated leucine zipper in earlier studies, suggesting that the point mutations used to stabilize the leucine zipper might also stabilize the dimer or the tetramer. These concepts are salient to potential future investigations.

An obvious next step is atomic resolution three-dimensional structural determination of cl-Par-4 via NMR spectroscopy, X-ray crystallography, and/or small-angle X-ray scattering (SAXS). Such studies of partially disordered proteins with possibly interconverting subpopulations and tendencies to aggregate can, however, be extremely difficult [53, 54]. The above knowledge of conditions favoring discreet conformations will be of assistance: structural analysis could be performed at high salt, at low pH, or using stabilizing mutations. These factors could also be combined. For instance, a multivariate analysis, including also divalent salts, may identify conditions under which moderate salt concentrations combined with moderately acidic pH may be sufficient to induce a well-folded dimer or tetramer structure. These combinatorial conditions may be more in line with conditions present within a cancer cell.

In addition, conformation equilibrium could be investigated via NMR spectroscopy. In this approach, a range of relaxation-based NMR experiments are combined to probe structural interconversions spanning the microsecond to second timescales [55]. This approach could also yield protein folding insights, such as the structure and dynamics of a critical but lowly populated nucleation seed conformation [56].

As with many IDPs, the folding of Par-4, cl-Par-4, or the SAC domain may be affected by micro-environment, including interaction with binding partners (for a review of disorder-to-order transitions, see [57]). These interactions may, for instance, drive cl-Par-4 into one of the above conformations. Figure 14 graphically

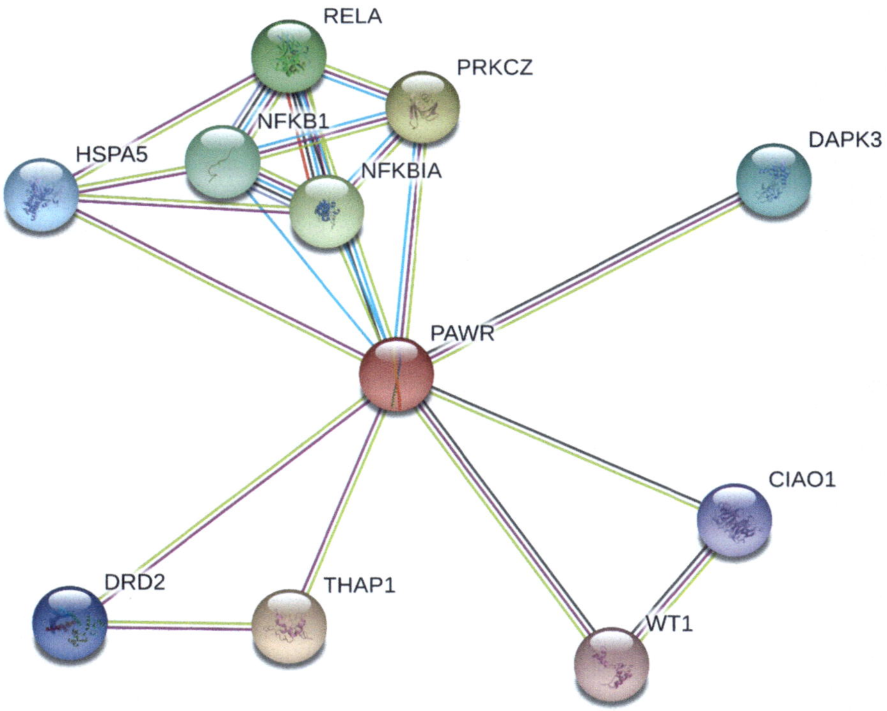

Fig. 14 Partial human Par-4 interactome. Database-derived interactome for human Par-4 using the STRING online tool [58]. Structural information regarding each protein is indicated by the cartoon within each bubble. Connecting colored lines indicate evidence of interaction from the following sources: teal, known from curated databases; magenta, experimentally determined; dark green, gene neighborhood; red, gene fusions; light green, text mining; black, co-expression; lavender, protein homology

presents a few of the many interactions attributed to Par-4 [58]. Future structural studies of protein complexes involving Par-4 may be integral in determining the fully functional conformation of the Par-4 molecule and its functional fragments. Another avenue would be in vivo NMR, wherein NMR-labeled Par-4 would be studied structurally within a relevant mammalian cell environment [59–64]. Full in vivo structural analysis may not be required: fingerprint analysis, such as identification of certain NMR peak patterns, could suffice to show that, for instance, Par-4 has a similar conformation in vivo as it does at low pH or high salt, in vitro.

Knowledge of environmental influences on Par-4 structure could also lead to the discovery of additional signaling pathways. To this point, most studies utilizing immunoprecipitation (IP) or similar approaches for identifying interactions with Par-4 have employed full-length Par-4 or fragments identified prior to 2012 (e.g., [65–69]). Similar studies could now be repeated with the more recently identified cl-Par-4 fragment. If different structures and accessibilities are present in cl-Par-4, then new interactions may be uncovered. These studies could potentially be performed using environments that favor the coiled coil dimer or the helical tetramer. This could not only provide additional opportunities to stabilize and detect new interactions but could also indicate which previously established interactions may preferentially interact with the dimer or the tetramer conformation. The results could help to determine whether the dimer, the tetramer, or both are active in apoptotic processes.

However, interactions involving partially disordered/conformationally unstable molecules such as Par-4 can be weak and transient and therefore difficult to detect via traditional methods. To solve this problem, in vivo proximity labeling approaches could be employed [70]. These approaches would biotinylate proteins that form complexes with cl-Par-4, even if that interaction is transient. Furthermore, this approach is selectively targetable to subcellular spaces such as the nucleus [71] and could be used in both healthy and tumor cells, to provide a comparison of the functional interactome of Par-4. Together, these structural and interaction studies would help to elucidate the role that Par-4 plays in apoptotic processes in cancer cells and could lead to new ways to therapeutically utilize specific Par-4 fragments or Par-4-based mimetics in cancer treatments.

References

1. Sells SF, Wood DP Jr, Joshi-Barve SS, Muthukumar S, Jacob RJ, Crist SA, Humphreys S, Rangnekar VM (1994) Commonality of the gene programs induced by effectors of apoptosis in androgen-dependent and -independent prostate cells. Cell Growth Differ 5:457–466
2. Barradas M, Monjas A, Diaz-Meco MT, Serrano M, Moscat J (1999) The downregulation of the pro-apoptotic protein Par-4 is critical for Ras-induced survival and tumor progression. EMBO J 18:6362–6369
3. Cook J, Krishnan S, Ananth S, Sells SF, Shi Y, Walther MM, Linehan WM, Sukhatme VP, Weinstein MH, Rangnekar VM (1999) Decreased expression of the pro-apoptotic protein Par-4 in renal cell carcinoma. Oncogene 18:1205–1208

4. Qiu SG, Krishnan S, el-Guendy, N. & Rangnekar, V. M. (1999) Negative regulation of Par-4 by oncogenic Ras is essential for cellular transformation. Oncogene 18:7115–7123
5. Boghaert ER, Sells SF, Walid AJ, Malone P, Williams NM, Weinstein MH, Strange R, Rangnekar VM (1997) Immunohistochemical analysis of the proapoptotic protein Par-4 in normal rat tissues. Cell Growth Differ 8:881–890
6. Duan W, Rangnekar VM, Mattson MP (1999) Prostate apoptosis response-4 production in synaptic compartments following apoptotic and excitotoxic insults: evidence for a pivotal role in mitochondrial dysfunction and neuronal degeneration. J Neurochem 72:2312–2322
7. Mattson MP, Duan W, Chan SL, Camandola S (1999) Par-4: an emerging pivotal player in neuronal apoptosis and neurodegenerative disorders. J Mol Neurosci 13:17–30
8. Duan W, Guo Z, Mattson MP (2000) Participation of par-4 in the degeneration of striatal neurons induced by metabolic compromise with 3-nitropropionic acid. Exp Neurol 165:1–11
9. Mattson MP, Duan W, Pedersen WA, Culmsee C (2001) Neurodegenerative disorders and ischemic brain diseases. Apoptosis 6:69–81
10. Guo Q, Fu W, Xie J, Luo H, Sells SF, Geddes JW, Bondada V, Rangnekar VM, Mattson MP (1998) Par-4 is a mediator of neuronal degeneration associated with the pathogenesis of Alzheimer disease [see comments]. Nat Med 4:957–962
11. Chaudhry P, Singh M, Parent S, Asselin E (2012) Prostate apoptosis response 4 (Par-4), a novel substrate of caspase-3 during apoptosis activation. Mol Cell Biol 32:826–839
12. El-Guendy N, Zhao Y, Gurumurthy S, Burikhanov R, Rangnekar VM (2003) Identification of a unique core domain of par-4 sufficient for selective apoptosis induction in cancer cells. Mol Cell Biol 23:5516–5525
13. Sells SF, Han SS, Muthukkumar S, Maddiwar N, Johnstone R, Boghaert E, Gillis D, Liu G, Nair P, Monnig S, Collini P, Mattson MP, Sukhatme VP, Zimmer SG, Wood DP Jr, McRoberts JW, Shi Y, Rangnekar VM (1997) Expression and function of the leucine zipper protein Par-4 in apoptosis. Mol Cel Biol 17:3823–3832
14. Libich DS, Schwalbe M, Kate S, Venugopal H, Claridge JK, Edwards PJ, Dutta K, Pascal SM (2009) Intrinsic disorder and coiled-coil formation in prostate apoptosis response factor 4. FEBS J 276:3710–3728
15. Park HH, Lo YC, Lin SC, Wang L, Yang JK, Wu H (2007) The death domain superfamily in intracellular signaling of apoptosis and inflammation. Annu Rev Immunol 25:561–586
16. Dutta K, Alexandrov A, Huang H, Pascal SM (2001) pH-induced folding of an apoptotic coiled coil. Protein Sci 10:2531–2540
17. Dutta K, Engler FA, Cotton L, Alexandrov A, Bedi GS, Colquhoun J, Pascal SM (2003) Stabilization of a pH-sensitive apoptosis-linked coiled coil through single point mutations. Protein Sci 12:257–265
18. Schwalbe M, Dutta K, Libich DS, Venugopal H, Claridge JK, Gell DA, Mackay JP, Edwards PJ, Pascal SM (2010) Two-state conformational equilibrium in the Par-4 leucine zipper domain. Proteins 78:2433–2449
19. Tiruttani Subhramanyam UK, Kubicek J, Eidhoff UB, Labahn J (2017) Structural basis for the regulatory interactions of proapoptotic Par-4. Cell Death Differ 24:1540–1547
20. Hebbar N, Burikhanov R, Shukla N, Qiu S, Zhao Y, Elenitoba-Johnson KSJ, Rangnekar VM (2017) A naturally generated decoy of the prostate apoptosis response-4 protein overcomes therapy resistance in tumors. Cancer Res 77:4039–4050
21. Thayyullathil F, Pallichankandy S, Rahman A, Kizhakkayil J, Chathoth S, Patel M, Galadari S (2013) Caspase-3 mediated release of SAC domain containing fragment from Par-4 is necessary for the sphingosine-induced apoptosis in Jurkat cells. J Mol Signal 8:2
22. Treude F, Kappes F, Fahrenkamp D, Muller-Newen G, Dajas-Bailador F, Kramer OH, Luscher B, Hartkamp J (2014) Caspase-8-mediated PAR-4 cleavage is required for TNFalpha-induced apoptosis. Oncotarget 5:2988–2998
23. Clark AM, Ponniah K, Warden MS, Raitt EM, Yawn AC, Pascal SM (2018) pH-induced folding of the caspase-cleaved Par-4 tumor suppressor: evidence of structure outside of the coiled coil domain. Biomol Ther 8

24. Clark AM, Ponniah K, Warden MS, Raitt EM, Smith BG, Pascal SM (2019) Tetramer formation by the caspase-activated fragment of the Par-4 tumor suppressor. FEBS J
25. Alexandrov A, Dutta K, Pascal SM (2001) MBP-fusion with a viral protease cleavage site: one-step cleavage/purification of insoluble proteins. BioTechniques 30:1194–1198
26. Zhou NE, Kay CM, Hodges RS (1994) The role of interhelical ionic interactions in controlling protein folding and stability. De novo designed synthetic two-stranded alpha-helical coiled-coils. J Mol Biol 237:500–512
27. Jelesarov I, Durr E, Thomas RM, Bosshard HR (1998) Salt effects on hydrophobic interaction and charge screening in the folding of a negatively charged peptide to a coiled coil (leucine zipper). Biochemistry 37:7539–7550
28. Krylov D, Barchi J, Vinson C (1998) Inter-helical interactions in the leucine zipper coiled coil dimer: pH and salt dependence of coupling energy between charged amino acids. J Mol Biol 279:959–972
29. Durr E, Jelesarov I, Bosshard HR (1999) Extremely fast folding of a very stable leucine zipper with a strengthened hydrophobic core and lacking electrostatic interactions between helices. Biochemistry 38:870–880
30. Krylov D, Mikhailenko I, Vinson C (1994) A thermodynamic scale for leucine zipper stability and dimerization specificity: e and g interhelical interactions. EMBO J 13:2849–2861
31. Kohn WD, Monera OD, Kay CM, Hodges RS (1995) The effects of interhelical electrostatic repulsions between glutamic acid residues in controlling the dimerization and stability of two-stranded alpha-helical coiled-coils. J Biol Chem 270:25495–25506
32. Kohn WD, Kay CM, Hodges RS (1997) Salt effects on protein stability: two-stranded alpha-helical coiled-coils containing inter- or intrahelical ion pairs. J Mol Biol 267:1039–1052
33. Muhle-Goll C, Gibson T, Schuck P, Schubert D, Nalis D, Nilges M, Pastore A (1994) The dimerization stability of the HLH-LZ transcription protein family is modulated by the leucine zippers: a CD and NMR study of TFEB and c-Myc. Biochemistry 33:11296–11306
34. Lavigne P, Kondejewski LH, Houston ME Jr, Sonnichsen FD, Lix B, Sykes BD, Hodges RS, Kay CM (1995) Preferential heterodimeric parallel coiled-coil formation by synthetic Max and c-Myc leucine zippers: a description of putative electrostatic interactions responsible for the specificity of heterodimerization. J Mol Biol 254:505–520
35. Ozeki S, Kato T, Holtzer ME, Holtzer A (1991) The kinetics of chain exchange in two-chain coiled coils: alpha- and beta beta-tropomyosin. Biopolymers 31:957–966
36. Wendt H, Berger C, Baici A, Thomas RM, Bosshard HR (1995) Kinetics of folding of leucine zipper domains. Biochemistry 34:4097–4107
37. Zitzewitz JA, Bilsel O, Luo J, Jones BE, Matthews CR (1995) Probing the folding mechanism of a leucine zipper peptide by stopped-flow circular dichroism spectroscopy. Biochemistry 34:12812–12819
38. Sosnick TR, Jackson S, Wilk RR, Englander SW, DeGrado WF (1996) The role of helix formation in the folding of a fully alpha-helical coiled coil. Proteins 24:427–432
39. Wendt H, Leder L, Harma H, Jelesarov I, Baici A, Bosshard HR (1997) Very rapid, ionic strength-dependent association and folding of a heterodimeric leucine zipper. Biochemistry 36:204–213
40. Marti DN, Jelesarov I, Bosshard HR (2000) Interhelical ion pairing in coiled coils: solution structure of a heterodimeric leucine zipper and determination of pKa values of Glu side chains. Biochemistry 39:12804–12818
41. Wishart DS, Sykes BD (1994) The 13C chemical-shift index: a simple method for the identification of protein secondary structure using 13C chemical-shft data. J Biomol NMR 4:171–180
42. Shekhtman A, Ghose R, Goger M, Cowburn D (2002) NMR structure determination and investigation using a reduced proton (REDPRO) labeling strategy for proteins. FEBS Lett 524:177–182
43. Wallace BA, Mao D (1984) Circular dichroism analyses of membrane proteins: an examination of differential light scattering and absorption flattening effects in large membrane vesicles and membrane sheets. Anal Biochem 142:317–328

44. Greenfield NJ (1996) Methods to estimate the conformation of proteins and polypeptides from circular dichroism data. Anal Biochem 235:1–10
45. Micsonai A, Wien F, Kernya L, Lee YH, Goto Y, Refregiers M, Kardos J (2015) Accurate secondary structure prediction and fold recognition for circular dichroism spectroscopy. Proc Natl Acad Sci U S A 112:E3095–E3103
46. Micsonai A, Wien F, Bulyaki E, Kun J, Moussong E, Lee YH, Goto Y, Refregiers M, Kardos J (2018) BeStSel: a web server for accurate protein secondary structure prediction and fold recognition from the circular dichroism spectra. Nucleic Acids Res 46:W315–W322
47. Whitmore L, Wallace BA (2004) DICHROWEB, an online server for protein secondary structure analyses from circular dichroism spectroscopic data. Nucleic Acids Res 32:W668–W673
48. Lagadic-Gossmann D, Huc L, Lecureur V (2004) Alterations of intracellular pH homeostasis in apoptosis: origins and roles. Cell Death Differ 11:953–961
49. Ishaque A, Al-Rubeai M (1999) Role of Ca2+, Mg2+ and K+ ions in determining apoptosis and extent of suppression afforded by bcl-2 during hybridoma cell culture. Apoptosis 4:335–355
50. Franklin RB, Costello LC (2009) The important role of the apoptotic effects of zinc in the development of cancers. J Cell Biochem 106:750–757
51. Bortner CD, Cidlowski JA (2014) Ion channels and apoptosis in cancer. Philos Trans R Soc Lond B Biol Sci 369:20130104
52. Pilchova I, Klacanova K, Tatarkova Z, Kaplan P, Racay P (2017) The involvement of Mg(2+) in regulation of cellular and mitochondrial functions. Oxidative Med Cell Longev 2017:6797460
53. Wright PE, Dyson HJ (2015) Intrinsically disordered proteins in cellular signalling and regulation. Nat Rev Mol Cell Biol 16:18–29
54. Bhowmick A, Brookes DH, Yost SR, Dyson HJ, Forman-Kay JD, Gunter D, Head-Gordon M, Hura GL, Pande VS, Wemmer DE, Wright PE, Head-Gordon T (2016) Finding our way in the dark proteome. J Am Chem Soc 138:9730–9742
55. Anthis NJ, Clore GM (2015) Visualizing transient dark states by NMR spectroscopy. Q Rev Biophys 48:35–116
56. Kotler SA, Tugarinov V, Schmidt T, Ceccon A, Libich DS, Ghirlando R, Schwieters CD, Clore GM (2019) Probing initial transient oligomerization events facilitating huntingtin fibril nucleation at atomic resolution by relaxation-based NMR. Proc Natl Acad Sci U S A 116:3562–3571
57. Mendoza-Espinosa P, Garcia-Gonzalez V, Moreno A, Castillo R, Mas-Oliva J (2009) Disorder-to-order conformational transitions in protein structure and its relationship to disease. Mol Cell Biochem 330:105–120
58. Szklarczyk D, Gable AL, Lyon D, Junge A, Wyder S, Huerta-Cepas J, Simonovic M, Doncheva NT, Morris JH, Bork P, Jensen LJ, Mering CV (2019) STRING v11: protein-protein association networks with increased coverage, supporting functional discovery in genome-wide experimental datasets. Nucleic Acids Res 47:D607–D613
59. Matecko-Burmann I, Burmann BM (2020) Recording in-cell NMR-spectra in living mammalian cells. Methods Mol Biol 2141:857–871
60. Gerez JA, Prymaczok NC, Riek R (2020) In-cell NMR of intrinsically disordered proteins in mammalian cells. Methods Mol Biol 2141:873–893
61. Bekei B, Rose HM, Herzig M, Stephanowitz H, Krause E, Selenko P (2012) In-cell NMR in mammalian cells: part 3. Methods Mol Biol 895:67–83
62. Barbieri L, Luchinat E, Banci L (2016) Characterization of proteins by in-cell NMR spectroscopy in cultured mammalian cells. Nat Protoc 11:1101–1111
63. Bekei B, Rose HM, Herzig M, Selenko P (2012) In-cell NMR in mammalian cells: part 2. Methods Mol Biol 895:55–66
64. Bekei B, Rose HM, Herzig M, Dose A, Schwarzer D, Selenko P (2012) In-cell NMR in mammalian cells: part 1. Methods Mol Biol 895:43–54
65. Diaz-Meco MT, Municio MM, Frutos S, Sanchez P, Lozano J, Sanz L, Moscat J (1996) The product of par-4, a gene induced during apoptosis, interacts selectively with the atypical isoforms of protein kinase C. Cell 86:777–786

66. Johnstone RW, See RH, Sells SF, Wang J, Muthukkumar S, Englert C, Haber DA, Licht JD, Sugrue SP, Roberts T, Rangnekar VM, Shi Y (1996) A novel repressor, par-4, modulates transcription and growth suppression functions of the Wilms' tumor suppressor WT1. Mol Cel Biol 16:6945–6956
67. Page G, Kogel D, Rangnekar V, Scheidtmann KH (1999) Interaction partners of Dlk/ZIP kinase: co-expression of Dlk/ZIP kinase and Par-4 results in cytoplasmic retention and apoptosis. Oncogene 18:7265–7273
68. Goswami A, Burikhanov R, de Thonel A, Fujita N, Goswami M, Zhao Y, Eriksson JE, Tsuruo T, Rangnekar VM (2005) Binding and phosphorylation of par-4 by akt is essential for cancer cell survival. Mol Cell 20:33–44
69. Goswami A, Qiu S, Dexheimer TS, Ranganathan P, Burikhanov R, Pommier Y, Rangnekar VM (2008) Par-4 binds to topoisomerase 1 and attenuates its DNA relaxation activity. Cancer Res 68:6190–6198
70. Roux KJ, Kim DI, Raida M, Burke B (2012) A promiscuous biotin ligase fusion protein identifies proximal and interacting proteins in mammalian cells. J Cell Biol 196:801–810
71. Rhee HW, Zou P, Udeshi ND, Martell JD, Mootha VK, Carr SA, Ting AY (2013) Proteomic mapping of mitochondria in living cells via spatially restricted enzymatic tagging. Science 339:1328–1331

Structural Analysis of Par-4 and Crystallographic Analysis of the Regulatory Domain

J. Labahn and U. K. Tiruttani Subhramanyam

Abstract Par-4 is a largely disordered 332 amino-acid proapoptotic protein with tumor suppressor activity that is part of extrinsic/death receptor-mediated and the intrinsic/mitochondrial apoptotic pathways. Par-4 contains the SAC domain (residues 137–195), which is sufficient to selectively induce apoptosis in cancer cells. A second domain of Par-4 (Par-4_{CC}, residues 254–332) was predicted to contain a leucine zipper domain and a nuclear export sequence.

The structural analysis of the par-4 protein showed for the SAC domain in NMR experiments a molten-globular-like state whereas the C-terminal domain of par-4 allowed crystallization and crystallographic structure determination at 3 Å resolution using MAD date obtained from Se-methionine-labeled protein. This domain forms a coiled-coil structure and consists of several subdomains and sites that govern folding, dimerization, and serve as a binding region for interaction partners. Three major subdomains were identified:

(i) The HDD domain appears to control folding, homodimerization, and interaction with Thap1. (ii) The leucine zipper domain is suggested to be relevant for forming higher hetero-oligomerization states in the cases of DAPK3, Amida, and AATF which interact with the C-terminal part of par-4 which includes the leucine zipper subdomain. (iii) A nuclear export sequence (NES) whose masking controls cellular localization is blocked by all four proteins in the oligomerization-dependent model for regulatory coiled-coiled interactions with par-4.

J. Labahn (✉)
Institut für Biologische Informationsprozesse, Strukturbiochemie (IBI-7), Forschungszentrum Jülich, Jülich, Germany

Institut für Physikalische Biologie, Heinrich-Heine-Universität Düsseldorf, Düsseldorf, Germany

Centre for Structural Systems Biology (CSSB), Deutsches Elektronen-Synchrotron (DESY), Hamburg, Germany
e-mail: j.labahn@fz-juelich.de

U. K. Tiruttani Subhramanyam
Institut für Biologische Informationsprozesse, Strukturbiochemie (IBI-7), Forschungszentrum Jülich, Jülich, Germany

Centre for Structural Systems Biology (CSSB), Deutsches Elektronen-Synchrotron (DESY), Hamburg, Germany

V. M. Rangnekar (ed.), *Tumor Suppressor Par-4*,
https://doi.org/10.1007/978-3-030-73572-2_4

Keywords Par-4 · Crystallography · Coiled-coil domain · Dimerization · Interaction models

1 Introduction

Par-4 is a 332 amino-acid proapoptotic protein with tumor suppressor activity which had been predicted to be largely disordered, respectively, natively unfolded [1]. Despite these discouraging predictions, we decided to investigate this protein structurally because of its involvement in important topics, e.g., apoptosis, cancer, and M. Alzheimer.

Par-4 interacts with many proteins and appears to be relevant in diseases that involve apoptotic processes. Its downregulation or deactivation has been linked to cancer [2] and neurodegenerative diseases [3–5] and the pathogenesis of HIV encephalitis [6]. It is specifically involved in both the extrinsic/death receptor-mediated and the intrinsic/mitochondrial pathways of apoptosis [7] and is conserved in vertebrates [8]. Par-4 contains the SAC domain (residues 137–195), which selectively induces apoptosis in cancer cells. It is sufficient for Fas pathway activation, inhibition of NF-κB activity, and apoptotic induction [9]. The latter two functions require its nuclear translocation and phosphorylation at Thr155 [7] by protein kinase A (PKA) [10] or death-associated protein-like kinase (Dlk/DAPK3) [11] and probably the nuclear localization sequence at residues 137–153 [9]. An Akt1 phosphorylation site between the SAC domain and the C-terminal region of Par-4 at Ser249 acts as a general pro-survival switch which is relevant in various cancers [2] and cellular localization [12].

A second regulatory domain of Par-4 (Par-4_{CC}, residues 254–332) was predicted to form a coiled coil [1] possibly containing leucine zipper domain (Par-4_{LZ}, residues 292–332) and a nuclear export sequence (Par-4_{NES}) [1, 7] relevant to sensitize cells to various apoptotic stimuli [13], but is not necessary for the direct induction of apoptosis [9]. This domain was shown to mediate the interaction with partner proteins such as PKCζ [14], WT1 [15], Akt1 [12], apoptosis antagonizing transcription factor (AATF)/Che-1 [16], DAPK3 [17], Amida [18], THAP1 [19], and p62 [20]. Par-4 specifically interacts with DNA-binding proteins like THAP1 (a promyelocytic leukemia nuclear bodies (PML-NBs)-associated protein), which contains a DNA-binding THAP domain [19], or WT1. The latter interaction transcriptionally downregulates Bcl-2 and thereby induces apoptosis [7]. The aforementioned features of Par-4 point to the C-terminal region of Par-4 as a regulatory domain, which interacts with various proteins forming an obscure signaling network.

Given this multitude of interaction partners, it appeared to be a distinct possibility that the largely disordered native state would be changed by modification, e.g., phosphorylation or physical interaction with these partners at different domains or sub-domains of Par-4.

2 Sample Preparation and Characterization

Being predicted to be a majorly disordered, the full-length purified Par-4 apparently does not fold properly and subsequently prone to aggregation. Hence the protein was separated into two functional domains, i.e., SAC domain (Par-4_{SAC}) and CC domain (Par-4_{CC}), which were investigated separately.

2.1 Cloning

The Par-4_{CC} construct was created using the Par-4 full-length plasmid [21], amplifying the DNA sequence corresponding to the C-terminal amino acids 240–332 of rat Par-4 (UniProt: Q62627) and introducing into pET-11a vector via NdeI and BamHI restriction sites along with an N-terminal Strep-tag II [22]. Par-4_{SAC} construct was made by introducing the residues corresponding to the region 137–195 along with N-terminal Strep-tag II into pET11a vector using the same restriction sites as above [23]. This strep-tagged Par-4_{SAC} construct was used as a template for the creation of its phosphorylation mimicking mutant. The threonine amino acid at 155th position (T155, numbering with respect to Par-4 protein sequence) was mutated to glutamic acid to get Par-$4_{SAC\text{-}T155E}$ mutant.

2.2 Overexpression and Purification

The Par-4_{CC} was overexpressed in RIL strain of Escherichia coli BL21(DE3) by 4 h continuous induction with 1 mM isopropyl β-d-1-thiogalactopyranoside (IPTG) at 37 °C [22]. The cell pellet after the bacterial cell culture harvest was resuspended in a lysis buffer (50 mM Tris buffer pH 7.5, 200 mM $MgCl_2$) along with protease inhibitor (Complete EDTA-free, Roche) and 15 μg ml − 1 deoxyribonuclease I (5 g of cell pellet in 50 ml of lysis buffer). The cell resuspension was subjected to a pressure of 103–152 MPa using an EmulsiFlex-C3 cell disruptor to achieve efficient cell lysis. The lysate was cleared by centrifugation at 57,000 × g for 45 min at 4 °C and the supernatant was loaded onto a gravity-flow column with pre-equilibrated Strep-Tactin Superflow high-capacity resin (equilibration buffer: 50 mM Tris, 200 mM $MgCl_2$, pH 7.5). The column was washed with five CV of equilibration buffer and the protein was eluted by lowering the pH with elution buffer (50 mM potassium phosphate pH 4). Concentrated protein fractions were filtered through a 0.22 μm filter and subsequently further purified using a pre-equilibrated HiLoad 16/60 Superdex 75 prep-grade column (SEC buffer: 50 mM potassium phosphate pH 4). Peak fractions were pooled, concentrated, and analyzed. The yield of the purified Par-4_{CC} was 1.5–2.0 mg per gram of cell pellet. The apparent molecular weight of the purified Par-4_{CC} on SDS–PAGE was 12.2 kDa, which is in good agreement with the value found by mass-spectrometric analysis, i.e., 12.26 kDa (Fig. 1). In SEC (size-exclusion chromatography) profile, the protein showed an apparent molecular weight of 95 kDa indicating an octamer (12.2 kDa × 8 = 97.6 kDa; data not shown). Considering the fact that the SEC separation is based on Stokes radii [24] and the oligomer is predicted to be an anisotropic molecule, there could be an offset in the estimated molecular weight of the oligomer, which was based on a calibration made using standard globular proteins. The pH during the affinity purification was lowered from pH 7.5 to pH 4.0 to obtain DNA-free Par-4_{CC}.

The Par-4_{CC} was labeled with selenomethionine to determine protein phases for solving its 3D structure. The Par-4_{CC} plasmid transformants of BL21(DE3) strain were grown as above to an OD_{600} of 0.8–1.2 and harvested at 4000 rpm for 15 min at 4 °C. The cell pellet was rinsed with SeMet minimal medium [25], resuspended gently in 20 ml of SeMet minimal medium, and inoculated into 750 ml of SeMet minimal medium with 200 μg/ml amp and 100 μg/ml L-SeMet. The culture was immediately induced with 1 mM IPTG and continued for 4 hrs. After harvesting, the cell pellet was used for purification of SeMet-Par-4_{CC} as outlined for the wild type. Selenomethionine incorporation was confirmed by mass spectroscopy where the mass difference between Par-4_{CC} and SeMet-Par-4_{CC} was found to be 144.5 Da (Fig. 1). It is in reasonable agreement with the theoretical value considering the three methionine residues per protein molecule replaced with selenomethionine, which should be 140.7 Da.

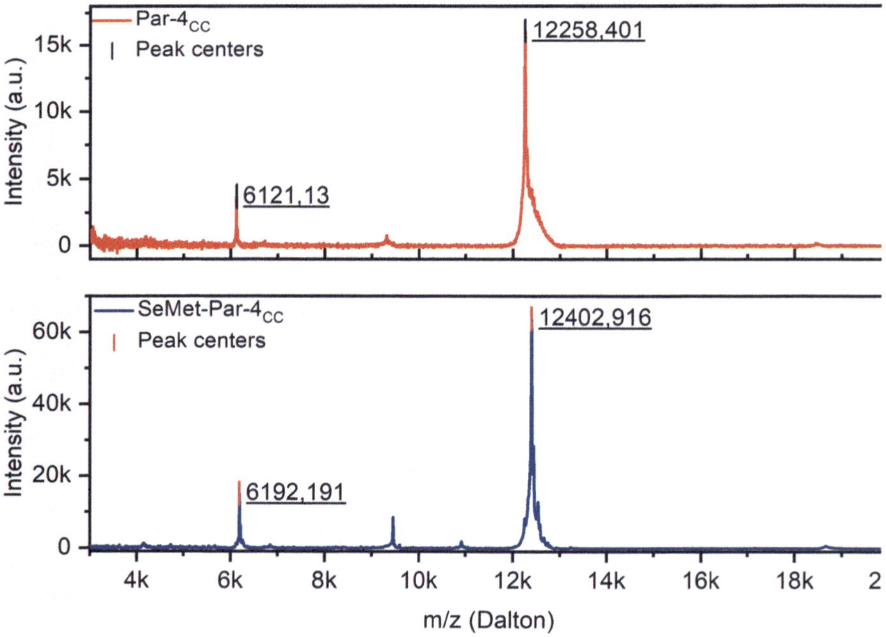

Fig. 1 Mass spectrometric analysis of Par-4_{CC} and SeMet-Par-4_{CC}

Par-4_{SAC} and its mutant were purified using the N-terminal Strep-tag as described above. The elution buffer was supplemented with 2.5 mM d-Desthiobiotin (in 50 mM Bicine, 100 mM NaCl pH 8.7 (SAC-buffer)). As the protein after the Strep column was found to contain DNA, a second purification step was conducted to separate DNA from the protein using an anion exchanger column (Macro-Prep High-Q, BioRad) equilibrated with SAC-buffer to bind DNA. The flow-through from the High-Q column was concentrated and further purified using a Superdex 75 prep-grade column (SEC buffer: 50 mM Bicine, 100 mM NaCl pH 8.7). The SEC profile for Par-4_{SAC} showed two peaks where both the peak samples were confirmed by mass spectroscopy to contain Par-4_{SAC} with a molecular mass of 8.1 kDa. The first SEC peak showed additionally another molecular mass peak (spectroscopy) at 16.3 kDa corresponding to the molecular weight of a dimer. Treatment of the Par-4_{SAC} sample with 100 mM DTT removed the disulfide bridge and a single peak corresponding to an apparent molecular weight of 18.2 kD in SEC (13.1 kDa in 20% SDS-page) was observed.

For nuclear magnetic resonance (NMR) spectroscopic analysis ^{15}N labeling of Par-4_{SAC} and Par-$4_{SAC\text{-}T155E}$ was required for which cell pellets from pre-grown cultures of respective E.coli BL21(DE3) transformants using 500 ml of DYT medium for each to OD_{600} of 0.8–1.0 were resuspended in equal volume of M9 minimal medium with $^{15}NH_4Cl$ (Cambridge Isotope Laboratories) and other supplements (9.1 g/l $NaH_2PO_4.2H_2O$; 3 g/l $KH_2PO_4.2H_2O$; 0.5 g/l NaCl; 1.0 g/l $^{15}NH_4Cl$; 100 μM $CaCl_2$ (1 M stock); 2 mM $MgSO_4$ (1 M stock); 10 μM Fe(III)citrate (10 mM stock); 4 g/l Glucose; 5 mg/l Thiamine hydrochloride; 0.2% (v/v) TS2 (Trace element stock solution [26]); 0.1% (v/v) Vitamin-cocktail stock solution (1 g/l Biotin; 1 g/l Choline chloride; 1 g/l Folic acid; 1 g/l Nicotinamide; 1 g/l Sodium-D-pantothenate; 1 g/l Pyridoxalhydrochlorid; 0.1 g/l Riboflavin dissolved in 20 mM Potassium Phosphate buffer, pH 7.4, filter sterilized and stored at −80 °C until use)). The resuspended cells were grown in a shaker incubator at 37 °C for 1 hr. before being induced with 1 mM IPTG for 4 hrs. Subsequently, cultures were harvested and ^{15}N labeled proteins were purified as mentioned earlier.

2.3 Sample Analysis by CD and NMR

The overall secondary structure content of the proteins was accessed by far-UV CD spectra (240 nm to 180 nm). The CD spectrum of the Par-4_{CC} showed a α-helical profile with double minima at 208 nm and 222 nm along with maxima near 190 nm (Fig. 2a). The MRE_{222}/MRE_{208} ratio of 1.095 indicated the coiled-coil conformation [27]. The far-UV CD spectra of the Par-4_{SAC} and its mutants showed a deep negative peak near 200 nm indicating its predominant unfolded state with a small shoulder near 222 nm indicating the presence of some residual secondary structure (Fig. 2a). The similar CD spectra of Par-$4_{SAC\text{-}T155E}$ showed small variations around 200 nm as well as 222 nm compared to the SAC domain (Fig. 2a).

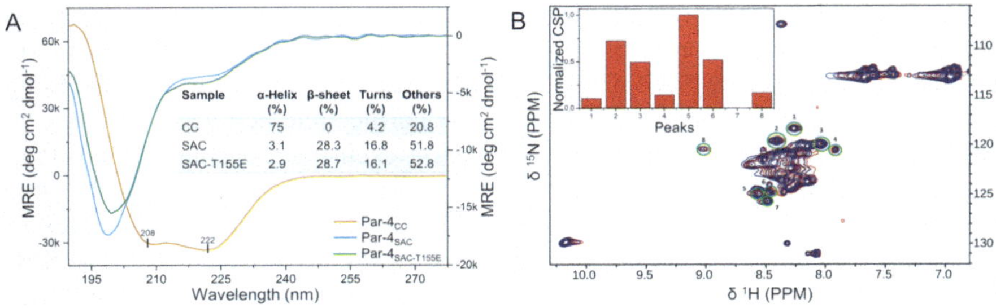

Fig. 2 (**a**) Far-UV CD analysis of Par-4_{CC}, Par-4_{SAC}, and Par-$4_{SAC\text{-}T155E}$. (**b**) 1H-15 N HSQC spectral overlay of Par-4_{SAC} and Par-$4_{SAC\text{-}T155E}$

The far-UV CD spectroscopic analysis of Par-4_{SAC} and its mutant indicated them to be majorly disordered and hence NMR spectroscopy was used to analyze these samples as it can provide further information on the state of disorder. The 2D ^{1}H-^{15}N HSQC (heteronuclear single quantum coherence) spectra of both proteins (Par-4_{SAC} and Par-$4_{SAC\text{-}T155E}$) showed poor peak dispersion indicating an unfolded state and their spectra-overlay showed some peak shifts (Fig. 2b). The normalized chemical shift perturbations (CSP, define as for the peak-shifts showed noticeable shifts with respect to the minimal shift, i.e., 0.25 (Fig. 2b), suggesting that structure induction can happen upon phosphorylation at T155.

2.4 *Protein Crystallization*

The sitting drop vapor diffusion approach was used for Par-4_{CC} and Par-4_{SAC} crystallization, which involves an aqueous drop containing the protein and the crystallization agents in a lower concentration than that required for the formation of crystals. In a sealed experiment the aqueous drop equilibrates against a reservoir buffer thereby gradually concentrating the ingredients in the sitting aqueous drop. During this process of diffusion, the crystallization trial experiments proceed through a range of conditions thereby conducting a self-screening process [28]. NeXtal Evolution μplates (96 wells) with 70 μl of crystallization buffer per well (reservoir) and different protein to reservoir buffer ratios (in sitting drop) were used for setting-up crystallization.

Freshly purified Par-4_{CC} at a concentration of 40–45 mg/ml was used for setting crystallization plates. An initial crystallization condition screening for the Par-4_{CC} was performed with commercially available crystallization screens (Hampton Research and Rigaku Reagents (Crystal Screen, Crystal Screen 2, Wizard Classic 1 and 2, and Wizard Cryo 1 and 2)). Crystal appeared with Crystal Screen 2 for the crystallization condition No. 17, i.e., 0.1 M sodium citrate tribasic dihydrate pH 5.6, 35% (v/v) tert-butanol. An improved crystallization condition was obtained by a grid screen (systematic variation for pH and tertiary butanol concentration) [29], which was 37–42% tert-butanol, 100 mM sodium citrate/citric acid pH 5.9 at 295.15 K with a protein concentration of 40 mg/ml.

Immediately after the addition of crystallization buffer protein precipitation was observed, but the sample usually cleared within an hour (Fig. 3). In some experiments, the protein remained in the precipitated form. Phase transformation of precipitated protein to a gel-like phase was observed when a 24-well plate with 1 ml reservoir buffer and 7–10 μl sitting-drop volume was used (Fig. 3). For the smaller sitting-drop volumes of 0.6–1.5 μl the gel-like phase was observed only infrequently. Crystals appeared randomly from the clear crystallization droplet as well as from the gel-like phase within a time range of 3 d to a couple of weeks. The protein mostly formed tetragonal bipyramidal crystals (Fig. 3).

Most crystals display a noticeable reduction in diffraction power within few seconds during data collection at room temperature in the X-ray beam at synchrotron sources. In such situations collection of a complete dataset will make the radiation damage even more severe, as the time of exposure to the X-ray beam will increase. Addition of cryoprotectants to crystals which is the standard procedure for cryo-cooling failed in the case of Par-4$_{CC}$ crystals. The crystals of Par-4$_{CC}$ were very sensitive because of its volatile crystallization buffer containing 37–42% tert-butanol, which tend to evaporate upon opening the experiment. Par-4$_{CC}$ crystals upon opening the experiment under the microscope could be seen moving very fast before turning into precipitate within seconds. Perfluoropolyether (PFO) or LV cryo oil (LVCO-1) was overlaid on the sitting drop immediately after opening the crystallization experiment to slow down the evaporation. Crystals were then picked and soaked in PFO before plunging into a nitrogen cryo-stream at 100 K.

Initial crystallization trials for Par-4$_{SAC}$ were made with different concentrations (5.7 mg/ml, 15 mg/ml and 33 mg/ml). Different screening suites from Qiagen

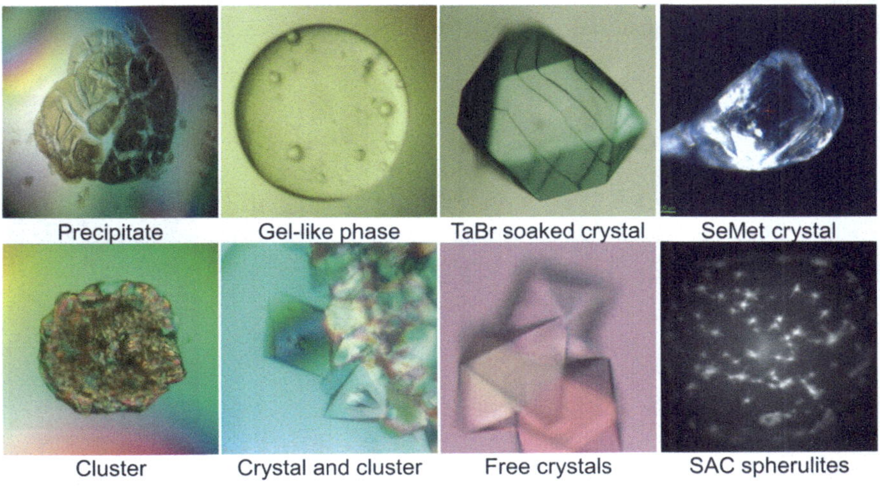

Fig. 3 Different phases of crystallization of C-terminal Par-4$_{CC}$ and Par-4$_{SAC}$

(JCSG core suites, PEGs II suite, Classics Lite suite, Classic II suite, pH clear II, cryos suite, comPAS suite, MPD suite, NeXtal PEG, NeXtal AmSO4 suite) were used. Precipitation of the protein was observed in most of the experiments with the PEGs II suite. Crystallization experiments were also made in the presence of DNA duplex of WT1 binding region on bcl-2 promoter (Bcl-2 WT1) [30] mixed in 1:1 molar ratio (SAC domain: Bcl2-WT1 DNA duplex). Spherulites were observed for the Par-4_{SAC} with Bcl-2 WTI duplex DNA (Fig. 3) suggesting that DNA binding may stabilize the Par-4_{SAC} structure to some extent. So far no investigated condition allowed the formation of Par-4_{SAC} crystals suitable for diffraction experiments.

3 Crystallographic Analysis

3.1 *Data Collection and Analysis*

Crystals were tested at different beamlines of the European Synchrotron Radiation Facility (ESRF, Grenoble, FRANCE) and PETRA, Germany. All measurements were performed at 100 K. Most Par-4_{CC} crystals diffracted to around 3.7 Å resolution. Par-4_{CC} X-ray diffraction images showed a distinct diffuse scattering pattern with stripes at ≈5.1 Å and widely spread diffuse scattering clouds at ≈10.5 Å at 90° to the former with eight-fold symmetry (Fig. 4) when the crystallographic four-fold axis (c axis) was oriented parallel to the beam axis.

Diffuse scattering occurs because of disturbances in the order of the crystal and causes the decrease of the Bragg intensities and the increase of non-Bragg scattering intensities, respectively [31]. The observed diffuse scattering pattern was similar to fiber diffraction patterns [32], where repetitive structural features in direction of the fiber axis appear in the meridional region and those perpendicular to the fiber axis appear in the equatorial region. The diffraction at ≈5.2 Å resolution in the

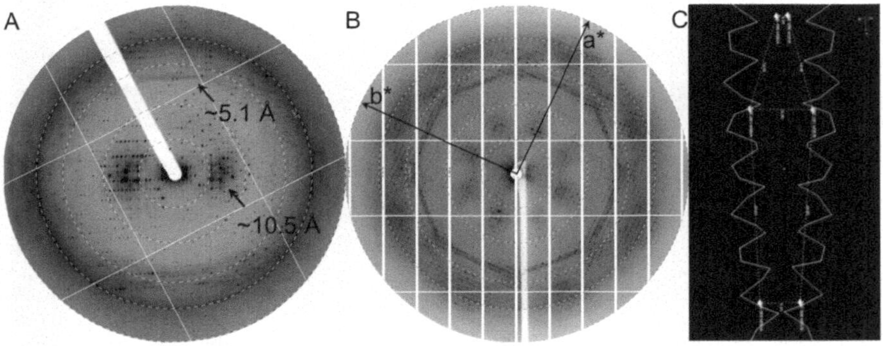

Fig. 4 (**a**) X-ray diffraction image of the crystal of C-terminal Par-4_{CC} showing diffuse scattering. (**b**) X-ray diffraction pattern when the crystal four-fold axis is pointed in the beam direction. (**c**) Found selenomethionine positions for a dimer

meridional region corresponded to the pitch of the helices in a coiled-coil structure, whereas the diffraction in the equatorial region at ≈10 Å resolution corresponded to the interhelical distance of the coiled-coil structure [27]. Therefore, it could be concluded from the mean resolutions of the meridional and equatorial diffuse scattering observed at ≈5.1 Å and ≈10.5 Å, respectively, that Par-4_{CC} had a disordered coiled-coil structure. The fiber diffraction-like pattern along a* and b* axes (Fig. 4b) indicated that these structures were oriented along the a and b axes. The pattern in the diagonal directions indicated the presence of similar structures at an angle of 45° to the first set.

Crystal orientation was determined based on test images. Auto-indexing that determines the unit-cell parameters and the orientation of the crystal was performed prior to the data collection with two diffraction images collected 90° apart using MOSFLM [34]. Complete datasets were collected with 0.5° rotation per image. The diffraction data displayed Laue class 4/mmm indicating point group 422 with possible primitive tetragonal spacegroups No. 89–96 [35]. Based on the observed extinction of axial Bragg spots, initial data were processed in the enantiomorphic spacegroups P43212 and P41212.

The phase problem is a well-known and a major hindrance in macromolecular crystallography, though methods are available to solve the issue either by heavy atom labeling or by using a similar structure [33]. From either, phases can be calculated that allow calculation of an electron density from the structure amplitudes (F_{hkl}) which in turn are calculated from the measured Bragg spot intensities for the respective Miller index (hkl) (Eq. 1).

$$\rho(xyz) = \frac{1}{V}\sum_{hkl}|F_{hkl}|\exp(i\alpha_{hkl})\exp[-2\pi i(hx+ky+lz)]\pi r^2 \tag{1}$$

In the case of the Par-4_{CC} heavy atom, labeling of the protein failed because of lacking isomorphy of the replacement, which can cause loss of diffraction and even damage the lattice observably [36] (Fig. 3: TaBr crystal).

Therefore, we employed selenomethionine replacement and multiwavelength anomalous dispersion (MAD). Anomalous scattering occurs when there exists a resonance between the incident X-ray waves and the electronic transitions from the bound atomic orbitals of the respective element [37]. It happens at the absorption edge. The respective anomalous scattering factor $f_{(anom.\,\theta)}$ consists of the normal scattering angle-dependent scattering factor $f_{(norm.\,\theta)}$, a wavelength-dependent real dispersive component $f'_{(disp.\,\lambda)}$ that modifies the normal scattering factor and an imaginary absorption component $f''_{(abs.\,\lambda)}$ that has 90° phase shift (Eq. 3).

$$f_{(anom.\theta.\lambda)} = f_{(norm.\theta)} + f'_{(disp.\lambda)} + f''_{(abs.\lambda)} \tag{2}$$

The presence of anomalous dispersion differences between centrosymmetric reflections breaks Friedel's law (i.e., $|F_{hkl}| = |F_{-h-k-l}|$; $\varphi_{hkl} = \varphi_{-h-k-l}$), which can be used to locate anomalous scatterers or heavy atom substructures and to break the two-fold phase ambiguity in isomorphous data.

MAD data were collected for the selenomethionine-labeled CC domain (SeMet-Par-4_{CC}) crystals on beamline ID23–1 at the European Synchrotron Radiation Facility (ESRF, Grenoble, France). A selenium K-edge fluorescence scan was performed with a SeMet- Par-4_{CC} crystal and f′ and f″ values were derived using the CHOOCH program [38]. Based on these spectra four wavelengths were selected: 0.9771 Å, high energy remote (he) wavelength; 0.9794 Å, absorption peak (pk) corresponding to f″ maxima (6.99 e-); 0.9795 Å, inflection point (ip) of the absorption curve corresponding to f′ minima (−10.87 e-); and 0.9825 Å, low energy remote (le) wavelength. Accordingly, a four-wavelength MAD dataset with 2.7 Å data resolution was collected with the same crystal.

POINTLESS, a program from CCP4 suite that determines the lattice with the highest possible symmetry compatible with the unit cell parameters from unmerged intensities [39], suggested the space group to be $P4_32_12$ for all the 4WL datasets that were integrated in P422 space groups. A solvent content of 47.7% was estimated using the MATTHEWS_COEF program [40], which corresponds to seven molecules per asymmetric unit in the space group $P4_32_12$.

Heavy atom substructure detection and initial phasing was performed by SHELXC/D/E [39] and also Phenix.AutoSol [41]. The SHELXD program, which uses the estimates for heavy atom structure factor (FA) from SHELXC, found 15 selenium sites in space group $P4_32_12$ (cc All/cc Weak: 52.5%/34.6%) within 100 trials [42]. Density modification was performed by using SHELXE which could also trace a Cα chain [43]. A reasonable Cα model could be built using SHELXE in space groups $P4_32_12$ with a correlation coefficient of the partial structure against native data of 39.31%. The Cα model consisted of 7 helices of which five were long (70–80 residues) and two were short (30–35 residues). The helices were paired forming three non-crystallographic dimers (two long and one short), whereas the remaining helix made a crystallographic dimer with a symmetry equivalent molecule. Phenix.AutoSol, which is an experimental phasing pipeline [41], was also used for finding heavy-atom sites and calculating experimental phases ad subsequently used for model building.

By model building and refinement (s.b) no additional electron density for the short helices could be found. Therefore, alternative space groups (symmetry reduced or non-isomorphic subgroups) [35] were investigated. In neither case, the densities for the short helices did improve.

3.2 Model Building and Refinement

Assignment of the protein sequence to the Cα models was performed manually using COOT, a graphical program in the CCP4 suite for building and refining the model. An initial assignment of the selenomethionine residues (corresponding to 285, 291, and 309 position of Par-4_{CC} protein sequence) was made for each helix based on the selenium substructure (Fig. 4c).

The sequence assigned model was refined using the PHENIX software suite [44] (Table 1). Several rounds of simulated annealing, energy minimization, and B-factor refinement were carried out before final refinement using phenix.refine [45]. Torsion-based NCS restraints, which allow local differences between NCS-related molecules, were used for all chains during the refinements. The refinement of the final structure of Par-4_{CC} of Par-4 includes residues 254–331 for chains A and C, 255–232 for chains B and E, 256–332 for chain D, and residues 255–294 for chains F and G.

These refinements are aimed to minimize the R-value which gives the relative agreement between the observed structure factor amplitudes and the model calculated amplitudes. Overfitting (error fitting) of the model was monitored by the free R-value which is calculated from a (free) subset of amplitudes (10%) that have not been used for refinement [46–48]. Finally, the structure of Par-4_{CC} was determined by MAD from a SeMet-Par-4_{CC} crystal at 3.0 Å resolution to a R_{work}/R_{free} of 24.54%/29.44% (Table 1) (PDB: 5FIY).

3.3 Structure of the Homodimer

The asymmetric unit Par-4_{CC} crystal consists of seven helical molecules (chains A, B, C, D, E, F, and G) forming three homodimeric structures (dimers AB, CD, and FG) (Fig. 5a). Another dimer whose two-fold symmetry coincides with a crystallographic two-fold symmetry forms a crystallographic dimer (dimer EE′). One of the non-crystallographic homodimers (dimer FG) displayed poor electron density for their C-terminal halves and did not allow model completion. This 3D structure of Par-4_{CC} confirmed the self-association of Par-4 via its C-terminal CC region [14]. In the crystal structure of Par-4_{CC} the dimers AB and FG are oriented along the a and b axes whereas the dimers CD and EE′ are oriented diagonal to the a and b axes. All seven helices are oriented approximately parallel to the ab plane (Fig. 5b). The arrangement of helices in the asymmetric unit supports the conclusions that were drawn from the diffuse scattering of the Par-4_{CC} crystals.

Residues corresponding to 257–330 in the Par-4_{CC} structure form a classical coiled-coil structure with two parallel right-handed helices forming a left-handed superhelix (Fig. 6b). Even the diffuse scattering in the diffraction images showed the characteristics of this feature (Fig. 4). Coiled-coil structure formation is guided by their heptad repeats in the sequence designated (abcdefg)n, with predominantly hydrophobic amino acids at positions a and d. A total of 11 heptad repeats could be assigned to residues 257–330 (Fig. 6b), with only four residues in the eleventh heptad (SOCKET [49]). It was shown that the presence of asparagine in position a of a heptad favors dimerization with parallel packing of helices [50, 51], which is confirmed by the Par-4_{CC} structure by having three asparagine residues at each position a of their respective heptads (Asn_{278}, Asn_{313}, and Asn_{320}). These residues in the crystallographic dimer have two rotamer conformations breaking the twofold non-crystallographic symmetry locally. Another characteristic feature for two-stranded

Table 1 Data collection statistics, phasing data, and refinement data

Data collection	SeMet-Par-4_{CC}							
Space group	P 4_32_12							
Cell dimensions: a, b, c (Å), α, β, γ (deg.)	114.99, 114.99, 121.81, 90.0, 90.0, 90.0							
	High-energy remote (λ1)		Peak (λ2)*		Inflection (λ3)		Low-energy remote (λ4)	
Wavelength (Å)	0.9771		0.9794		0.9795		0.9825	
Resolution (Å)	41.81–3.0		41.81–3.0		41.81–3.0		41.81–3.0	
R_{merge} [a]	0.084 (0.284)		0.096 (0.319)		0.096 (0.354)		0.081 (0.305)	
I/σ(I)[a]	15.9 (6.1)		17.4 (5.6)		14.8 (5.1)		24.4 (7.5)	
Total observations[a]	152,595 (28283)		152,969 (28368)		152,789 (28286)		152,261 (28257)	
Unique reflections[a]	16,918 (3002)		16,918 (3002)		16,917 (3002)		16,912 (3002)	
CC(1/2)[a]	0.998 (0.981)		0.997 (0.974)		0.996 (0.968)		0.997 (0.977)	
Completeness (%)[a]	99.7 (100.0)							
Redundancy[a]	9.0 (9.4)							
Phasing (AutoSol in PHENIX)								
Bragg spacing limit (Å)	Total	10.22	6.41	4.99	4.23	3.73	3.37	3.10
Phasing power (r.m.s. anomalous FH/E)								
λ1	0.6	1.8	2.1	1.7	1.2	0.8	0.5	0.3
λ2	0.6	1.3	1.8	1.7	1.4	0.9	0.6	0.4
λ3	0.5	0.9	1.4	1.4	1.1	0.6	0.4	0.3
λ4	0.5	1.0	1.1	1.2	1.0	0.7	0.4	0.3
Phasing power (r.m.s. dispersive FH/E)								
λ1–λ2	0.1	0.3	0.3	0.3	0.2	0.2	0.1	0.1
λ1–λ3	0.2	0.5	0.5	0.5	0.3	0.2	0.2	0.1
λ2–λ3	0.1	0.4	0.5	0.4	0.3	0.2	0.1	0.1
λ2–λ4	0.1	0.2	0.2	0.2	0.2	0.1	0.1	0.1
λ3–λ4	0.2	0.6	0.7	0.6	0.5	0.3	0.2	0.1
Mean figure of merit	0.39 (after SOLVE), 0.72 (after RESOLVE)							
Refinement								
Resolution (Å)	41.81–3.0							
No. of reflections	16,869							
Rwork/Rfree	24.54/29.44							
No. of atoms	3958							
Protein	3913							
Potassium ions	18							
Water	27							
Mean B-factors ($Å^2$)	75.20							
Protein	75.20							
Potassium ions	97.60							

(continued)

Table 1 (continued)

Data collection	SeMet-Par-4_{CC}
Water	54.90
R.M.S. deviations	
Bond length (Å)	0.005
Bond angle (deg.)	0.810

[a]values in parentheses correspond to the highest-resolution shell (3.21–3.0 Å)
bdataset used for refinement
$R_{merge} = \Sigma_{hkl}\Sigma_j|I_{hkl,j} - \langle I_{hkl}\rangle| / \Sigma_{hkl}\Sigma_j I_{hkl,j}$.
CC(1/2) = correlation coefficient between random half data sets.

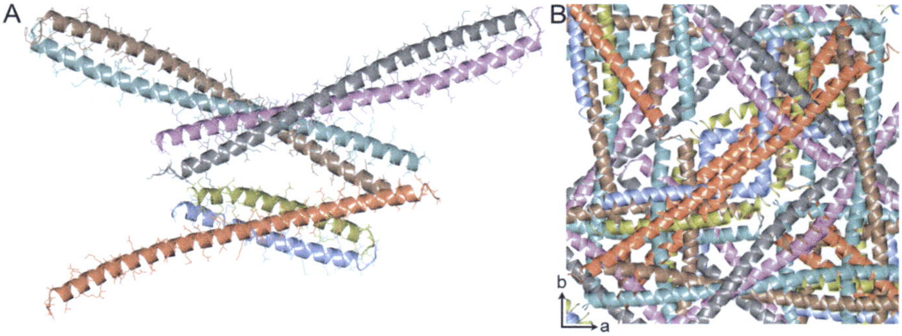

Fig. 5 (**a**) Asymmetric unit showing seven molecules of C-terminal domain of Par-4. (**b**) Molecular packing in the ab plane with dimers along a axis, b axis, and diagonal to the ab plane [60]

parallel coiled coils [52] that could be seen in the Par-4_{CC} structure is that the Cα–Cβ bonds of the residues at positions a and d of each heptad are oriented parallel and perpendicular, respectively, to the peptide bonds facing each of them in the opposite helix. The side chains at position a and d in the Par-4_{CC} structure (Fig. 6b) face the interhelical space, which are predominantly hydrophobic (73%) and constitute the hydrophobic dimer interface. The Par-4_{CC} structure also includes leucine zipper (Par-4_{LZ}) (Fig. 6B) where the position d of their heptads is predominantly occupied by leucine and is known to dimerize via coiled-coil interaction [53, 54]. The last six heptads of the Par-4_{CC} belong to the Par-4_{LZ} which features four hydrophobic and two polar residues at position a (Ile292, Ile299, Leu306, Asn313, Asn320, Val327), and exclusively leucine residues at position d (Leu295, Leu302, Leu316, Leu323, Leu330), with the exception of the eighth heptad containing a methionine (Met309) (Fig. 6b).

Coiled-coil structures are known to be stabilized via interhelical salt bridges with their α-helices being stabilized by intrahelical salt bridges [55–57]. The Par-4_{CC} structure displays both interhelical and intrahelical salt bridges belonging to g–e′, d–e′, and g–c types (Fig. 6b). The interhelical interacting pairs Arg263-Glu268′, Glu270-Arg275′, and Glu312-Lys317′ belong to the g–e′ type, whereas

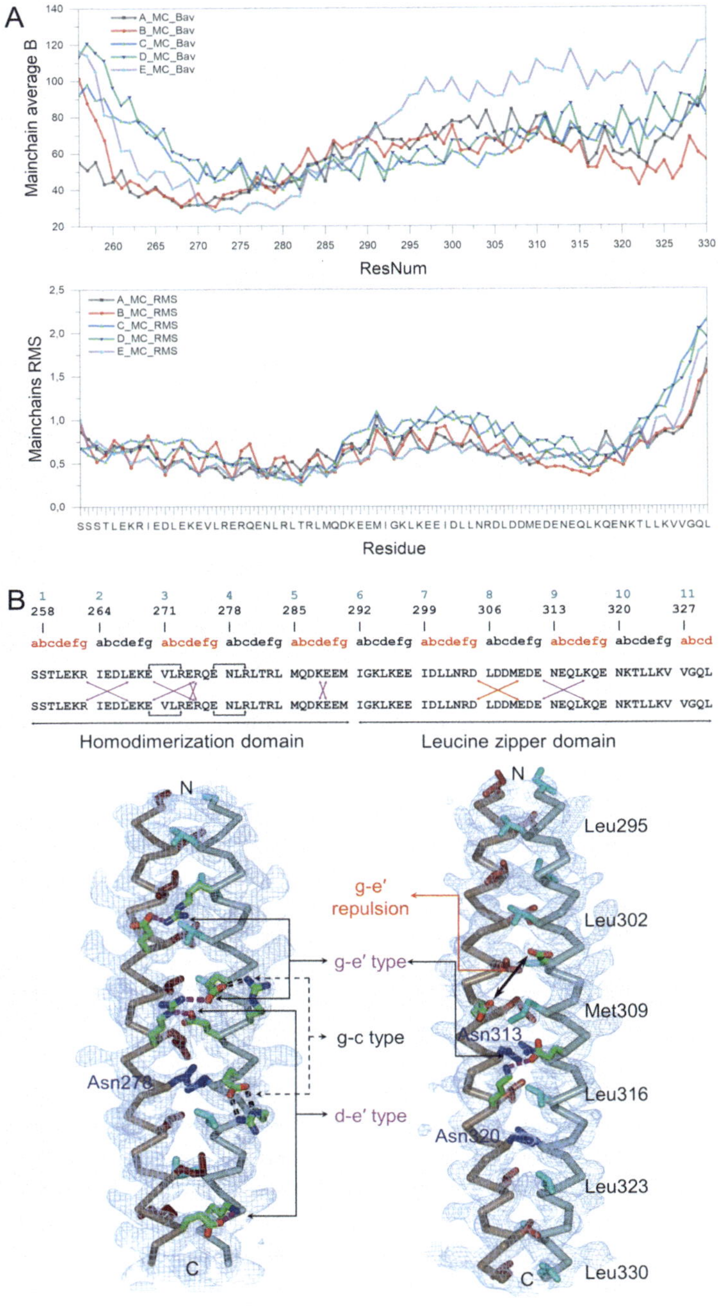

Fig. 6 (**a**) Mainchain B-factors and RMS deviations of the chains A, B, C, D, and E. (**b**) Homodimerization and leucine zipper domain of the C-terminal domain of Par-4 [60]

Glu274-Arg275′ and Lys288-Glu289′ belong to the d–e′ type (Fig. 6b). And the intrahelical salt bridges between Glu270-Arg273 and Glu277-Arg280 belong to the g–c type (Fig. 6b).

Residues from 256–291 belonging to the N-terminus of all the helices display systematically lower disorder (Fig. 6a) than the C-terminal halves (residues 292–330), which resulted in relatively poor map quality for the C-terminal parts of the molecules. The helices A, B, C, D, and E are relatively less disordered compared to the helices (F and G) with their backbone average B-value being ~65 Å^2 (mean of A, B, C, D, and E (residues 256–330)) and ~ 108 Å^2 (mean of G and F (residues 256–294)), respectively. Most of the earlier mentioned salt bridges are localized in the N-terminal half of the structure while Par-4_{LZ} has only one salt bridge. This resulted in the weakening of side-chain electron density and increasing temperature factors from around residue 290. Par-4_{LZ} depends apparently on the hydrophobic interface for the stability in the homodimeric coiled-coil structure. These observations clearly indicated that the N-terminal half of the Par-4_{CC} constitutes the homodimerization subdomain (Par-4_{HDD}).

All the long helices in the asymmetric unit are similar with RMS for their main-chains (residues 256–330) in the range of 0.25–2.15 Å for all five chains compared to all other chains (i.e., A to B, C, D and E; B to A, C, D and E; C to A, B, D and E; D to A, B, C and E; E to A, B, C and D). This range is lower for the residues 260–291 than for C-terminal residues 292–330, which is 0.25–1.09 Å and 0.36–2.15 Å, respectively (Fig. 6a). Dimers AB and CD displayed a better-defined electron densities compared to the crystallographic and the short dimers and the reason for which is apparently the interaction of Arg304 of chains B and C with Asp308 and Asp 311 of chains C and B, respectively. Chains A and B display a better temperature factors overall compared to other chains, while that for chains C and D got effected esp. in their N-terminus because of being neighbored by highly disordered short helices F and G (Fig. 6a). The slight variations in atomic coordinates for the different molecules did not affect the interactions between the molecules that are described below.

4 Functional Implications

The crystallographic characterization of Par-4_{CC} delivered structural information for this protein. But only a further analysis by molecular docking with the relevant parts of interaction partners allowed gaining some insight in the role of the subdomain of the protein and their functional relevance.

4.1 Regulatory Interactions of Par-4$_{CC}$

The C-terminal region of Par-4 had been shown as an active domain involving interactions with many of its interacting partners. Few of the known binding proteins were selected and docking analysis was performed for parallel helical pairing with Par-4$_{CC}$ to understand the mechanism of interaction. Regions in the sequence of AATF, Amida, DAPK3, and THAP1 that showed high coiled-coil probability [58], which also overlapped with their reported binding regions with Par-4, were modeled [59] and used for docking to monomeric Par-4$_{CC}$ structure [60]. This analysis showed that AATF, Amida, and DAPK3 interact with the C-terminal Par-4$_{LZ}$ while THAP1 showed interaction in the N-terminal Par-4$_{HDD}$ region with six interhelical salt bridges (Fig. 7). THAP1 interaction with the stability equivalent to the homodimeric Par-4$_{CC}$ structure considering the number of interhelical salt bridges clearly explains Par-4's colocalization with THAP1 in the PML-nuclear bodies [19]. The DAPK3 model that was used for this docking analysis did not include its own LZ domain as its deletion mutant interacts with Par-4 [17]. The interaction of AATF and DAPK3, which are also known to interact with each other [61], with Par-4$_{CC}$ in the same region, i.e., Par-4$_{LZ}$ (Fig. 7) suggests a competition between them for Par-4.

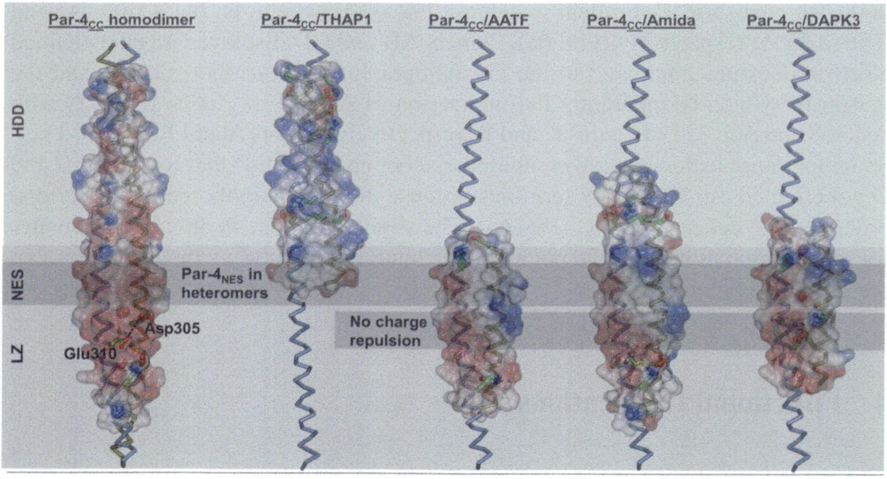

Interacting region	Par-4: 262-299 THAP1: 147-184	Par-4 : 291-320 AATF: 270-299	Par-4: 280-322 Amida: 84-126	Par-4: 289-316 DAPK3: 366-393
Salt-bridges	Arg263 - Glu153 Glu268 - Arg148 Glu270 - Arg160 Glu274 - Arg160 Glu289 - Arg169 Lys288 - Glu174	Lys296-Glu270 Lys317-Glu295	Glu289-Arg88 Lys288-Glu93 Lys317-Glu116 Glu312-Arg117	Lys296-Asp372 **Asp305**-Arg387 **Glu310**-Arg382

Fig. 7 Docking analysis of the C-terminal domain of Par-4 with four of its interacting partners [60]

The interaction model of Par-4_{LZ}/Amida is systematically supported by the earlier reported data where two of the Par-4_{LZ} residues (Leu295 and Leu316) in the interacting region were shown to influence their interaction [18]. It was shown before that the Par-4_{LZ} stability is pH-dependent and is attributed to the repulsive interaction between Asp305 and Glu310′ (g–e′ type) [62, 63]. This destabilizing g–e′ type charge repulsion of the Par-4_{CC} homodimer was not seen in the three Par-4_{LZ} heteromeric interacting models (Fig. 7). Furthermore, these residues (Asp305 and Glu310) of Par-4_{LZ} were stabilized by forming salt bridges in the DAPK3 interacting model (Fig. 7). Overall, these results substantially suggest charge as an important factor for specific heteromerization of Par-4_{CC}.

4.2 *Trigger Motif for Homodimerization*

The homodimerization of two-stranded parallel coiled-coil structure was described to depend on a trigger sequence, which is a necessary autonomous helical folding unit [64]. The consensus motif for this trigger sequence was derived, to be xxLExchxcxccx, where x, h, and c are any (x), a hydrophobic (h), or a charged (c) residue, respectively [64]. In Par this consensus sequence is found twice (conserved residues in bold) within the Par-4_{HDD} subdomain: 258-ST**LEKR**-**IED**L**EK**E-270 and 265-ED**LEKE**-**V**L**RER**QE-277. This subdomain exhibits the least disorder (lowest temperature factors) in the Par-4_{CC} structure (Fig. 6a), which suggests that Par-4_{HDD} being the most stable part of the structure could trigger the homodimeric coiled-coil interaction even before the full folding of the domain could occur. A heteromeric interaction model of Par-4_{CC}/THAP1 also showed such a trigger motif (HQ**LE**QQ-**VEK**L**RK**K) in the interaction region of THAP1, which directly interacts with the second trigger consensus sequence of Par-4_{HDD}. Therefore, the annotation as homodimerization domain may be overly restrictive. In contrast, the Par-4_{CC} interacting proteins AATF, Amida, and DAPK3 do not contain such a trigger sequence and exhibit dimerization only with Par-4_{LZ} in the interaction models (Fig. 7).

In the ninth and tenth heptads of Par-4_{LZ}, Asn (Asn313 and Asn320) (Fig. 6b) is located at positions replacing Leu which allows energetic discrimination between the different possible assemblies of the coil-coil structure and provide specificity in the dimeric coiled-coil structure formation at the cost of a destabilizing effect by the replacement of Leu by Asn [51]. Par-4_{HDD} also contains such an Asn (Asn278). Difference in the stability or apparent disorder of the structures of Par-4_{HDD} and Par-4_{LZ} could be explained at least partially by the different number of destabilizing Asn residues in these subdomains.

Further stabilization of the homodimer of Par-4_{HDD} is provided by intrahelical salt bridges which stabilize monomeric α-helices, and interhelical salt bridges which stabilize the dimer (Fig. 6b). The majority of these stabilizing salt bridges are located within the trigger sequences in the Par-4_{HDD} subdomain. This is also the case for THAP1 in the model of the THAP1/Par-4 heterodimer (Fig. 7). The similarity of salt-bridge formation with the Par-4_{HDD} homodimer (Fig. 7) implies a preferential heterodimerization via the HDD subdomains.

4.3 Preference for Heterodimerization in Par-4_{LZ} Region

Par-4_{LZ} was shown to exhibit environment-dependent structural interconversion between a partially ordered monomer and a predominantly dimeric coiled-coil state [27]. Despite the formation of a homodimeric state assisted by the N-terminal segment (Par-4_{HDD}) of Par-4_{CC}, the presence of g–e′ type charge repulsion between Asp305-Glu310′ is likely causing instability in the Par-4_{LZ} dimerization of the homodimeric structure of Par-4_{CC}. Clearly, the structure of Par-4_{LZ} is only weakly stabilized by the interaction at the hydrophobic interface in the homodimer, which suggests that more stable interactions involving the Par-4_{LZ} domain may be formed with other proteins. This is supported by our analysis of heterodimeric-interaction models, where Par-4_{LZ} showed 2–4 interhelical salt bridges and no charge repulsions with AATF, Amida, and DAPK3 (Fig. 7). The interaction model with DAPK3 displays optimal heteromeric interaction, where stabilizing interactions between Par-4_{Asp305}-DAPK3_{Arg387} and Par-4_{Glu310}-DAPK3_{Arg382} replaced the destabilizing repulsive g–e′ interaction of the Par-4_{CC} homodimer (Fig. 7).

Though the Par-4_{LZ} homodimer has one stabilizing interhelical salt bridge (Glu312-Lys317′), homodimerization appears disfavored by the g–e′ type repulsion and the presence of Asn313 and Asn320 in the helical interface (Fig. 6b). More favorable interactions with regard to charge repulsion and salt bridges are seen in the heterodimeric models of Par-4_{LZ}. Therefore, the structure of the Par-4_{CC} and the interaction models point at Par-4 providing preferential interactions with signaling proteins by either Par-4_{LZ} or Par-4_{HDD} with the majority of important interactions being provided by Par-4_{HDD}. This conclusion is supported by observations that the Par-4_{LZ} domain is not folded under physiological conditions and inherently instable [62], but still appeared in coiled-coil structure within Par-4 [1].

4.4 Regulation of Complex Formation by Par-4_{CC}

The structure of Par-4_{CC} and the docking studies suggested two different binding modes for Par-4_{CC} interaction with other proteins: THAP1 binds to Par-4_{HDD}, whereas AATF, Amida, and DAPK3 bind Par-4_{LZ}. The binding of THAP1 to Par-4_{HDD} is therefore similar to the homodimerization of Par-4 and is expected to be heterodimeric. The other three proteins do not bind preferentially to Par-4_{HDD} but to Par-4_{LZ}, and are therefore not necessarily disrupting Par-4_{CC} homodimerization completely. Therefore, for these, protein heterotetrameric assemblies that consist of homodimers of the respective proteins may form at least transiently.

For example, the interaction of Par-4 and DAPK3 is mediated by the predicted coiled-coil domain within the interacting region of DAPK3 and Par-4_{LZ} (Fig. 7), whereas the LZ domain of DAPK3 is needed for its homodimerization and kinase activation [65] and therefore can remain relatively undisturbed by heterotetramerization. The putative complex of Par-4 and DAPK3 is a tetramer with intertwined homodimers [60]. Such a quaternary structure would require bending of the Par-4 helices between the Par-4_{HDD} and the Par-4_{LZ} subdomains near residue 290. But this is the region where the Par4_{CC} structure shows disorder for molecules F and G, whereas in Par-4_{CC} molecules A-E high temperature factors appear. A closer inspection of the predicted coil-coil region (residues 366–393) in the DAPK3 homodimer shows that there is a repulsive g–e′ interaction between Arg382 and Arg387′, which would be replaced by attractive interactions in the suggested heterotetramer [60]. The DAPK3 structure does not contain the residues adjacent to its terminal LZ. Therefore, it remains unclear if DAPK3 interacts with Par-4_{HDD} and its preceding segment which connects to the Par-4_{SAC}.

4.5 *Masking of the Par-4_{NES} by Dimerization*

Par-4 is sensitivity to selective inhibitors of nuclear export [66], indicating its involvement in nuclear export sequence (NES)-dependent nuclear export and the Par-4_{LZ} was claimed to contain an NES [7]. Analyzing the Par-4_{CC} sequence indicated the residues 295–301 (**L**KEE**I**D**L**) to be a potential NES (NetNES [67] and ValidNES [68]). NES consensus consists of hydrophobic residues (shown in bold), which are essential for binding to the hydrophobic pockets of exportin-1/XPO1/CRM1 [69]. In the Par-4_{CC} homodimeric structure hydrophobic surface formed by Par-4_{NES} is masked by dimerization (Fig. 8), suggesting a mechanism to escape CRM1-mediated nuclear export by being inaccessible and thereby suggesting a mechanism for Par-4 nuclear localization. Moreover, most of the structures of NES display higher temperature factors [69], which is also true for the predicted Par-4_{NES} region.

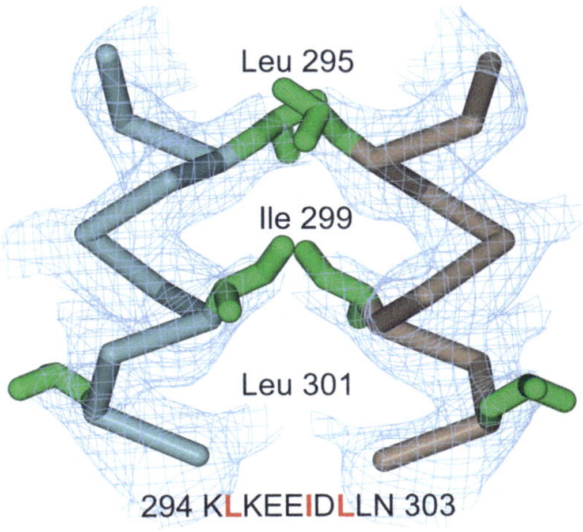

Fig. 8 Hydrophobic residues of the NES consensus in the homodimeric interface of Par-4_{CC}

Par-4 is known to appear in different localizations in different cell types [9], but the regulatory mechanism of its nuclear import/export is not described. However, the nuclear export by exportin-1 requires the NES to be accessible. Our models show masking of the Par-4_{NES} for the Par-4_{CC} homodimeric structure as well as the heteromeric interaction models (Fig. 6). Regulation of nuclear localization by oligomerization and NES masking was already suggested for p53 [70]. For the Par-4-mediated recruitment of DAPK3 to the actin cytoskeleton in the cytoplasm [17] was also suggested to depend on the Par-4_{NES} [2]. However, in the Par-4/DAPK3 complex (Fig. 7) the Par-4_{NES} would be masked and disallow translocation to the cytoplasm. Therefore, the interaction of Par-4 with DAPK3 would need to occur in the cytoplasm.

5 Conclusions and Future Perspectives

Despite large efforts invested into the structural analysis of Par-4, the difficulties in preparing a structured state for this protein allowed so far only the determination of about one-third of the total structure (residues 240–332).

For the N-terminal part of Par-4 (residues 1–136) especially residues 1 to 53 there is currently no evidence for any kind of structure that this segment may adopt under the right conditions.

For the SAC domain (residues 137–195) we made some progress in preparing a state that may eventually lead to a structure, but currently our results indicate a β-structured molten globular-like state, where clearly more stabilization is needed than provided by phosphorylation at Thr155. An obvious route is the preparation of complexes with either other proteins or DNA.

Currently, we are investigating the N-terminal elongation of the C-terminal domain (residues 137–322) which includes the SAC domain under the working hypothesis that the coiled-coil domain may induce structure or stabilization in the SAC domain. But between these domains a further control domain (residues 196–239) is located that contains the Akt kinase phosphorylation switch at Ser249 whose role for structure induction and stability requires investigation. Specifically, the question if Akt phosphorylation induces ternary structure reorganization needs to be addressed. Therefore, experiments with CD to investigate secondary structure composition, dynamic light scattering to monitor changes of compactness as an indicator for ternary structure changes, and binding experiments for complex formation which address functional attribution are required for sample characterization before a 3D-analysis is attempted.

References

1. Libich DS, Schwalbe M, Kate S, Venugopal H, Claridge JK, Edwards PJB, Dutta K, Pascal SM (2009) Intrinsic disorder and coiled-coil formation in prostate apoptosis response factor 4. FEBS J 276:3710–3728. https://doi.org/10.1111/j.1742-4658.2009.07087.x
2. Shrestha-Bhattarai T, Rangnekar VM (2010) Cancer-selective apoptotic effects of extracellular and intracellular Par-4. Oncogene 29:3873–3880. https://doi.org/10.1038/onc.2010.141
3. Duan W, Guo Z, Mattson MP (2000) Participation of Par-4 in the degeneration of striatal neurons induced by metabolic compromise with 3-Nitropropionic acid. Exp Neurol 165:1–11. https://doi.org/10.1006/exnr.2000.7434
4. Guo Q, Fu W, Xie J, Luo H, Sells SF, Geddes JW, Bondada V, Rangnekar VM, Mattson MP (1998) Par-4 is a mediator of neuronal degeneration associated with the pathogenesis of Alzheimer disease. Nat Med 4:957–962. https://doi.org/10.1038/nm0898-957
5. Pedersen WA, Luo H, Kruman I, Kasarskis E, Mattson MP (2000) The prostate apoptosis response-4 protein participates in motor neuron degeneration in amyotrophic lateral sclerosis. FASEB J 14:913–924. https://doi.org/10.1096/fasebj.14.7.913
6. Kruman II, Nath A, Maragos WF, Chan SL, Jones M, Rangnekar VM, Jakel RJ, Mattson MP (1999) Evidence that Par-4 participates in the pathogenesis of HIV encephalitis. Am J Pathol 155:39–46. https://doi.org/10.1016/S0002-9440(10)65096-1
7. Ranganathan P, Rangnekar VM (2005) Regulation of Cancer cell survival by Par-4. Ann N Y Acad Sci 1059:76–85. https://doi.org/10.1196/annals.1339.046
8. Boghaert ER, Sells SF, Walid AJ, Malone P, Williams NM, Weinstein MH, Strange R, Rangnekar VM (1997) Immunohistochemical analysis of the proapoptotic protein Par-4 in normal rat tissues. Cell Growth Differ Mol Biol J Am Assoc Cancer Res 8:881–890
9. El-Guendy N, Zhao Y, Gurumurthy S, Burikhanov R, Rangnekar VM (2003) Identification of a unique Core domain of Par-4 sufficient for selective apoptosis induction in Cancer cells. Mol Cell Biol 23:5516–5525. https://doi.org/10.1128/MCB.23.16.5516-5525.2003
10. Gurumurthy S, Goswami A, Vasudevan KM, Rangnekar VM (2005) Phosphorylation of Par-4 by protein kinase a is critical for apoptosis. Mol Cell Biol 25:1146–1161. https://doi.org/10.1128/MCB.25.3.1146-1161.2005
11. Boosen M, Vetterkind S, Kubicek J, Scheidtmann K-H, Illenberger S, Preuss U (2009) Par-4 is an essential downstream target of DAP-like kinase (Dlk) in Dlk/Par-4–mediated apoptosis. Mol Biol Cell 20:4010–4020. https://doi.org/10.1091/mbc.e09-02-0173
12. Goswami A, Burikhanov R, de Thonel A, Fujita N, Goswami M, Zhao Y, Eriksson JE, Tsuruo T, Rangnekar VM (2005) Binding and phosphorylation of Par-4 by Akt is essential for Cancer cell survival. Mol Cell 20:33–44. https://doi.org/10.1016/j.molcel.2005.08.016
13. Zhao Y, Rangnekar VM (2008) Apoptosis and tumor resistance conferred by Par-4. Cancer Biol Ther 7:1867–1874. https://doi.org/10.4161/cbt.7.12.6945
14. Díaz-Meco MT, Municio MM, Frutos S, Sanchez P, Lozano J, Sanz L, Moscat J (1996) The product of par-4, a gene induced during apoptosis, interacts selectively with the atypical isoforms of protein kinase C. Cell 86:777–786. https://doi.org/10.1016/S0092-8674(00)80152-X
15. Johnstone RW, See RH, Sells SF, Wang J, Muthukkumar S, Englert C, Haber DA, Licht JD, Sugrue SP, Roberts T et al (1996) A novel repressor, par-4, modulates transcription and growth suppression functions of the Wilms' tumor suppressor WT1. Mol Cell Biol 16:6945–6956. https://doi.org/10.1128/mcb.16.12.6945
16. Guo Q, Xie J (2004) AATF inhibits aberrant production of amyloid beta peptide 1-42 by interacting directly with Par-4. J Biol Chem 279:4596–4603. https://doi.org/10.1074/jbc.M309811200
17. Page G, Kögel D, Rangnekar V, Scheidtmann KH (1999) Interaction partners of Dlk/ZIP kinase: co-expression of Dlk/ZIP kinase and Par-4 results in cytoplasmic retention and apoptosis. Oncogene 18:7265–7273. https://doi.org/10.1038/sj.onc.1203170

18. Boosen M, Vetterkind S, Koplin A, Illenberger S, Preuss U (2005) Par-4-mediated recruitment of Amida to the actin cytoskeleton leads to the induction of apoptosis. Exp Cell Res 311:177–191. https://doi.org/10.1016/j.yexcr.2005.09.010
19. Roussigne M, Cayrol C, Clouaire T, Amalric F, Girard J-P (2003) THAP1 is a nuclear proapoptotic factor that links prostate-apoptosis-response-4 (Par-4) to PML nuclear bodies. Oncogene 22:2432–2442. https://doi.org/10.1038/sj.onc.1206271
20. Chang S, Kim JH, Shin J (2002) p62 forms a ternary complex with PKCzeta and PAR-4 and antagonizes PAR-4-induced PKCzeta inhibition. FEBS Lett 510:57–61. https://doi.org/10.1016/s0014-5793(01)03224-0
21. Eidhoff UB Heterologe Expression, Kristallisation und Untersuchungen zur Struktur. Dissertation, Heinrich Heine University Düsseldorf
22. Tiruttani Subhramanyam UK, Kubicek J, Eidhoff UB, Labahn J (2014) Cloning, expression, purification, crystallization and preliminary crystallographic analysis of the C-terminal domain of Par-4 (PAWR). Acta Crystallogr Sect F Struct Biol Commun 70:1224–1227. https://doi.org/10.1107/S2053230X14014691
23. U. K. Tiruttani, S. *Analysis of functional domains of Par-4*; Schriften des Forschungszentrums Jülich Reihe Gesundheit; Forschungszentrum Jülich: Jülich, 2012; ISBN 978-3-89336-766-5
24. Hong P, Koza S, Bouvier ESP (2012) Size-exclusion chromatography for the analysis of protein biotherapeutics and their aggregates. J Liq Chromatogr Relat Technol 35:2923–2950. https://doi.org/10.1080/10826076.2012.743724
25. Guerrero SA, Hecht HJ, Hofmann B, Biebl H, Singh M (2001) Production of selenomethionine-labelled proteins using simplified culture conditions and generally applicable host/vector systems. Appl Microbiol Biotechnol 56:718–723. https://doi.org/10.1007/s002530100690
26. Meyer O, Schlegel HG (1983) Biology of aerobic carbon monoxide-oxidizing bacteria. Annu Rev Microbiol 37:277–310. https://doi.org/10.1146/annurev.mi.37.100183.001425
27. Schwalbe M, Dutta K, Libich DS, Venugopal H, Claridge JK, Gell DA, Mackay JP, Edwards PJB, Pascal SM (2010) Two-state conformational equilibrium in the Par-4 leucine zipper domain. Proteins 78:2433–2449. https://doi.org/10.1002/prot.22752
28. Chayen NE (1998) Comparative studies of protein crystallization by vapour-diffusion and microbatch techniques. Acta Crystallogr D Biol Crystallogr 54:8–15. https://doi.org/10.1107/s0907444997005374
29. Kubicek J, Untersuchungen zur Funktion und Kristallisation von heterolog exprimiertem Par-4, Amida und DLK. Dissertation, Heinrich Heine University Düsseldorf
30. Cheema SK, Mishra SK, Rangnekar VM, Tari AM, Kumar R, Lopez-Berestein G (2003) Par-4 transcriptionally regulates Bcl-2 through a WT1-binding site on the bcl-2 promoter. J Biol Chem 278:19995–20005. https://doi.org/10.1074/jbc.M205865200
31. Welberry TR, Heerdegen AP, Goldstone DC, Taylor IA (2011) Diffuse scattering resulting from macromolecular frustration. Acta Crystallogr B 67:516–524. https://doi.org/10.1107/S0108768111037542
32. Chandrasekaran R, Stubbs G (2006) Fibre diffraction. In International Tables for Crystallography. Am Cancer Soc:444–450. ISBN 978-0-470-68575-4
33. Taylor GL (2010) Introduction to phasing. Acta Crystallogr D Biol Crystallogr 66:325–338. https://doi.org/10.1107/S0907444910006694
34. Battye TGG, Kontogiannis L, Johnson O, Powell HR, Leslie AGW (2011) iMOSFLM: a new graphical interface for diffraction-image processing with MOSFLM. Acta Crystallogr D Biol Crystallogr 67:271–281. https://doi.org/10.1107/S0907444910048675
35. (International Tables) Volume A home page Available online: https://it.iucr.org/A/. Accessed on 30 Nov 2020
36. Ke H (1997) [25] Overview of isomorphous replacement phasing. In Methods in Enzymology; macromolecular crystallography Part A. Academic Press Vol. 276, pp. 448–461
37. Hendrickson WA (1991) Determination of macromolecular structures from anomalous diffraction of synchrotron radiation. Science 254:51–58. https://doi.org/10.1126/science.1925561

38. Evans G, Pettifer RF (2001) CHOOCH: a program for deriving anomalous-scattering factors from X-ray fluorescence spectra. J Appl Crystallogr 34:82–86. https://doi.org/10.1107/S0021889800014655
39. Evans P (2006) Scaling and assessment of data quality. Acta Crystallogr D Biol Crystallogr 62:72–82. https://doi.org/10.1107/S0907444905036693
40. Matthews BW (1968) Solvent content of protein crystals. J Mol Biol 33:491–497. https://doi.org/10.1016/0022-2836(68)90205-2
41. Terwilliger TC, Adams PD, Read RJ, McCoy AJ, Moriarty NW, Grosse-Kunstleve RW, Afonine PV, Zwart PH, Hung L-W (2009) Decision-making in structure solution using Bayesian estimates of map quality: the PHENIX AutoSol wizard. Acta Crystallogr D Biol Crystallogr 65:582–601. https://doi.org/10.1107/S0907444909012098
42. Schneider TR, Sheldrick GM (2002) Substructure solution with SHELXD. Acta Crystallogr D Biol Crystallogr 58:1772–1779. https://doi.org/10.1107/S0907444902011678
43. Sheldrick GM (2010) Experimental phasing with SHELXC/D/E: combining chain tracing with density modification. Acta Crystallogr D Biol Crystallogr 66:479–485. https://doi.org/10.1107/S0907444909038360
44. Adams PD, Afonine PV, Bunkóczi G, Chen VB, Davis IW, Echols N, Headd JJ, Hung L-W, Kapral GJ, Grosse-Kunstleve RW et al (2010) PHENIX: a comprehensive python-based system for macromolecular structure solution. Acta Crystallogr D Biol Crystallogr 66:213–221. https://doi.org/10.1107/S0907444909052925
45. Afonine PV, Grosse-Kunstleve RW, Echols N, Headd JJ, Moriarty NW, Mustyakimov M, Terwilliger TC, Urzhumtsev A, Zwart PH, Adams PD (2012) Towards automated crystallographic structure refinement with phenix.refine. Acta Crystallogr D Biol Crystallogr *68*:352–367. https://doi.org/10.1107/S0907444912001308
46. Kleywegt GJ, Alwyn Jones T (1997) [11] Model building and refinement practice. In *Methods in Enzymology*; macromolecular crystallography part B; Academic Press, Vol. 277, pp. 208–230
47. Brünger AT (1992) Free R value: a novel statistical quantity for assessing the accuracy of crystal structures. Nature 355:472–475. https://doi.org/10.1038/355472a0
48. Brünger AT (1993) Assessment of phase accuracy by cross validation: the free R value. Methods and applications. Acta Crystallogr D Biol Crystallogr 49:24–36. https://doi.org/10.1107/S0907444992007352
49. Walshaw J, Woolfson DN (2001) SOCKET: a program for identifying and analysing coiled-coil motifs within protein structures11Edited by J. Thornton J Mol Biol 307:1427–1450. https://doi.org/10.1006/jmbi.2001.4545
50. Lee DL, Ivaninskii S, Burkhard P, Hodges RS (2003) Unique stabilizing interactions identified in the two-stranded α-helical coiled-coil: crystal structure of a cortexillin I/GCN4 hybrid coiled-coil peptide. Protein Sci 12:1395–1405. https://doi.org/10.1110/ps.0241403
51. Harbury PB, Zhang T, Kim PS, Alber T (1993) A switch between two-, three-, and four-stranded coiled coils in GCN4 leucine zipper mutants. Science 262:1401–1407. https://doi.org/10.1126/science.8248779
52. Lupas AN, Gruber M (2005) The structure of α-helical coiled coils. In *Advances in Protein Chemistry*; fibrous proteins: coiled-coils, Collagen and Elastomers; Academic Press, Vol. 70, pp. 37–38
53. Krylov D, Vinson CR (2001) Leucine Zipper. In eLS; American Cancer Society, ISBN 978-0-470-01590-2
54. Rasmussen R, Benvegnu D, O'Shea EK, Kim PS, Alber T (1991) X-ray scattering indicates that the leucine zipper is a coiled coil. Proc Natl Acad Sci 88:561–564. https://doi.org/10.1073/pnas.88.2.561
55. Meier M, Lustig A, Aebi U, Burkhard P (2002) Removing an Interhelical salt bridge abolishes coiled-coil formation in a de novo designed peptide. J Struct Biol 137:65–72. https://doi.org/10.1006/jsbi.2002.4467
56. Burkhard P, Ivaninskii S, Lustig A (2002) Improving coiled-coil stability by optimizing ionic interactions. J Mol Biol 318:901–910. https://doi.org/10.1016/S0022-2836(02)00114-6

57. Kammerer RA, Jaravine VA, Frank S, Schulthess T, Landwehr R, Lustig A, Garcı´a-Echeverrı´a C, Alexandrescu AT, Engel J, Steinmetz MO (2001) An Intrahelical Salt Bridge within the Trigger Site Stabilizes the GCN4 Leucine Zipper. J Biol Chem *276*:13685–13688. https://doi.org/10.1074/jbc.M010492200
58. Lupas A, Dyke MV, Stock J (1991) Predicting coiled coils from protein sequences. Science 252:1162–1164. https://doi.org/10.1126/science.252.5009.1162
59. Arnold K, Bordoli L, Kopp J, Schwede T (2006) The SWISS-MODEL workspace: a web-based environment for protein structure homology modelling. Bioinformatics 22:195–201. https://doi.org/10.1093/bioinformatics/bti770
60. Tiruttani Subhramanyam UK, Kubicek J, Eidhoff UB, Labahn J (2017) Structural basis for the regulatory interactions of proapoptotic Par-4. Cell Death Differ 24:1540–1547. https://doi.org/10.1038/cdd.2017.76
61. Page G, Lödige I, Kögel D, Scheidtmann KH (1999) AATF, a novel transcription factor that interacts with Dlk/ZIP kinase and interferes with apoptosis 1. FEBS Lett 462:187–191. https://doi.org/10.1016/S0014-5793(99)01529-X
62. Dutta K, Alexandrov A, Huang H, Pascal SM (2001) pH-induced folding of an apoptotic coiled coil. Protein Sci 10:2531–2540. https://doi.org/10.1110/ps.28801
63. Dutta K, Engler FA, Cotton L, Alexandrov A, Bedi GS, Colquhoun J, Pascal SM (2003) Stabilization of a pH-sensitive apoptosis-linked coiled coil through single point mutations. Protein Sci 12:257–265. https://doi.org/10.1110/ps.0223903
64. Kammerer RA, Schulthess T, Landwehr R, Lustig A, Engel J, Aebi U, Steinmetz MO (1998) An autonomous folding unit mediates the assembly of two-stranded coiled coils. Proc Natl Acad Sci 95:13419–13424. https://doi.org/10.1073/pnas.95.23.13419
65. Kawai T, Matsumoto M, Takeda K, Sanjo H, Akira S (1998) ZIP kinase, a novel serine/threonine kinase which mediates apoptosis. Mol Cell Biol 18:1642–1651. https://doi.org/10.1128/MCB.18.3.1642
66. Azmi AS, Aboukameel A, Bao B, Sarkar FH, Philip PA, Kauffman M, Shacham S, Mohammad RM (2013) Selective inhibitors of nuclear export block pancreatic Cancer cell proliferation and reduce tumor growth in mice. Gastroenterology 144:447–456. https://doi.org/10.1053/j.gastro.2012.10.036
67. la Cour T, Kiemer L, Mølgaard A, Gupta R, Skriver K, Brunak S (2004) Analysis and prediction of leucine-rich nuclear export signals. Protein Eng Des Sel 17:527–536. https://doi.org/10.1093/protein/gzh062
68. Fu S-C, Huang H-C, Horton P, Juan H-F (2013) ValidNESs: a database of validated leucine-rich nuclear export signals. Nucleic Acids Res 41:D338–D343. https://doi.org/10.1093/nar/gks936
69. Xu D, Farmer A, Collett G, Grishin NV, Chook YM (2012) Sequence and structural analyses of nuclear export signals in the NESdb database. Mol Biol Cell 23:3677–3693. https://doi.org/10.1091/mbc.e12-01-0046
70. Stommel JM, Marchenko ND, Jimenez GS, Moll UM, Hope TJ, Wahl GM (1999) A leucine-rich nuclear export signal in the p53 tetramerization domain: regulation of subcellular localization and p53 activity by NES masking. EMBO J 18:1660–1672. https://doi.org/10.1093/emboj/18.6.1660

Regulation of Par-4 by Ubiquitinases

Nikhil Hebbar and Vivek M. Rangnekar

Abstract The ubiquitin-proteasome system (UPS) is a complex cellular machine that is responsible for maintaining normal homeostasis by targeting misfolded and damaged proteins for degradation. It is characterized by a multienzyme cascade that catalyzes the addition of ubiquitin molecules onto the target proteins. A ubiquitin-activating enzyme E1, a conjugation enzyme E2, and a multi-domain ligase E3 form the complex triad that controls the ubiquitylation of substrate proteins. The substrates for the UPS play important roles in cell signaling pathways, and, in several diseases, linked to neurodegeneration and cancer also, E3 ligases that are part of the UPS enzyme cascade can be dysregulated and function as oncogenes. E3 ligases promote ubiquitination of tumor suppressor proteins leading to their degradation by the 26S proteasome. The RING E3 ligase subfamily contains evolutionarily conserved F-box and SOCS-box proteins that recognize and bind substrates through a characteristic SPRY domain. The RING E3 ligases like Fbxo45 specifically recognize the tumor suppressor protein Par-4 via a consensus VASA-like ELNNNL binding motif, which is located within the pro-apoptotic PAF domain of Par-4. In therapy-sensitive tumors, the PAF domain undergoes caspase-induced cleavage and acts as a decoy substrate for Fbxo45, thereby rescuing Par-4 from ubiquitination. The degradation of Par-4 leads to tumorigenesis, drug resistance, and poor overall survival in several cancers. Greater understanding of E3 ligase interactions with tumor suppressor proteins like Par-4 can drive the development of novel therapeutic approaches to better regulate the UPS in cancer and improve survival outcomes in patients.

Keywords Ubiquitinases · E3 ligases · F-box · Fbxo45 · TRIM-21 · PAF · Par-4 · COP1 degron · Skp1-Cullin-F-box subfamily · RING

N. Hebbar (✉)
Department of Bone Marrow Transplantation and Cellular Therapy, St. Jude Children's Research Hospital, Memphis, TN, USA
e-mail: nikhil.hebbar@stjude.org

V. M. Rangnekar
Associate Director, Markey Cancer Center, University of Kentucky, Lexington, KY, USA

V. M. Rangnekar (ed.), *Tumor Suppressor Par-4*,
https://doi.org/10.1007/978-3-030-73572-2_5

1 The Ubiquitin-Proteasome System (UPS)

The vast majority of functional as well as structural components of a cell are comprised of proteins and protein complexes. They are responsible for driving and regulating a multitude of cellular processes that are crucial for maintaining homeostasis, cell division, and apoptosis [1]. The functional characteristics of cellular proteins are highly dependent on their three-dimensional structure which is derived from the amino acid composition [2, 3]. Although favorable thermodynamics instructs the folding of proteins into their tertiary or quaternary structures, the protein folding process itself is extremely bioenergetically demanding. This is due to the employment of an intricate network of chaperone proteins that can detect regions of hydrophobicity within fully or partially unfolded proteins.

The chaperones use the energy derived from ATP hydrolysis to guide the protein folding process [4–6]. Despite the tight regulation and energy expenditure, proteins are sometimes misfolded or get damaged. Several factors such as reactive oxygen species, DNA damage, exposure to toxic chemicals, or drugs can lead to incorrect protein folding or can damage correctly folded proteins structures. These aberrations can result in loss of protein function and sometimes cause proteins to aggregate which can be toxic to the cell [7, 8]. Misfolding and aggregation of proteins result in a number of debilitating neurodegenerative diseases like Parkinson's disease and Alzheimer's disease, and it has also been implicated in diabetes and cancer development and progression [9–11]. In order to prevent and control the accumulation of misfolded proteins and protein aggregates, cells have developed two important mechanisms of quality control, the ubiquitin-proteasome system (UPS) and autophagy which involves lysosome-mediated proteolysis [7].

The UPS is comprised of a multienzyme cascade that works sequentially to tag misfolded or aberrant proteins with ubiquitin moieties which allows them to be identified and targets them for degradation by the proteasomal machinery. Central to this process is a small and evolutionarily highly conserved protein called ubiquitin which acts as a proteolytic signal moiety [12]. The ubiquitin-activating enzyme (E1), ubiquitin-conjugating enzyme (E2), and ubiquitin ligases (E3) are responsible for catalyzing the formation of an isopeptide bond between the C-terminus of ubiquitin and a lysine residue on either side of the substrate or another ubiquitin moiety [13–15]. A single molecule of ubiquitin contains seven Lys residues, thus allowing proteins to be ubiquitylated in multiple homo- or heterotypic linkage combinations. The length of the ubiquitin chains can also vary allowing for much diversity in the ubiquitin signals [16]. Once a substrate has been tagged with a ubiquitin molecule, the Lys-48 (K48) residue serves as linkage site for the addition of more ubiquitin moieties. Polyubiquitination of substrates through the K48 residue renders them as targets for proteasome-mediated degradation, while Lys-63-linked ubiquitin chains signal transcription factor nuclear factor-κB (NF-κB) activation or play a role in the regulation of DNA repair [17–19]. The 26S proteasome degrades the polyubiquitinated protein and releases the ubiquitin molecules which get recycled for tagging

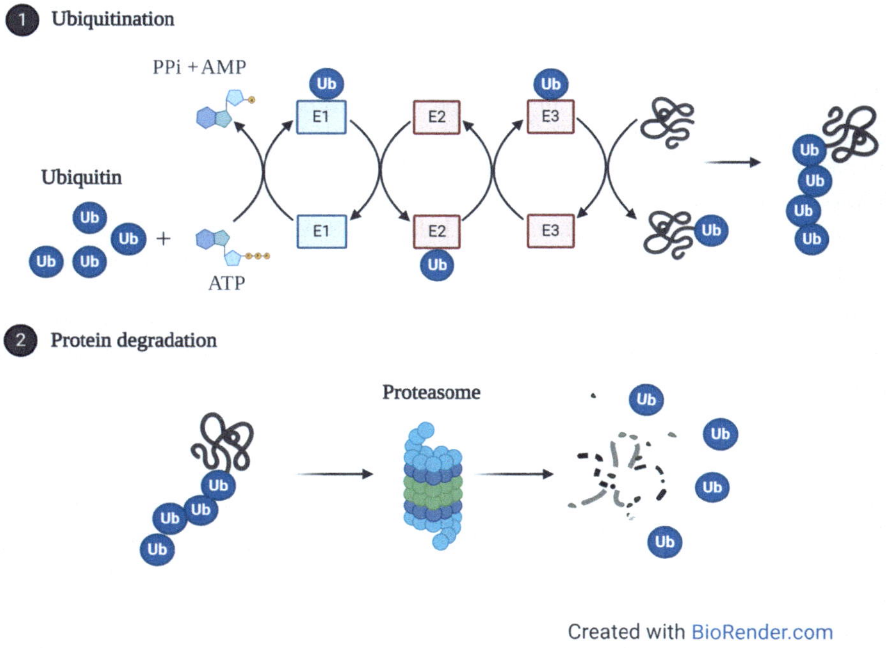

Fig. 1 Overview of the ubiquitin-proteasome system
(1) Sequence of ubiquitin activation by the enzyme cascade beginning with the activation of ubiquitin molecules by the E1-activating enzyme followed by the transfer of the activated ubiquitin molecule onto the E2-conjugating enzyme. Lastly, the ubiquitin moiety is transferred from the E2 to the E3 ligase which adds the ubiquitin moiety to specific lysine residues on a substrate protein. The E3-bound substrate goes through multiple rounds of ubiquitination. (2) The polyubiquitinated substrate is then degraded into free amino acids by the 26S proteasome, and the ubiquitin moieties are recycled by isopeptidases

other substrates [20, 21]. The process of ubiquitination can be reversed by a family of isopeptidase enzymes known as deubiquitinases (DUBs) [22, 23] (Fig. 1).

Here, we shall focus on the role and the key components of the ubiquitin-proteasome system in the regulation of cellular proteins like prostate apoptosis response-4 (Par-4).

2 Ubiquitin-Activating Enzyme (E1)

The ubiquitin-activating enzyme (E1) also known as UBA1 is the first of a three-part enzymatic cascade that regulates the process of ubiquitination. Its primary role in the activation of ubiquitin moieties is critical for maintaining protein homeostasis. There are two known isoforms of UBA1: UBA1a which is localized primarily in the nucleus and UBA1b which is predominantly found in the cytoplasm [24, 25].

The UBA1 gene displays a high degree of evolutionary conservation, and it has been classified as essential gene in yeast [26]. In addition to UBA1, other enzymes like the FAT10-actiavting E1 enzyme UBA6 have the ability to activate ubiquitin [27]; however, the difference in the expression levels of UBA1 as compared to UBA6 is more than tenfold. This observation clearly demonstrates that UBA1 is responsible for the majority of E2 charging [28] and underscores the importance of UBA1 in the regulating the homeostatic balance of cellular proteins.

In order to successfully maintain this balance, UBA1 facilitates several cellular processes. It is responsible for the activation and transfer of the ubiquitin-like protein neural precursor cell-expressed developmentally downregulated protein 8 (UBL NEDD8). This is especially important under cellular stress conditions resulting from decreased ubiquitin levels, where Uba1 can conjugate proteins with a combination of ubiquitin and UBL NEDD8 [29]. Another interesting function of UBA1 is its atypical role in the autophagy pathway where it facilitates Atg-8-dependent autophagy by circumventing the Atg7 (E1) and Atg7 (E2) enzymes that typically regulate the autophagic pathway [30], thereby uncovering an interesting interdependence of the UPS and autophagic pathway. UBA1 also plays a role in fundamental cellular processes such a cell cycle progression UBA 1 through the regulation of key proteins such as histones and p53 involved in this process [31, 32]. Studies using temperature-sensitive loss-of-function mutants of UBA1 showed that UBA1 loss can downregulate degradation of crucial cell cycle proteins due to a decrease in ubiquitylation. This can have severe consequences leading to cell cycle arrest [33, 34]. The subcellular localization of UBA1 also varies depending on the phase of the cell cycle, with UBA1 being restricted to the nucleus during the G1 and G2 phases, while it is found in both nuclear and cytoplasmic compartments during the other mitotic phases [35].

The role of UBA1 or the E1 as the first enzyme in the UPS cascade involves the activation of ubiquitin in two-step reaction. The first step being the C-terminal adenylation of ubiquitin and the second involving the formation of a thioester bond between a catalytic cysteine of the E1 enzyme and the C-terminus of ubiquitin [36, 37]. This E1-Ub intermediate results in the recruitment of the E2-conjugating enzyme, thus allowing the cascade to proceed. Since the E1 enzyme plays a critical role in the initiation of the UPS pathway, its structure and the function role of its various residues have been extensively studied. Many of the seminal studies that allowed the elucidation of the structure of the E1 (Uba1) enzyme were performed using yeast [38–41]. However, more recently the crystal structure of the human E1-ubiquitin complex has been identified, and it provides some interesting insights into the structural similarities and overall conservation between the yeast and human E1 enzyme [42].

The discovery of the structure of Uba1 has allowed for a greater understanding of its catalytic ability and its engagement with other enzymes of the UPS pathway. The structural studies on Uba1 have shown that it is comprised of multiple different domains and that it assumes several conformations to perform it catalytic functions. At its core Uba1 contains two adenylation domains, one that is active (AAD) and a second inactive adenylation domain (IAD) that come together to form a pseudodimer which provides the enzyme with structural rigidity. It is the AAD that is responsible for ubiquitin, ATP, and Mg^{2+} binding and adenylation potential of the E1 enzyme [38]. The cysteine domain in Uba1 is comprised of two cysteine half domains, the first (FCCH) and second cysteine catalytic half domain (SCCH). The SCCH is the domain that harbors the catalytic cysteine. The process involving thioester bond formation has been revealed through studies of the E1 enzyme for the Ub-like modifier SUMO in which the adenylation active site is disassembled and the SCCH domain undergoes a rotation of approximately 130°. This allows the transfer of the catalytic cysteine and other components necessary for catalyzing the formation thioester bond, from the SCCH domain into the active site [43, 44]. The rotational feature of SCCH is also seen on the yeast ortholog of Uba1 albeit at a slightly smaller angle of 106° [41]. The C-terminal end of Uba1 contains the ubiquitin fold domain (UFD) which is responsible for bringing the catalytic cysteines in the E1 and E2 enzymes within proximity of each other, thereby allowing transfer of the thioester. The UFD achieves this by recruiting the E2 enzymes in a "distal" conformation and rotating them to a more "proximal" conformation bringing it close to the E1 catalytic site [45] (Fig. 2). Although a number of different E2 enzymes interact with the various orthologs of Uba1, most of the Ub-E2 sequences at the E1-E2 interaction sites do not display a high degree of conservation; however, it has been observed that the N-terminal helix at the core of an E2 enzyme contains a three-residue basic motif crucial for its binding to E1 [45, 46]. These studies provide evidence that the variability with the sequence and structure of the interaction motifs between Uba1 and E2 provides a large degree of flexibility and diverse binding affinities between different E1-E2 complexes.

The central role Uba1 in the regulation of protein homeostasis also means that Uba1 dysfunction or dysregulation leads to debilitating neurodegenerative disease. In *Drosophila*, mutations causing even low level impairment of Uba1 function result in a neurodegenerative disease like phenotype [47]. Several preclinical and clinical studies in humans have shown that a cluster of mutations found in exon15 of UBA1 can give rise to a rare disease called X-linked spinal muscular atrophy (XL-SMA)

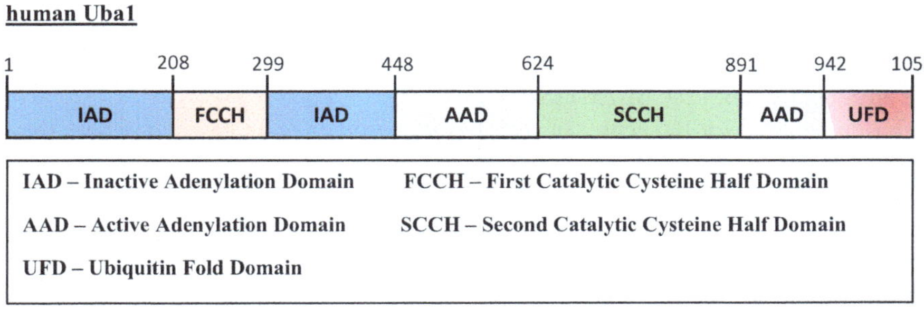

Fig. 2 Domain structure of human E1 enzyme Uba1
Domain organization of the human ubiquitin-activating enzyme E1 (Uba1). The IAD domain is responsible for ubiquitin and ATP binding, while the FCCH and SCCH domains catalyze the thioester bond formation required for activation of the ubiquitin molecule

[48, 49]. In addition to SMA, downregulation or inhibition of Uba1 has been known to play a role in Huntington's disease through polyQ protein aggregation [50], and it gets sequestered in disease-associated protein aggregates in Parkinson's disease [51].

The importance of the E1 enzyme Uba1 as a gatekeeper in the UPS pathway and deep understanding of its structure and function have made it an attractive target for therapeutic intervention. As mentioned earlier, the structure and mechanism of the E1 enzymes display a high degree of plasticity to accommodate the diversity of E2 enzymes. This has posed significant challenges in the development of small molecule inhibitors. However, small molecule inhibitors like pevonedistat or TAK-924 which inhibits the E1 for the ubiquitin-like molecule NEDD8 have been developed [52]. More recently small molecules that inhibit E1 enzymes have shown to be an effective approach to cancer treatment. TAK 243, a cell-active inhibitor of the ubiquitin-activating enzyme E1, has shown potent antitumor preclinical efficacy in a wide range of tumor models [53]. With new agents being developed that specifically target E1 enzymes, it allows for many opportunities and effective tools to study the role of E1 enzymes in normal protein turnover as well as in disease.

3 Ubiquitin-Conjugating Enzyme (E2)

The second stage of the process of ubiquitination involves a conjugation step that is facilitated by a family of E2 ubiquitin-conjugating enzymes. E2 enzymes have been found across all eukaryotic organisms, indicating the importance of these enzymes in normal cellular function. The number of E2 enzymes varies depending on the species but is generally proportional to the evolutionary hierarchy of the organism. In lower organisms like yeast, 15 E2 enzymes have been found, whereas in higher organisms like humans, 35 active E2 enzymes have been discovered. The higher number of enzymes has evolved to meet the needs of increased complexity and specialized function of the higher eukaryotes. Interestingly, despite the high degree of variability in the number of E2 enzymes across species, there exists a small subgroup of E2 enzymes that is highly conserved and is found to be identical across all eukaryotes; some prominent examples of these are SUMO E2 Ubc9(UBE2I) and the NEDD8 E2 Ubc12(UBE2M) [54].

E2 enzymes lie at the center of the ubiquitination enzyme cascade, where they interact with the E1 enzymes which activate and transfer a ubiquitin moiety to the active site cysteine of E2. This is followed by the dissociation of the E2 from E1. Once the E2 has dissociated from the E1 enzyme, it then binds to an E3 ubiquitin ligase which is responsible for the recruitment of the substrate. Depending on the type of E3 ligase, it either transfers the ubiquitin molecule to a Cys residue in the HECT domain of the HECT family of E3 ligase or in the case of E3 ligases containing a RING domain or U-box motif, the E2 is activated, and the substrate is recruited within close proximity allowing it to transfer the ubiquitin to a lysine residue in the substrate. Finally, the ubiquitin is transferred to the substrate, and the E2 is released from the E3 ligase and is available for further conjugation [20, 55].

Although E2 enzymes have been considered primarily as "ubiquitin carriers," they play diverse roles that are crucial for regulating the ubiquitin chain length and efficiency of the entire process of ubiquitin loading and assembly. In order to efficiently fulfill these roles, E2 enzymes have evolved into many specialized subclasses. Based on phylogenetic analyses, human E2 enzymes fall into 17 distinct subfamilies [54]. At the core of all active E2 enzymes lies a ubiquitin-conjugating domain (UBC) which is equipped with a catalytic Cys residue and is responsible for interacting with E1 enzymes. The sequences of UBC domains belonging to different E2s are very homologous and assume similar conformations containing 4 α-helices, an anti-parallel β-sheet made up of four strands, and a short 3_{10}-helix. A short loop that connects α-helix 2 with α-helix 3 combined with a long loop that is near the active site form a pocket that houses the evolutionarily conserved active site Cys residue. The UBC domain can also be found in another group of proteins called ubiquitin E2 variant (UEV) proteins; however, unlike the E2-UBCs, these domains do not possess an active site cysteine [56–58].

As mentioned above, the first phase of ubiquitin conjugation reaction involves an E1-E2 interaction. In order to successfully conjugate ubiquitin, the E2 enzyme must be able to differentiate ubiquitin from a structurally similar but unrelated group of proteins called ubiquitin-like modifiers (UBLs). Much like ubiquitin, these ubiquitin-like modifiers are linked to specific Lys residues on substrate proteins via the catalytic activity of certain E1, E2, and E3 enzymes. The tagging of substrates with UBLs regulates the localization and activation of substrate proteins [59]. The difference between ubiquitin and UBLs is clearly demonstrated by the regulation of proliferating cell nuclear antigen (PCNA), an auxiliary protein that plays a role in eukaryotic replication. When the UBL small ubiquitin-like modifier (SUMO) is bound to a specific Lys residue of PCNA, it prevents sister chromatid recombination, instead monoubiquitylation of the same Lys residue facilitates translesion synthesis DNA repair [18], thus exemplifying the need for E2 enzymes to discern ubiquitin from UBLs.

Even though the E1 and E2 enzymes that recognize UBLs are structurally quite comparable to those that are responsible for binding ubiquitin, the ubiquitin-binding E2 enzymes exclusively interact with the two E1 enzymes involved in the ubiquitin cascade. Usually, it is only when a cognate E1 enzyme is loaded with its modifier that the E2 enzyme will bind to it with high affinity [14, 60]. A few elegant studies analyzing the structure of E1-E2 complexes have shown that an E1 enzyme charged with ubiquitin or a UBL undergoes a change in conformation, thereby allowing access to cryptic E2-binding sites making it highly conducive for the formation of E1-E2 complexes [38, 43, 61]. The ubiquitin fold domain (UFD) of E1 that contains a negatively charged groove becomes exposed by these conformational changes and can be recognized by two Lys residues in the α-helix 1 of ubiquitin E2 enzymes that are highly conserved in all ubiquitin E2 enzymes but are not found in E2 enzymes that recognize the UBL small ubiquitin-like modifier (SUMO) and neuronal precursor cell-expressed developmentally downregulated protein 8 (NEDD8). In addition to the UBC domain, sequences that are either extensions or appendices to the N-terminus of E2 enzymes also play a significant role in modulating the specificity of E1 binding. For example, the E2 UBE2M that is specific for NEDD8 has extensions at its N-terminus that are required for reducing its affinity to the E1 for ubiquitin, but they also provide stability to its interactions with its cognate E1 [62, 63].

In the second phase of ubiquitin conjugation, the E2 enzymes interact with the E3 ligases to complete the process of substrate ubiquitylation. E2 enzymes can engage with a number of different E3s. Some of these E2-E3 pairs have been successfully characterized: in yeast, the Skp1-Cullin-F-box protein (SCF) and the E2 cell division cycle 34 (Cdc34); in humans, the human anaphase-promoting complex or cyclosome (APC/C) and the E2s UBE2C and UBE2S [64, 65]; the endoplasmic reticulum (ER)-resident E3 gp78 and the E2 UBE2G2 [66, 67]; and the human TNF receptor-associated factor 6 (TRAF6) and the heterodimeric E2 UBE2N-UBE2V1 [68], to name a few. It is important to understand the dynamics of E1-E3 pairing because depending on which E3 is chosen by a specific E2, the outcome of substrate ubiquitylation can vary greatly.

The surface of E2s has an N-terminal α-helix 1 and two loops (L1 and L2) that are used for recognizing their corresponding E3s. Minor differences in the sequences of these motifs can alter the specificity of E3 binding [69, 70]. Differences in the amino acids involved in recognizing an E3 allow a single E2 enzyme to recognize various E3s. For example, the residues Arg6 and Lys10 in UBE2N are necessary to recognize the RING domain of the E3 TRAF6, while the interaction of UBE2N with the C-terminal U-box of HSC70-interacting protein (CHIP) is mediated by its connection to the Arg7 and Lys10 residues [71, 72].

Just like the N-terminal extensions in the E2 enzyme modulate its E1 binding properties, certain added features that are specific to E2s can improve the binding affinity to E3 ligases. For example, E2 Rad6 in yeast binds to a Rad18 motif that is not part of the RING domain [73], while the C-terminus of Cdc34 employs a negative charge to facilitate its binding with its E3 SCF [74]. Similarly, the human UBE2G2 binds it E3 gp78 via an additional G2BR motif by using two residues Leu163 and Leu165, thereby greatly enhancing the E2-E3 binding affinity [67, 75, 76]. A single E2 can employ multiple binding sites that can act synergistically to improve the affinity of a given E2-E3 interaction. This can be seen in the case of UBE2G2 and its interaction with the RING domain of its E3 gp78, where addition of G2BR peptide in *trans* increases the affinity of the interaction [76].

The binding between and E2 and E3 enzyme is typically quite weak with low micromolar dissociation constants. Since most E2s rely on using binding surfaces that overlap for E1 and E3 enzymes, the weak nature of E2-E3 binding likely provides an advantage during the formation of ubiquitin chains. Due to the overlap between binding surfaces, an E2 enzyme that is bound to cognate E3 has to undergo multiple iterations of binding and dissociations, while a ubiquitin chain is being assembled and cannot be simultaneously recharged by its corresponding E1. It is only after the E2 has completely dissociated from the E3 that it gets recharged by the E1 enzyme and is maintained in a ubiquitin charged steady state in cells [27, 77].

E2 enzymes are vital to the process of ubiquitin chain assembly and are comprised of an initiation step, where the first ubiquitin molecule being transferred a Lys residue on the substrate, followed by an elongation step, in which additional ubiquitin molecules are added to substrate-linked ubiquitin. After the initial ubiquitin has been added, the subsequent ubiquitin molecules can be added to Lys residues on either the substrate or on the initial ubiquitin itself, a decision that is made by the E2 enzyme. This can have profound effects on the function and fate of the ubiquitinated substrate. The initiation and elongation steps are also regulated by distinct E2s. One such example is that of heterodimeric E3 breast cancer type 1 susceptibility protein (BRCA1)-associated RING domain 1 (BARD1). A number of E2s like UBE2W and UBE2E2 control the ubiquitin chain initiation step, while the elongation step is facilitated by the heterodimer UBE2N-UBE2V1 and UBE2K [78].

In addition to ubiquitin chain initiation and elongation, the E2 enzymes also control the processivity of ubiquitin chain formation. The processivity of ubiquitylation is defined as the number of molecules of ubiquitin that are added to a growing chain in one round of a substrate being bound to the E3 [79]. This is extremely important in the physiological context because the 26S proteasome has substrate receptors that require a minimum of four ubiquitin moieties to be attached to the substrate in order to be recognized [80]. The processivity needs to be maintained by the E2 for the substrates to be degraded efficiently and in a timely manner.

4 E3 Ubiquitin Ligases

The E3 ubiquitin ligases are the last piece of the enzymatic cascade that catalyzes the formation of a covalent attachment of ubiquitin of wide array of substrates. The ubiquitin moiety is first activated by the E1-activating enzyme and attached to an E2 enzyme through a thioester bond. Subsequently, the binding of an E3 enzyme to the E2-Ub complex allows for the formation of an isopeptide bond between the amino group at the free N-terminal end or the ε-amino group of a lysine side chain of the substrate and the carboxyl group at the C-terminus of ubiquitin. Since ubiquitin molecules possess seven Lys residues, the E3 can repetitively catalyze multiple reactions using the different lysines to form polyubiquitin chains.

As discussed earlier, the ubiquitination pattern of the substrate has a profound impact on its interactions with various downstream effectors, which each pattern resulting in distinct signals. So far, 2 E1 and 35 active E2 enzymes have been discovered in humans; however, the library of E3 enzymes stands at an impressive 600 enzymes [27, 54, 81], underscoring the importance of these enzymes and their role in the coordination of the ubiquitin cascade and linking the ubiquitin molecules to various substrates. In this chapter, we shall discuss three import subtypes of E3 ligases: HECT (homologous to E6-associated protein C-terminus) domain E3 ligases, U-box E3 ligases, and RING E3 ligases (Fig. 3).

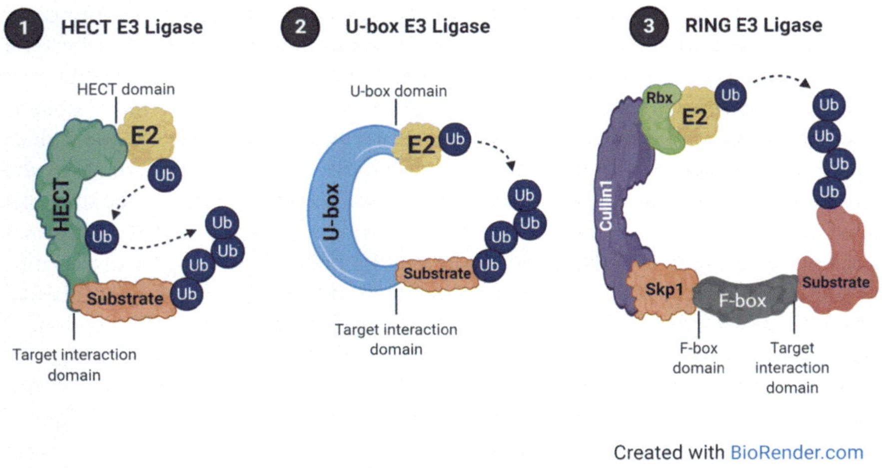

Fig. 3 E3 Ubiquitin ligases
(1) HECT E3 ligases have a conserved HECT domain that possesses intrinsic ligase activity. They catalyze the transfer of ubiquitin from the E2 to the substrate through the formation of thioester intermediate. (2) U-box E3 ligases interact with E2 enzymes through their U-box domain, and they catalyze direct transfer of ubiquitin to the substrate without the formation of a thioester bond. (3) SCF (Skp1-Cullin1-F-box) RING E3 ubiquitin ligase. It consists of a scaffold protein Cullin1 and the Skp1-F-box fusion protein that recruits the substrate proteins and a RING domain containing protein Rbx1 which activates the E2 for the transfer of ubiquitin

4.1 HECT E3 ligases

A number of studies in the 1990s found that a number of proteins share a domain that is approximately 350 amino acids in the C-terminus of E6-associated protein (E6AP). It was discovered that the homologous to E6AP C-terminus (HECT) could facilitate polyubiquitin chain formation through its ability to catalyze the linkage of a ubiquitin C-terminus with a thioester [82]. These findings led to the classification of these enzymes as HECT E3 ubiquitin ligases. Twenty-eight HECT E3 ligases have been found in humans, and they range from relatively small 80 kDa to much larger 500 kDa proteins [82–84].

HECT E3s are multi-domain enzymes with each domain serving a distinct role. The catalytic function the HECT E3 ligases resides within the HECT domain near the C-terminal region, whereas several domains and extensions at the N-terminal region are tasked with binding the substrate to be ubiquitinated. The unique motif found in the N-terminal extensions has led to the division of HECT E3 ligases into three broad subgroups: the WW domain containing Nedd4/Nedd4-like E3s, the HECT and RCC1-like domain containing HERC E3s, and the third group of “other HECT E3s” that is comprised of enzymes containing a combination of other domains [83, 84].

The HECT domains catalyze the formation of the ubiquitin-thioester linkage by interacting with the cognate E2s. These interactions are primarily carried out with the UbcH7 and UbcH5 E2 subfamily of enzymes. The HECT domain is made up of an N-terminal lobe that controls E2 binding, while its catalytic cysteine is found in the C-terminal lobe. The two lobes of the HECT domain are quite flexible owing to the presence of a hinge region between them. This flexible linker allows the catalytic cysteine and the E2-Ub thioester to be favorably positioned, thus facilitating the transthioesterification of the HECT domain to ubiquitin. The distance between the N- and C-terminal lobes of the HECT domain heavily relies on the ubiquitin-loading state of the E2 enzymes and noncovalent interactions between the E2-Ub complex and the two lobes of the HECT domain. Studies with the NEDD4–2/NEDD4L and ubiquitin-loaded UbcH5B have shown that the distance between the catalytic cysteine residues of E2s and HECT domains gets significantly reduced when the E2 is loaded with a ubiquitin moiety [85], while the distance is significantly greater in the absence of bound ubiquitin [86, 87]. This proximity is crucial for transthioesterification to occur.

There are two levels of regulation of HECT E3 activity. The first level involves protein interaction motifs at the N-terminal of the HECT that control substrate binding. Several regulatory auxiliary/adaptor proteins that bind these N-terminal interaction motifs on the cognate E3s have a facilitatory or inhibitory effect on substrate binding. For example, members of the α-arrestin protein family (Arrdc) bind to WW domains of Nedd4 family E3s and facilitate the process of ubiquitination of substrates [88, 89]. On the other hand, hormone-induced phosphorylation of Nedd4–2 results in its binding to the 14–3-3 proteins which inhibit the interaction of Nedd4–2 substrates like the epithelial sodium channel subunits [90].

The next level of regulation comes from the HRECT domain which is responsible for catalyzing the ubiquitination of the substrate. To ensure that the process and timing of ubiquitination of a substrate protein are tightly controlled, several inter- and/or intramolecular interactions regulate the activity of HECT domains of HECT E3 ligases. The enzymes SUMRF2 and Itch belonging to the Nedd4 family of E3s are examples of HECT E3 ligases regulated by such interactions. Studies have shown that the HECT domain interaction between SMURF2 and its cognate E2 UbcH7 is rather weak in comparison to other HECT domains. In order to improve the efficiency of the interaction of the HECT domain of SMURF2 and UbcH7, both the proteins need to be bound to an adaptor protein SMAD7 [91]. NMR studies have shown that a C2 domain within SMURF2 masks the HECT domain, thereby blocking its catalytic cysteine access to UbcH7. SMAD7 prevents this inhibitory interaction allowing the ubiquitination reaction to proceed [92]. Under normal conditions, the Nedd4 E3 ligase Itch is maintained in an inactive state due to its HECT domain interaction with its WW domain. In response to stimuli, specific serine and threonine residues at the N-terminus get phosphorylated. This causes a change in conformation that disrupts the inhibitory WW-HECT interaction resulting in the activation of Itch [93].

HECT E3 ligases control the ubiquitination of many important protein, thereby regulating important physiological processes like inflammatory signaling [94], the epithelial sodium channel ENaC [95], and the water channel AQP2 [96]. Considering its role in regulation of cellular function required for the maintain homeostasis, it is not surprising to see that dysregulation of the HECT E3 ligases leads to the development of many disease states that affect the nervous system and cancer. The HECT E3 ligase E6AP associates with the DNA binding of domain of the tumor suppressor p53, while its N-terminal can simultaneously bind the HPV protein E6 and drive HPV-induced cervical cancer [97]. Mutation in the C-terminal region of the epithelial sodium channel ENaC can inhibit its ability to bind the HECT E3 ligase Nedd4–2, thus preventing its ubiquitination and proteasomal degradation. Increased cellular levels of ENaC results in the development of an autosomal dominant disorder known as Liddle's syndrome that is characterized by sodium retention and hypertension [98]. Another HECT ligase HUWE1 has been shown to play a role in the carcinogenesis of different tumors. It is known to target pro-apoptotic proteins such a p53 and Miz1 for degradation and also activate the oncogene c-myc through the ubiquitination of K63 residues [99–101]. The role of HUWE1 in tumorigenesis is also supported by evidence showing that it is inhibited by the tumor suppressor p14ARF [100]. WWP2 which belongs to the NEDD4-like protein family of HECT E3 ligases is known to bind and ubiquitinate the tumor suppressor PTEN, thereby promoting its degradation, leading to a pro-tumorigenic effect. Treatment of cells with the chemotherapeutic agent doxorubicin has been shown to inhibit WWP2-PTEN binding and stabilize the levels of PTEN within cells [102]. Various observations that have confirmed the role of HECT E3 ligases in disease have led to much interest in the therapeutic targeting of these enzymes. A number of inhibitors have been developed over the years that target different components of HECT E3 ligases. Heclin is a small molecule inhibitor that works by targeting the active site cysteine and inducing a conformation change, inhibiting the catalytic activity of the enzyme [103]. Antidepressant drugs like clomipramine have been used to inhibit the trans-thiolation step, preventing the auto-ubiquitylation of the HECT E ligase ITCH [104].

4.2 U-Box Ligases

Typically, the process of ubiquitination is carried out by the E1, E2, and E3 enzymes; however, the process of multiubiquitylation of substrates involves a higher degree of complexity that require specialized conjugation factors [105]. The conjugation factors, classified as E4 enzymes, have the ability to catalyze the elongation of ubiquitin chain in concert with the traditional E1, E2, and E3 enzymes. The E4 enzymes have a highly conserved UFD2-homology or U-box domain at their C-terminus. Structurally, the U-box domain bears significant similarity to the RING finger domain which is a characteristic feature of RING finger E3 ligases [106].

Studies in yeast have shown that UFD2 in combination with the E1, E2, and E3 enzymes catalyzes the addition of ubiquitin moieties to substrate that already have 1 to 3 molecules of ubiquitin attached. This results in the formation of multiubiquitylated substrates that are targeted proteasome mediated degradation. In mammalian systems, many of the U-box proteins such as UFD2a have been shown to possess E3 ligase-like activity and function in a manner that is similar to that of the RING finger proteins [107]. However, in the case of polyglutamine diseases like spinocerebellar ataxia type 3 (SCA3), pathological forms of ataxin-3 require to undergo polyubiquitination in order to be cleared by the proteasome. This requires a more unique enzyme-substrate reaction that is catalyzed by the mammalian UFD2a, where it acts as a ubiquitin chain assembly factor (E4) rather than a typical E3 ligase [108]. Some U-box proteins like the UbcM4/UbcH7-interacting protein 5 (UIP5) contain both a U-box domain and a RING finger domain. It is interesting to note that the ubiquitin ligase activity of UIP5 is primarily dependent on its U-box domain and not the RING domain. This has been confirmed by mutating residues which show that ability of UIP5 to interact with its cognate E2 is mediated by the U-box domain [109].

Proteins containing the U-box domains play very significant physiological roles by maintaining cellular homeostasis. One example of such a U-box-containing protein is the carboxyl terminus of Hsc70-interacting protein (CHIP) [110]. It contains three N-terminal tetratricopeptide (TPR) domains which bind the chaperones Hsc70 and Hsp90 and a C-terminal U-box domain which possesses an E3 ligase-like activity. CHIP has also been shown to work in conjunction with the RING finger E3 ligase Parkin to catalyze the polyubiquitination of unfolded Pael receptor (Pael-R). The dopaminergic neurons under ER stress conditions accumulate unfolded Pael-R and undergo degradation leading to Parkinson's disease. Studies have shown that the ubiquitin chain assembly activity of CHIP is required in addition to the E3 activity of Parkin for the multiubiquitination of Pael-R for it to be efficiently targeted for degradation [111].

Human cyclophilin-60 (hCyp-60), also known as CYC4 that belongs to the cyclophilin class, is believed to play the role of a molecular chaperone. It contains an N-terminal U-box domain and a cyclophilin-like domain with peptidyl-prolyl cis/trans isomerase activity at its C-terminal end. Cyclophilin-60 is found in the nucleus and performs an E3 ligase-like function with its cognate E2 Ubc3 (Cdc34), and it is believed that its peptidyl-prolyl cis/trans isomerase activity assists in the ubiquitination of nuclear proteins [112]. Other U-box-containing proteins have also been described in various studies. U-box type E3 is also known to mediate unusual types of ubiquitin linkages that signal proteins toward specific functions. For example, substrates tagged with polyubiquitin chains linked at Lys63 are primed for their functions in DNA repair, resistance to cellular stress, and ribosomal modifications regulated by the cell cycle [113–115].

4.3 RING E3 Ligases

The third and probably one of the largest classes of E3 ligases is the RING E3 ligases class with approximately 600 RING E3s having been identified in humans. Studies have shown that the genetic code for the RING domain involves 300 genes in human [81]. As the name suggests, they contain a signature Really Interesting New Gene (RING) domain that plays a role in the recruitment of E2-Ub the subsequent transfer of ubiquitin to the substrate [13, 116]. Structural analysis of the RING domain has revealed that at its core it contains highly conserved residues of cysteine and histidine that are responsible for binding two zinc atoms. This cross-braced binding interaction, which is different from zinc fingers, is crucial for maintaining the structural integrity of the RING domain and provides a globular structure for protein-protein interaction [117, 118].

A polypeptide sequence within the RING domain is typically where the majority of E3 substrate binding sites are found. However, many of the RING domain ligases occur in the form of protein complexes in which the substrate recognition function is carried out by a separate protein. Prominent examples of these auxiliary proteins are the Rbx1/Roc1 and Rbx2/Roc2 proteins that are part of extremely diverse SCF and Cullin-RING ubiquitin ligases (CRL) with a variety of substrate-recognition subunits such as the F-box, SOCS-box, VHL box, and BTB domains [119]. A number of RING proteins are also known to have signaling functions, of note being SH2, SH3, FHA, ankyrin repeats, and PDZ. RING proteins are classified into several subfamilies based on homology and the composition of domains, with the TRIM/RBCC subfamily being the largest with approximately 76 members in humans. The presence of a B-box and a coiled coil defines the members of the TRIM/RBCC subfamily [120, 121], while the RBR/TRIAD which is the second largest subfamily with 14 representatives is characterized by an in-between RING (IBR) and two RING domains [122, 123].

The ubiquitin ligase activity of RING domains is elicited through their direct binding to their cognate E2 ubiquitin-conjugating enzymes. Early studies with Rbx1/Roc1, c-Cbl, and AO7 and discovery of the crystal structure of the Cbl-RING domain bound to the E2 UbcH7 were seminal to the understanding of how RING ligases function through their association with the E2 enzymes [69]. The critical residues that are required for the E2-RING E3 binding have been discovered through structural and mutational studies. The surface of the RING E3 has a shallow cleft which is made up of loop regions containing two RING domain zinc coordination sites that are connected by a central helix. This shallow cleft region serves as a binding site for the E2s [69]. The zinc-binding residues in the RING domain are essential for the E2-RING interaction as well as its E3 ligase function. This has been clearly demonstrated by mutational analyses with RING domain of c-Cbl and Cnot4 where perturbation of the zinc-binding residues results in abrogation of its ability to bind its cognate E2 and loss of its E3 ligase activity [124, 125].

RING E3s can be classified as bi-substrate enzymes where a ubiquitin linked to an E2 via a thioester linkage (E2-Ub) is the first and the lysine residue on the target protein acts as the second substrate. Once the ubiquitin has been transferred, the discharged E2 becomes the first product of the RING E3 ligase, while the newly formed ubiquitinated substrate becomes the second product. The RING E3s can also polyubiquitinate substrates by different mechanisms. One model being sequential ubiquitination, where the RING E3 adds ubiquitin molecules to the substrates after the dissociation of the E2 after each round of ubiquitin transfer, and alternate model being the substrate transfer of complete ubiquitin chains that have been pre-assembled on the E2 platform.

The mechanism of ubiquitin transfer by RING E3 ligases is quite different when compared to the previously described HECT E3 ligases. In the case of HECT E3s, the ubiquitin molecule is first transferred from the E2-Ub complex to conserved cysteines within the HECT from a thioester-linked intermediate and is subsequently transferred onto the substrate. However, in the case of RING E3 ligases, the active cysteine residues within the RING domain are not exposed to allow ubiquitin binding, suggesting that the transfer of the ubiquitin moiety from the E2-Ub complex to the substrate happens in a single direct transfer reaction [126].

The regulation of RING E3 activity occurs through multiple mechanisms in the cell. The phosphorylation of substrate is an important modification that modulates the ubiquitination process. An example of this would be Skp1-Cullin-F-box (SCF) and other Cullin-RING ligases whose substrates often need to be phosphorylated to ensure binding to the RING E3s and get ubiquitinated. In the case of the anaphase promoting complex/cyclosome (APC/C)-Cdc20, phosphorylation ensures tighter binding to Cdc-20 which acts as a substrate activator and enhances E3 ligase activity [127].

Autoubiquitination of components of the RING E3 ligase complex is another important mechanism of regulation its activity. Prominent examples include the ubiquitination of the F-box protein that is part of the multi-subunit SCF complex. The E2 enzyme that is bound to the RING domain catalyzed the ubiquitination of the F-box protein targeting it for degradation. The presence of a substrate bound to the SCF complex renders the F-box protein ubiquitination sites inaccessible to the E2 which allows preservation of the F-box protein when the substrate needs to be ubiquitinated [128]. Another important example is the regulation of p53, where the RING finger E3 ligase Mdm2 catalyzes the ubiquitination of p53 and targets it for proteasomal degradation. Mdm2 can also mediate its own ubiquitination via its zinc-coordinated RING finger domain. This negative regulation through autoubiquitination is completely abolished by mutations in the zinc-coordinating residues [129].

5 F-Box Proteins

F-box proteins are an important part of the Cullin-RING E3 Ligases (CRLs) and are found as a part of Skp1-Cullin-F-Box protein (SCF)-type RING E3 ligase complex [130]. The F-box proteins are typically made up of 50 amino acid residues, and there are approximately 38 F-box proteins in humans. They have a multitude of secondary motifs like zinc fingers, leucine zippers, proline-rich regions, tetratricopeptide (TPR) repeats, and cyclin domains. In humans, the N-terminal end of the protein is where the F-box motif is situated, while the C-terminal end typically contains leucine-rich repeats (LRRs) or WD repeats. This structural organization led to the development of nomenclature of the F-box proteins, where those with

LRRs are known as FBXL and those with WD repeats being called FBXW. The F-box proteins with an alternative or no secondary motif are called FBXO proteins [131].

The characteristic F-box motif allows Skp1 to bind to the F-box proteins forming a Skp1-F-box protein dimer that is capable of docking onto the CRL scaffold protein Cullin 1 (Cul1). Being part of the SCF E3 ligase complex, the main function of the F-box protein subunit is to recognize and bind the substrate which brings it in proximity to Skp1 and facilitating the ubiquitin transfer by the E3 ligase onto the substrate [132]. The mechanisms by which F-box proteins recognize substrates varies; however, substrate phosphorylation is one of the most commonly found modes of recognition. In the case of the FBXW7, phosphorylation of the highly conserved CDC4 phospho-degron sequence 4-X-pThr (or pSer)-Pro-Pro-X-pSer (or pThr, Glu or Asp) (where X is any amino acid) is essential for the recognition and subsequent ubiquitination of its substrates [133, 134]. In some cases, modifications like glycosylation or addition N-linked high-mannose oligosaccharides are required for recognition by certain F-box proteins [135, 136].

5.1 *F-Box Proteins in Cancer*

F-box proteins have a wide range of substrates including proteins involved in the cell cycle, cell proliferation, and apoptosis and are important in the regulation of normal cellular homeostasis. The substrates recognized by F-box proteins are involved in various pathways like AKT/PI3K, epithelial to mesenchymal transition (EMT), AMPK/mTOR, NF-κB, and others that are often dysregulated in cancer and are known to promote tumor progression, increased cell growth, invasion, and metastasis [137–141].

All three families of F-box proteins are known to play a role in cancer development and progression. Some members of the FBXL family, like Skp2 (FBXL1), function as oncogenes. It is responsible for the degradation of cell cycle regulators such as the cyclin-dependent kinase inhibitors p27, p21, and p57 [142–144]. Skp2 is responsible for the ubiquitination and degradation of the tumor suppressor FOXO1 [145]; similarly its role in promoting the phosphorylation and ubiquitination of the tumor suppressor PDCD4 results in the development of breast tumors that are resistant to treatment with radiation and have worse prognosis [146]. The degradation of these substrates occurs through the ubiquitination of Lys48 residues; however, Skp2 is also known to activate prosurvival proteins like Akt through the ubiquitination of the Lys63 residues [147].

Another member of the FBXL family that is known to have both tumor-suppressive and tumor-promoting functions is FBXL2. Studies in leukemia and gastric and lung cancers have shown that FBXL2 mediates the ubiquitination and degradation of proteins like cyclin D2 and cyclin D3 that drive cell cycle progression. It also recognizes the phosphorylated form of the transcription factor forkhead box 1 (FOXM1) that drives the transcription of the cell cycle regulators cyclin B1 and cyclin D1 and targets it for degradation.

In these studies, FBXL2 performs tumor-suppressive functions and inhibits cancer progression [148–150]. However, studies with PTEN-null tumors identified that PTEN competes with FBXL2 for binding its substrate with inositol 1,4,5-trisphosphate receptor IP3R3 which is important for Ca^{2+}-dependent apoptosis. These studies have shown that FBXL2 has tumor-promoting functions in PTEN-null tumors and that pharmacological inhibition of FBXL2 has a pro-apoptotic effect and sensitizes the tumors to photodynamic therapy [151].

Among members of the FBXW family, two proteins have been rather well characterized. The first being β-TrCP which has been shown to play a role in the activation of the NF-κB pathway by recognizing phosphorylated forms of IκBα [152]. It also recognizes cell division cycle 25A (CDC25A) when it is hyperphosphorylated by the activated checkpoint kinase-1 (CHK1) and 2 (CHK2) under conditions of DNA damage [153].

The second member of this family, FBXW7, is known to be mutated and downregulated in different types of cancers like T cell acute lymphatic leukemia and cholangiocarcinoma [154]. In chronic lymphocytic leukemia (CLL), mutations in FBXW7 prevent the efficient binding of NOTCH1, causing its cleaved form to accumulate and resulting in increased c-Myc and HIF1α levels [155].

The third family of F-box proteins is the FBXO family which contains many members that have been shown to play different roles in tumor promotion as well as tumor suppression. In this chapter, we shall focus on one of the key members of this family, FBXO45.

5.2 *Fbxo45*

Fbxo45 was first characterized as an estrogen-induced protein. The coding region of human and mouse FBXO45 is made up of three exons, and its promoter contains multiple consensus binding sites for the estrogen receptor. Studies showed that exposure of the breast cancer cell line MCF-7 cells to 17β-estradiol resulted in rapid induction of Fbxo45 mRNA transcription [156]. Fbxo45 has been known to play diverse roles ranging from nervous system development to cancer. Studies have shown that Fbxo45 binds to protein associated with Myc (PAM) and its murine equivalent Phr1 which is crucial for neuronal development. Also, mice lacking Fbxo45 did not survive after birth due to severe respiratory distress [157]. In contrast to the role of F-box proteins targeting proteins for degradation, Fbxo45 has been shown to interact with N-cadherin and promote neuronal differentiation by actually suppressing its degradation via proteolysis [158]. Another interesting role for Fbxo45 was uncovered by a genome-wide association study which showed that its regulation of synapse maturation directly correlated with the expression of emotion [159].

Fbxo45 is known to play important roles in tumorigenesis and tumor progression in some cases, while it has tumor-suppressive effects in others. Gene expression analysis based on the TCGA and Gene Expression Omnibus datasets have shown that Fbxo45 is overexpressed in squamous cell lung carcinoma (SCLC) when to compared to surrounding normal tissues expression. The elevated gene expression data also correlates with poor survival outcomes in these patients [160]. Another mechanism by which Fbxo45 improves cancer cell survival is by mediating the ubiquitination of the tumor suppressor protein Par-4. The degradation of Par-4 inhibits apoptosis of cancer cells [161].

However, gastric cancer patients who had lower levels of Fbxo45 expression had worse prognosis, an increase in cancer progression, and poor overall survival as compared to those who had higher levels of Fbxo45 [162]. Fbxo45 is also known to inhibit the process of EMT where Fbxo45 complexes with the E3 ubiquitin ligase SIAH1, and together they ubiquitinate one of the key EMT transcriptional factors Zeb1 and target it for proteasomal degradation [163]. The substrates of Fbxo45 are varied with many of them being involved in cell division and apoptosis. Prominent examples include the transcription factor p73 which belongs to the p53 family of proteins. Fbxo45 as part of the SCF-type RING E3 ligase complex recognizes and ubiquitylates p73 and targets it for degradation. In the case of DNA damage, Fbxo45 is downregulated by the cell to allow for an increase in p73 levels [164].

6 Ubiquitin-Proteasome System and Tumor Suppressor Proteins

In order to prevent the malignant transformation of cells, the levels of oncogenic and tumor suppressor proteins are always under tight regulation. Often, this tight control is mediated by the ubiquitin-proteasome system which can modify these proteins to target them for degradation. One of the tumor suppressor proteins that is often mutated or deleted in various cancers is the phosphatase and tensin homolog (PTEN) which acts as a PI3K antagonist, among other critical cellular functions. Proteasomal degradation of PTEN to control its cellular concentrations is mediated by the HECT-type E3 ligase NEDD4–1 [165].

The RING E3 ligase Mdm2 is responsible for the ubiquitylation of a number of important tumor suppressor proteins like p53, pRb, as well as p73. Mdm2 has been known to function as an oncogene and is amplified in sarcomas with wild-type p53 [166]. In a number of tumors, the overexpression of Mdm2 leads to increased degradation of these tumor suppressors [167].

Another example is of the F-box protein Skp2 which is known to recognize proteins associated with tumor suppression. Several proteins like $p27^{Kip1}$, p130, c-Myc, and $p57^{Kip2}$ act as substrates for Skp2-mediated ubiquitination [143, 144, 168, 169]. One of the tumor suppressor proteins involved in the inhibition of cyclin D1 expression and regulating G1/S progression is a member of the Tob/BTG family called

Tob1. The ubiquitin-dependent proteasomal degradation of Tob1 is facilitated by the SCF^{Skp2} E3 ligase. Overexpression of Skp2 in tumors elevates Cyclin D1 levels through the continuous degradation of Tob1 which leads to uncontrolled cell cycle progression and promotes tumor growth [170].

6.1 Role of Ubiquitination in Prostate Apoptosis Response-4 (Par-4) Regulation

Par-4 is a tumor suppressor protein that was initially identified in a yeast two-hybrid screen. It is considered to be an immediate early apoptotic gene that despite its early discovery in prostate cancer has been later found to be expressed ubiquitously in several different types of tissue [171, 172]. Studies have shown that the nuclear translocation of Par-4 induces apoptosis in various cancer types, and it is able to override the prosurvival signals from proteins like Bcl-xl and Bcl2. It can exert its pro-apoptotic effects to compensate for the lack of p53 and PTEN in some tumors. Consistent with its classification as a tumor suppressor gene, tumors are known to downregulate Par-4 but a variety of mechanisms such as point mutations in exon 3 of PAR-4 resulting in the formation of a premature stop codon or through PAR-4 gene silencing by promoter hypermethylation [173]. Another important mechanism of Par-4 regulation is through the ubiquitin-proteasome system, more specifically through the ubiquitin RING E3 ligases that recognize Par-4 as a substrate protein. Among the RING E3 ligases, the Skp1-Cullin-F-box (SCF) and Elongin-C-Cullin-SOCS-box (ECS) subfamilies have been rather well characterized. The F-box and SOCS-box proteins within these enzyme complexes serve as sites for substrate recognition and interaction. A signature SPRY motif within some F-box and SOCS-box proteins allows them to specifically recognize and ubiquitinate Par-4 and target it for proteasome mediated degradation [174].

6.2 Structure and Domains of Par-4

Par-4 is a 340 amino acid protein that is made up of multiple domains. A high degree of conservation can be seen in the key domains of Par-4 across the rat, mouse, and human genomes. Par-4 contains two nuclear localization sequences (NLS1 and NLS2) and leucine zipper (LZ) domain at its C-terminal end. The LZ domain allows Par-4 to interact and with a number of different binding partners like Wilm's tumor protein (WT1) [175] and ζPKC [176] by forming homo- and heterodimers. The LZ domain is also involved in the negative regulation of Par-4 by mediating its interaction with Akt1 causing Par-4 inactivation through the phosphorylation of specific serine residues [177]. Two important domains that are central to the pro-apoptotic function of Par-4 have been identified.

The selective for apoptosis (SAC) domain is unique to Par-4 and as the name suggests confers Par-4 with the selective ability to specifically kill tumor cells but not normal cells. This selectivity comes from the PKA-mediated phosphorylation of a Thr residue within the SAC domain. The elevated PKA levels of cancer cells compared to normal cells drive the activation of Par-4 in a cancer-specific manner [178].

The second pro-apoptotic domain that was more recently identified was the Par-4 amino terminal fragment (PAF). Studies have shown that treatment of heterogeneous tumors with targeted therapy or chemotherapeutic agents results in the activation of the caspase cascade in the therapy-sensitive tumor cells. This drives the caspase-mediated cleavage of Par-4 at the D131 site resulting in the formation and release of PAF into the cytoplasm. The shedding of PAF into the tumor microenvironment induces apoptosis in therapy-sensitive as well as therapy-resistant tumor cells [179].

Within PAF and hence also within Par-4 lies a very interesting and highly conserved domain called the VASA-like domain. The Glu-Leu-Asn-Asn-Asn-Leu (ELNNNL) containing VASA-like domain is also highly conserved like the other functional domains of Par-4. It was identified as a recognition site for the protein SSB-1 which is part of the ECS subfamily of E3 ligases and contains a central SPRY domain [180].

Another putative domain that may regulate Par-4 levels through ubiquitination is the sequence EPDGVPE which bears significant homology to the COP1 target consensus site [D/E](x)xxVP[D/E] [181]. Next, we shall discuss the role of these E3 ligases in the ubiquitination and degradation of Par-4.

6.3 Fbxo45-Mediated Degradation of Par-4

Fbxo45 is part of the F-box family of proteins that forms a multi-domain complex with the Skp1-Cullin-F-box (SCF) subfamily of RING E3 ligases. Interestingly, the central SPRY domain that imparts the protein recognition and binding properties is commonly found among the SOCS-box and B-box families of E3 ligases [180, 182]. But, in the case of the F-box protein family, the SPRY binding domain is found exclusively in Fbxo45. Studies using a yeast two-hybrid system confirmed that interaction of Fbxo45 with Skp1 is mediated by the SPRY domain resulting in the formation of an atypical multicomponent E3 ligase. LC-MS/MS analysis of Fbxo45 immunoprecipitation samples revealed that Par-4 was one of the binding partners for the E3 ligase.

Further studies with using deletion mutagenesis showed that the SPRY domain within Fbxo45 was responsible for its ability to recognize and bind Par-4. The sequence motif that is recognized by the SPRY domain has been elucidated for the *Drosophila* orthologue GUSTAVUS which binds the DEAD-box RNA helicase VASA (DINNNN) and Par-4 (ELNNNL) [183]. The interaction between the Fbxo45 SPRY domain and Par-4 also occurs at this signature VASA-like domain (^{68}ELNNNL^{72}) within the N-terminal region of Par-4. This was confirmed by using a Par-4 VASA mutant in which the ELNNNL sequence was replaced by ALANAL residues. The mutation in the VASA domain of Par-4 completely abrogated its interaction with Fbxo45 which prevented the ubiquitination and subsequent proteasome-mediated degradation of Par-4 Fbxo45-mediated degradation of the tumor-suppressor Par-4 regulates cancer cell survival [161].

6.4 Rescue of Par-4 Degradation by the Par-4 Amino-Terminal Fragment (PAF)

In heterogenous tumors, Par-4 in the therapy-resistant cell compartment gets degraded by the SPRY domain containing Fbxo45 E3 ligases. The VASA-like (ELNNNL) domain that is recognized by the Fbxo45 SPRY domain located at the N-terminal end of Par-4 and lies within the pro-apoptotic PAF domain. The 131aa PAF domain contains a D131 caspase cleavage site. In therapy-sensitive tumor cell compartment of heterogenous tumors, Par-4 gets cleaved in response to therapy-induced caspase activation at the D131 site releasing the PAF domain. Studies have shown that treatment of therapy-resistant tumors using recombinant PAF protein or ectopic expression of the PAF domain results in stabilization of full-length Par-4. Since PAF contains the same VASA-like domain as Par-4, it can act as a decoy and compete with Par-4 as a substrate for Fbxo45 binding. This allows PAF to rescue Par-4 from being ubiquitinated by Fbxo45 and prevents its degradation. The resulting increase in Par-4 levels translates into responsiveness to therapy in previously resistant tumor cells as well as better overall survival in vivo [179] (Fig. 4).

6.5 Regulation of Par-4 by Other E3 Ligases

One of the E3 ligases that plays a role in Par-4 regulation is the RING E3 ligase tripartite motif containing-21 (TRIM21) which is comprised of multiple domains like a B-box domain, a coiled-coil dimerization domain, an N-terminal RING domain, and C-terminal PRY-SPRY domain [184]. The specificity of substrate binding is controlled by the highly conserved PRY-SPRY domain which is comprised of two elements called PRY and SPRY, respectively [185]. A number of studies have identified the role of TRIM21 in the regulation of innate immune signaling through the ubiquitination of the DNA sensor DDX41 [186] and transcription factors of the IRF protein family that are activated by toll-like receptors (TLRs) [187].

In addition to regulating the innate immunity, TRIM21 is also known to promote tumorigenesis by targeting pro-apoptotic proteins such a p53 [188] and Par-4 [189]. It was identified that PRY-SPRY domain is necessary for Par-4 binding. Experiments with deletion mutants of TRIM-21 showed that only ΔPRY-SPRY mutants were unable to recognize and bind Par-4. The study showed that elevated TRIM21

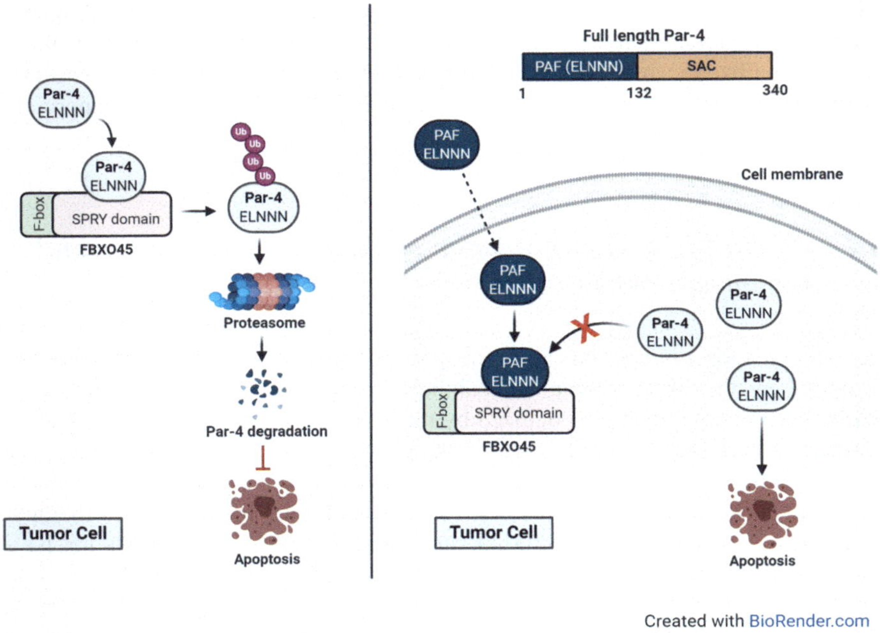

Fig. 4 PAF-mediated rescue of Par-4 degradation
The SPRY domain of Fbxo45 binds to the VASA-like ELNNN region and targets Par-4 for proteasomal degradation in tumor cells, allowing them to evade Par-4-mediated apoptosis. Caspase activation in response to therapy induces Par-4 cleavage and releases the PAF domain. The VASA-like ELNNN region resides within PAF, thereby enabling it to act as a decoy substrate for Fbxo45 and protects Par-4 from being degraded, allowing it induce apoptosis in tumor cells.

expression in pancreatic cancer resulted in cisplatin resistance. In cisplatin-treated tumors, the elevated TRIM21 levels are responsible for the ubiquitination of Par-4 and targeting it for proteasomal degradation. This contributes to therapeutic resistance and worse prognosis in pancreatic cancer patients [189].

Another mechanism of Par-4 regulation by E3 ligases might be through a putative degron EPDGVPE that is found close to the C-terminal region of the PAF domain. This degron displays sequence similarity to the consensus binding site [D/E](x)xxVP[D/E] of the constitutive photomorphogenic 1 (COP1) E3 ligase. COP1, also known as RING finger and WD repeat domain protein 2 (RFWD2), belongs to the RING E3 ligase family of proteins [181]. COP1 can function as an oncogene, and it is often overexpressed in several different types of cancer, with over 80% of breast adenocarcinomas and approximately 70% of pancreatic cancers showing elevated COP1 levels [190, 191]. The high level of expression combined with its ability to putatively bind and target key tumor suppressor proteins such as p53 makes a strong case for Par-4 also being a substrate for the COP1 E3 ligase. COP1 has been shown to promote leukemogenesis in a murine model which identified the WD40 domain in COP1 as essential for substrate recognition [192]. It is likely the WD40 domain within COP1 also recognizes the putative EPDGVPE degron within Par-4 and targets it for degradation.

7 Conclusions and Future Perspectives

In conclusion, the role of Par-4 regulation by the ubiquitin-proteasome system is complex and involves multiple E3 ligases. As a result, Par-4 is often degraded in tumor cells leading to therapeutic resistance. Targeting the UPS with proteasomal inhibitors like first-generation bortezomib or subsequent generations like carfilzomib, oprozomib, and ixazomib to prevent the degradation of tumor suppressor proteins has been fraught with challenges associated with toxicity. However, using inhibitors to target the UPS has come a long way over since bortezomib was first discovered and several new targeting strategies have been developed.

Highly selective agents known as ubiquitin variants (UbVs) can potently target individual components of the UPS. They have been developed using a combination of Phage display and protein engineering technology. UbVs have been successfully used to target deubiquitinases (DUBs) like UPS7 resulting in stabilization of p53 though enhanced Mdm2 ubiquitination [193]. UbVs have also been effectively used to target E3 ligases. Studies with Skp1-Cul1-F-box (SCF) E3 ligase family have shown that UbVs block sites on Skp1 and F-box proteins that are essential for binding to Cullin1, thereby inhibiting activity of the enzyme. In the context of Par-4, developing UbVs to bind and target Fbxo45 might be an effective approach to disrupt its ligase function and prevent Par-4 from being degraded. Abrogation of Fbxo45-Par-4 binding by the decoy function of the PAF domain provides evidence to support this line of investigation.

More recently the role of the immune system in cancer treatment and regulation has been an area of intense investigation, particularly the role of ubiquitination in tumor immunology and the tumor microenvironment. For example, in activated T cells, polyubiquitination of the programmed cell death protein 1 (PD-1) by the E3 ligase Fbxo38 for proteasomal degradation promoting tumor development [194]. Conversely, cancer cells overexpress PD-L1 (the ligand for PD1) which inhibits T cell function. Studies with the COP9 signalosome 5 (CSN5) which catalyzes the deneddylation of Cullin-RING ubiquitin E3 ligases can also prevent PD-L1 from being ubiquitinated and stabilize PD-L1 in cancer cells [195]. It is interesting to note that, both these examples bear resemblance to the Fbxo45-mediated degradation of Par-4 and the putative role of COP1 in targeting Par-4 for proteasomal degradation on tumor cells.

Another interesting approach involves proteolysis-targeting chimeras (PROTACs), which function by the regulating components of the UPS by controlling their degradation in a targeted manner instead of inhibiting their activity and expression. PROTAC's are heterodimers that are composed of E3 Ub ligase binder that is chemically linked to a protein-targeting warhead. A stable interaction between the target protein and the E3 ligase is mediated by the PROTAC which leads to the ubiquitination and proteasomal degradation of the target protein. Although this technology provides many advantages over conventional small molecule inhibitors, PROTACs have only been used for a limited number of E3 ligase targets [196, 197]. For future studies, an interesting avenue of PROTAC research could involve targeting specific proteins like Fbxo45 that are part of E3 ligase complexes. Since Fbxo45 is often dysregulated in cancer and neurodegenerative disease, using PROTACs could provide a promising therapeutic option and improve treatment outcomes.

With a deeper understanding of the molecular and structural basis of Par-4 binding with its interacting partners in the UPS, combined with the advent of newer forms of targeted therapy and CRISPR/Cas9-based genome editing technologies, we can make significant progress in improving therapeutic response with safer toxicity profiles and better overall survival outcomes of cancer patients.

References

1. Alberts B (2008) Molecular biology of the cell, 5th edn. Garland Science, New York
2. Dobson CM, Sali A, Karplus M (1998) Protein folding: a perspective from theory and experiment. Angew Chem Int Ed Engl 37(7):868–893
3. Berg JM, Tymoczko JL, Stryer L (2012) Biochemistry, 7th edn. Freeman
4. Jahn TR, Radford SE (2005) The Yin and Yang of protein folding. FEBS J 272(23):5962–5970
5. Hartl FU, Hayer-Hartl M (2009) Converging concepts of protein folding in vitro and in vivo. Nat Struct Mol Biol 16(6):574–581
6. Hartl FU (1996) Molecular chaperones in cellular protein folding. Nature 381(6583):571–579
7. Tyedmers J, Mogk A, Bukau B (2010) Cellular strategies for controlling protein aggregation. Nat Rev Mol Cell Biol 11(11):777–788
8. Hartl FU, Bracher A, Hayer-Hartl M (2011) Molecular chaperones in protein folding and proteostasis. Nature 475(7356):324–332
9. Chiti F, Dobson CM (2006) Protein misfolding, functional amyloid, and human disease. Annu Rev Biochem 75:333–366
10. Ross CA, Poirier MA (2004) Protein aggregation and neurodegenerative disease. Nat Med 10(Suppl):S10–S17
11. Kayed R, Head E, Thompson JL, McIntire TM, Milton SC, Cotman CW et al (2003) Common structure of soluble amyloid oligomers implies common mechanism of pathogenesis. Science 300(5618):486–489
12. Dikic I, Wakatsuki S, Walters KJ (2009) Ubiquitin-binding domains – from structures to functions. Nat Rev Mol Cell Biol 10(10):659–671
13. Deshaies RJ, Joazeiro CA (2009) RING domain E3 ubiquitin ligases. Annu Rev Biochem 78:399–434
14. Schulman BA, Harper JW (2009) Ubiquitin-like protein activation by E1 enzymes: the apex for downstream signalling pathways. Nat Rev Mol Cell Biol 10(5):319–331
15. Ye Y, Rape M (2009) Building ubiquitin chains: E2 enzymes at work. Nat Rev Mol Cell Biol 10(11):755–764
16. Ikeda F, Dikic I (2008) Atypical ubiquitin chains: new molecular signals. 'Protein Modifications: Beyond the Usual Suspects' review series. EMBO Rep 9(6):536–542
17. Tokunaga F, Sakata S, Saeki Y, Satomi Y, Kirisako T, Kamei K et al (2009) Involvement of linear polyubiquitylation of NEMO in NF-kappaB activation. Nat Cell Biol 11(2):123–132
18. Hoege C, Pfander B, Moldovan GL, Pyrowolakis G, Jentsch S (2002) RAD6-dependent DNA repair is linked to modification of PCNA by ubiquitin and SUMO. Nature 419(6903):135–141

19. Sims JJ, Cohen RE (2009) Linkage-specific avidity defines the lysine 63-linked polyubiquitin-binding preference of rap80. Mol Cell 33(6):775–783
20. Pickart CM, Fushman D (2004) Polyubiquitin chains: polymeric protein signals. Curr Opin Chem Biol 8(6):610–616
21. Xu P, Duong DM, Seyfried NT, Cheng D, Xie Y, Robert J et al (2009) Quantitative proteomics reveals the function of unconventional ubiquitin chains in proteasomal degradation. Cell 137(1):133–145
22. Nijman SM, Luna-Vargas MP, Velds A, Brummelkamp TR, Dirac AM, Sixma TK et al (2005) A genomic and functional inventory of deubiquitinating enzymes. Cell 123(5):773–786
23. Lindner HA (2007) Deubiquitination in virus infection. Virology 362(2):245–256
24. Sugaya K, Ishihara Y, Inoue S (2015) Nuclear localization of ubiquitin-activating enzyme Uba1 is characterized in its mammalian temperature-sensitive mutant. Genes Cells 20(8):659–666
25. Handley-Gearhart PM, Stephen AG, Trausch-Azar JS, Ciechanover A, Schwartz AL (1994) Human ubiquitin-activating enzyme, E1. Indication of potential nuclear and cytoplasmic subpopulations using epitope-tagged cDNA constructs. J Biol Chem 269(52):33171–33178
26. McGrath JP, Jentsch S, Varshavsky A (1991) UBA 1: an essential yeast gene encoding ubiquitin-activating enzyme. EMBO J 10(1):227–236
27. Jin J, Li X, Gygi SP, Harper JW (2007) Dual E1 activation systems for ubiquitin differentially regulate E2 enzyme charging. Nature 447(7148):1135–1138
28. Finley D, Ciechanover A, Varshavsky A (1984) Thermolability of ubiquitin-activating enzyme from the mammalian cell cycle mutant ts85. Cell 37(1):43–55
29. Leidecker O, Matic I, Mahata B, Pion E, Xirodimas DP (2012) The ubiquitin E1 enzyme Ube1 mediates NEDD8 activation under diverse stress conditions. Cell Cycle 11(6):1142–1150
30. Chang TK, Shravage BV, Hayes SD, Powers CM, Simin RT, Wade Harper J et al (2013) Uba1 functions in Atg7- and Atg3-independent autophagy. Nat Cell Biol 15(9):1067–1078
31. Kitagaki J, Yang Y, Saavedra JE, Colburn NH, Keefer LK, Perantoni AO (2009) Nitric oxide prodrug JS-K inhibits ubiquitin E1 and kills tumor cells retaining wild-type p53. Oncogene 28(4):619–624
32. Joo HY, Zhai L, Yang C, Nie S, Erdjument-Bromage H, Tempst P et al (2007) Regulation of cell cycle progression and gene expression by H2A deubiquitination. Nature 449(7165):1068–1072
33. Ghaboosi N, Deshaies RJ (2007) A conditional yeast E1 mutant blocks the ubiquitin-proteasome pathway and reveals a role for ubiquitin conjugates in targeting Rad23 to the proteasome. Mol Biol Cell 18(5):1953–1963
34. Sugaya K, Ishihara Y, Inoue S, Tsuji H (2014) Characterization of ubiquitin-activating enzyme Uba1 in the nucleus by its mammalian temperature-sensitive mutant. PLoS One 9(5):e96666
35. Grenfell SJ, Trausch-Azar JS, Handley-Gearhart PM, Ciechanover A, Schwartz AL (1994) Nuclear localization of the ubiquitin-activating enzyme, E1, is cell-cycle-dependent. Biochem J 300(Pt 3):701–708
36. Haas AL, Rose IA (1982) The mechanism of ubiquitin activating enzyme. A kinetic and equilibrium analysis. J Biol Chem 257(17):10329–10337
37. Haas AL, Warms JV, Hershko A, Rose IA (1982) Ubiquitin-activating enzyme. Mechanism and role in protein-ubiquitin conjugation. J Biol Chem 257(5):2543–2548
38. Lee I, Schindelin H (2008) Structural insights into E1-catalyzed ubiquitin activation and transfer to conjugating enzymes. Cell 134(2):268–278
39. Lv Z, Rickman KA, Yuan L, Williams K, Selvam SP, Woosley AN et al (2017) S. pombe Uba1-Ubc15 Structure Reveals a Novel Regulatory Mechanism of Ubiquitin E2 Activity. Mol Cell 65(4):699–714. e6
40. Schafer A, Kuhn M, Schindelin H (2014) Structure of the ubiquitin-activating enzyme loaded with two ubiquitin molecules. Acta Crystallogr D Biol Crystallogr 70(Pt 5):1311–1320

41. Lv Z, Yuan L, Atkison JH, Aldana-Masangkay G, Chen Y, Olsen SK (2017) Domain alternation and active site remodeling are conserved structural features of ubiquitin E1. J Biol Chem 292(29):12089–12099
42. Lv Z, Williams KM, Yuan L, Atkison JH, Olsen SK (2018) Crystal structure of a human ubiquitin E1-ubiquitin complex reveals conserved functional elements essential for activity. J Biol Chem 293(47):18337–18352
43. Lois LM, Lima CD (2005) Structures of the SUMO E1 provide mechanistic insights into SUMO activation and E2 recruitment to E1. EMBO J 24(3):439–451
44. Olsen SK, Capili AD, Lu X, Tan DS, Lima CD (2010) Active site remodelling accompanies thioester bond formation in the SUMO E1. Nature 463(7283):906–912
45. Olsen SK, Lima CD (2013) Structure of a ubiquitin E1-E2 complex: insights to E1-E2 thioester transfer. Mol Cell 49(5):884–896
46. Tokgoz Z, Siepmann TJ, Streich F Jr, Kumar B, Klein JM, Haas AL (2012) E1-E2 interactions in ubiquitin and Nedd8 ligation pathways. J Biol Chem 287(1):311–321
47. Pfleger CM, Harvey KF, Yan H, Hariharan IK (2007) Mutation of the gene encoding the ubiquitin activating enzyme uba1 causes tissue overgrowth in Drosophila. Fly (Austin) 1(2):95–105
48. Dlamini N, Josifova DJ, Paine SM, Wraige E, Pitt M, Murphy AJ et al (2013) Clinical and neuropathological features of X-linked spinal muscular atrophy (SMAX2) associated with a novel mutation in the UBA1 gene. Neuromuscul Disord 23(5):391–398
49. Jedrzejowska M, Jakubowska-Pietkiewicz E, Kostera-Pruszczyk A (2015) X-linked spinal muscular atrophy (SMAX2) caused by de novo c.1731C>T substitution in the UBA1 gene. Neuromuscul Disord 25(8):661–666
50. Wade BE, Wang CE, Yan S, Bhat K, Huang B, Li S et al (2014) Ubiquitin-activating enzyme activity contributes to differential accumulation of mutant huntingtin in brain and peripheral tissues. J Neurosci 34(25):8411–8422
51. McNaught KS, Shashidharan P, Perl DP, Jenner P, Olanow CW (2002) Aggresome-related biogenesis of Lewy bodies. Eur J Neurosci 16(11):2136–2148
52. Brownell JE, Sintchak MD, Gavin JM, Liao H, Bruzzese FJ, Bump NJ et al (2010) Substrate-assisted inhibition of ubiquitin-like protein-activating enzymes: the NEDD8 E1 inhibitor MLN4924 forms a NEDD8-AMP mimetic in situ. Mol Cell 37(1):102–111
53. Hyer ML, Milhollen MA, Ciavarri J, Fleming P, Traore T, Sappal D et al (2018) A small-molecule inhibitor of the ubiquitin activating enzyme for cancer treatment. Nat Med 24(2):186–193
54. Michelle C, Vourc'h P, Mignon L, Andres CR (2009) What was the set of ubiquitin and ubiquitin-like conjugating enzymes in the eukaryote common ancestor? J Mol Evol 68(6):616–628
55. Scheffner M, Nuber U, Huibregtse JM (1995) Protein ubiquitination involving an E1-E2-E3 enzyme ubiquitin thioester cascade. Nature 373(6509):81–83
56. Lin Y, Hwang WC, Basavappa R (2002) Structural and functional analysis of the human mitotic-specific ubiquitin-conjugating enzyme, UbcH10. J Biol Chem 277(24):21913–21921
57. Eddins MJ, Carlile CM, Gomez KM, Pickart CM, Wolberger C (2006) Mms2-Ubc13 covalently bound to ubiquitin reveals the structural basis of linkage-specific polyubiquitin chain formation. Nat Struct Mol Biol 13(10):915–920
58. Ozkan E, Yu H, Deisenhofer J (2005) Mechanistic insight into the allosteric activation of a ubiquitin-conjugating enzyme by RING-type ubiquitin ligases. Proc Natl Acad Sci U S A 102(52):18890–18895
59. Kerscher O, Felberbaum R, Hochstrasser M (2006) Modification of proteins by ubiquitin and ubiquitin-like proteins. Annu Rev Cell Dev Biol 22:159–180
60. Haas AL, Bright PM, Jackson VE (1988) Functional diversity among putative E2 isozymes in the mechanism of ubiquitin-histone ligation. J Biol Chem 263(26):13268–13275
61. Huang DT, Hunt HW, Zhuang M, Ohi MD, Holton JM, Schulman BA (2007) Basis for a ubiquitin-like protein thioester switch toggling E1-E2 affinity. Nature 445(7126):394–398

62. Huang DT, Miller DW, Mathew R, Cassell R, Holton JM, Roussel MF et al (2004) A unique E1-E2 interaction required for optimal conjugation of the ubiquitin-like protein NEDD8. Nat Struct Mol Biol 11(10):927–935
63. Huang DT, Zhuang M, Ayrault O, Schulman BA (2008) Identification of conjugation specificity determinants unmasks vestigial preference for ubiquitin within the NEDD8 E2. Nat Struct Mol Biol 15(3):280–287
64. Rape M, Kirschner MW (2004) Autonomous regulation of the anaphase-promoting complex couples mitosis to S-phase entry. Nature 432(7017):588–595
65. Williamson A, Wickliffe KE, Mellone BG, Song L, Karpen GH, Rape M (2009) Identification of a physiological E2 module for the human anaphase-promoting complex. Proc Natl Acad Sci U S A 106(43):18213–18218
66. Fang S, Ferrone M, Yang C, Jensen JP, Tiwari S, Weissman AM (2001) The tumor autocrine motility factor receptor, gp78, is a ubiquitin protein ligase implicated in degradation from the endoplasmic reticulum. Proc Natl Acad Sci U S A 98(25):14422–14427
67. Chen B, Mariano J, Tsai YC, Chan AH, Cohen M, Weissman AM (2006) The activity of a human endoplasmic reticulum-associated degradation E3, gp78, requires its Cue domain, RING finger, and an E2-binding site. Proc Natl Acad Sci U S A 103(2):341–346
68. Deng L, Wang C, Spencer E, Yang L, Braun A, You J et al (2000) Activation of the IkappaB kinase complex by TRAF6 requires a dimeric ubiquitin-conjugating enzyme complex and a unique polyubiquitin chain. Cell 103(2):351–361
69. Zheng N, Wang P, Jeffrey PD, Pavletich NP (2000) Structure of a c-Cbl-UbcH7 complex: RING domain function in ubiquitin-protein ligases. Cell 102(4):533–539
70. Dominguez C, Bonvin AM, Winkler GS, van Schaik FM, Timmers HT, Boelens R (2004) Structural model of the UbcH5B/CNOT4 complex revealed by combining NMR, mutagenesis, and docking approaches. Structure 12(4):633–644
71. Zhang M, Windheim M, Roe SM, Peggie M, Cohen P, Prodromou C et al (2005) Chaperoned ubiquitylation--crystal structures of the CHIP U box E3 ubiquitin ligase and a CHIP-Ubc13-Uev1a complex. Mol Cell 20(4):525–538
72. Yin Q, Lin SC, Lamothe B, Lu M, Lo YC, Hura G et al (2009) E2 interaction and dimerization in the crystal structure of TRAF6. Nat Struct Mol Biol 16(6):658–666
73. Bailly V, Prakash S, Prakash L (1997) Domains required for dimerization of yeast Rad6 ubiquitin-conjugating enzyme and Rad18 DNA binding protein. Mol Cell Biol 17(8):4536–4543
74. Kolman CJ, Toth J, Gonda DK (1992) Identification of a portable determinant of cell cycle function within the carboxyl-terminal domain of the yeast CDC34 (UBC3) ubiquitin conjugating (E2) enzyme. EMBO J 11(8):3081–3090
75. Li W, Tu D, Li L, Wollert T, Ghirlando R, Brunger AT et al (2009) Mechanistic insights into active site-associated polyubiquitination by the ubiquitin-conjugating enzyme Ube2g2. Proc Natl Acad Sci U S A 106(10):3722–3727
76. Das R, Mariano J, Tsai YC, Kalathur RC, Kostova Z, Li J et al (2009) Allosteric activation of E2-RING finger-mediated ubiquitylation by a structurally defined specific E2-binding region of gp78. Mol Cell 34(6):674–685
77. Soucy TA, Smith PG, Milhollen MA, Berger AJ, Gavin JM, Adhikari S et al (2009) An inhibitor of NEDD8-activating enzyme as a new approach to treat cancer. Nature 458(7239):732–736
78. Christensen DE, Brzovic PS, Klevit RE (2007) E2-BRCA1 RING interactions dictate synthesis of mono- or specific polyubiquitin chain linkages. Nat Struct Mol Biol 14(10):941–948
79. Hochstrasser M (2006) Lingering mysteries of ubiquitin-chain assembly. Cell 124(1):27–34
80. Thrower JS, Hoffman L, Rechsteiner M, Pickart CM (2000) Recognition of the polyubiquitin proteolytic signal. EMBO J 19(1):94–102
81. Li W, Bengtson MH, Ulbrich A, Matsuda A, Reddy VA, Orth A et al (2008) Genome-wide and functional annotation of human E3 ubiquitin ligases identifies MULAN, a mitochondrial E3 that regulates the organelle's dynamics and signaling. PLoS One 3(1):e1487

82. Huibregtse JM, Scheffner M, Beaudenon S, Howley PM (1995) A family of proteins structurally and functionally related to the E6-AP ubiquitin-protein ligase. Proc Natl Acad Sci U S A 92(11):5249
83. Scheffner M, Staub O (2007) HECT E3s and human disease. BMC Biochem 8(Suppl 1):S6
84. Rotin D, Kumar S (2009) Physiological functions of the HECT family of ubiquitin ligases. Nat Rev Mol Cell Biol 10(6):398–409
85. Kamadurai HB, Souphron J, Scott DC, Duda DM, Miller DJ, Stringer D et al (2009) Insights into ubiquitin transfer cascades from a structure of a UbcH5B approximately ubiquitin-HECT(NEDD4L) complex. Mol Cell 36(6):1095–1102
86. Huang L, Kinnucan E, Wang G, Beaudenon S, Howley PM, Huibregtse JM et al (1999) Structure of an E6AP-UbcH7 complex: insights into ubiquitination by the E2-E3 enzyme cascade. Science 286(5443):1321–1326
87. Verdecia MA, Joazeiro CA, Wells NJ, Ferrer JL, Bowman ME, Hunter T et al (2003) Conformational flexibility underlies ubiquitin ligation mediated by the WWP1 HECT domain E3 ligase. Mol Cell 11(1):249–259
88. Shea FF, Rowell JL, Li Y, Chang TH, Alvarez CE (2012) Mammalian alpha arrestins link activated seven transmembrane receptors to Nedd4 family e3 ubiquitin ligases and interact with beta arrestins. PLoS One 7(12):e50557
89. Shearwin-Whyatt L, Dalton HE, Foot N, Kumar S (2006) Regulation of functional diversity within the Nedd4 family by accessory and adaptor proteins. BioEssays 28(6):617–628
90. Ichimura T, Yamamura H, Sasamoto K, Tominaga Y, Taoka M, Kakiuchi K et al (2005) 14-3-3 proteins modulate the expression of epithelial Na+ channels by phosphorylation-dependent interaction with Nedd4-2 ubiquitin ligase. J Biol Chem 280(13):13187–13194
91. Ogunjimi AA, Briant DJ, Pece-Barbara N, Le Roy C, Di Guglielmo GM, Kavsak P et al (2005) Regulation of Smurf2 ubiquitin ligase activity by anchoring the E2 to the HECT domain. Mol Cell 19(3):297–308
92. Wiesner S, Ogunjimi AA, Wang HR, Rotin D, Sicheri F, Wrana JL et al (2007) Autoinhibition of the HECT-type ubiquitin ligase Smurf2 through its C2 domain. Cell 130(4):651–662
93. Gallagher E, Gao M, Liu YC, Karin M (2006) Activation of the E3 ubiquitin ligase Itch through a phosphorylation-induced conformational change. Proc Natl Acad Sci U S A 103(6):1717–1722
94. Grimsey NJ, Narala R, Rada CC, Mehta S, Stephens BS, Kufareva I et al (2018) A Tyrosine Switch on NEDD4-2 E3 Ligase Transmits GPCR Inflammatory Signaling. Cell Rep 24(12):3312–3323. e5
95. Konstas AA, Shearwin-Whyatt LM, Fotia AB, Degger B, Riccardi D, Cook DI et al (2002) Regulation of the epithelial sodium channel by N4WBP5A, a novel Nedd4/Nedd4-2-interacting protein. J Biol Chem 277(33):29406–29416
96. Trimpert C, Wesche D, de Groot T, Pimentel Rodriguez MM, Wong V, van den Berg DTM et al (2017) NDFIP allows NEDD4/NEDD4L-induced AQP2 ubiquitination and degradation. PLoS One 12(9):e0183774
97. Beaudenon S, Huibregtse JM (2008) HPV E6, E6AP and cervical cancer. BMC Biochem 9(Suppl 1):S4
98. Lifton RP (1995) Genetic determinants of human hypertension. Proc Natl Acad Sci U S A 92(19):8545–8551
99. Adhikary S, Marinoni F, Hock A, Hulleman E, Popov N, Beier R et al (2005) The ubiquitin ligase HectH9 regulates transcriptional activation by Myc and is essential for tumor cell proliferation. Cell 123(3):409–421
100. Chen D, Kon N, Li M, Zhang W, Qin J, Gu W (2005) ARF-BP1/Mule is a critical mediator of the ARF tumor suppressor. Cell 121(7):1071–1083
101. Yang Y, Do H, Tian X, Zhang C, Liu X, Dada LA et al (2010) E3 ubiquitin ligase Mule ubiquitinates Miz1 and is required for TNFalpha-induced JNK activation. Proc Natl Acad Sci U S A 107(30):13444–13449

102. Maddika S, Kavela S, Rani N, Palicharla VR, Pokorny JL, Sarkaria JN et al (2011) WWP2 is an E3 ubiquitin ligase for PTEN. Nat Cell Biol 13(6):728–733
103. Mund T, Lewis MJ, Maslen S, Pelham HR (2014) Peptide and small molecule inhibitors of HECT-type ubiquitin ligases. Proc Natl Acad Sci U S A 111(47):16736–16741
104. Rossi M, Rotblat B, Ansell K, Amelio I, Caraglia M, Misso G et al (2014) High throughput screening for inhibitors of the HECT ubiquitin E3 ligase ITCH identifies antidepressant drugs as regulators of autophagy. Cell Death Dis 5:e1203
105. Koegl M, Hoppe T, Schlenker S, Ulrich HD, Mayer TU, Jentsch S (1999) A novel ubiquitination factor, E4, is involved in multiubiquitin chain assembly. Cell 96(5):635–644
106. Aravind L, Koonin EV (2000) The U box is a modified RING finger – a common domain in ubiquitination. Curr Biol 10(4):R132–R134
107. Hatakeyama S, Yada M, Matsumoto M, Ishida N, Nakayama KI (2001) U box proteins as a new family of ubiquitin-protein ligases. J Biol Chem 276(35):33111–33120
108. Matsumoto M, Yada M, Hatakeyama S, Ishimoto H, Tanimura T, Tsuji S et al (2004) Molecular clearance of ataxin-3 is regulated by a mammalian E4. EMBO J 23(3):659–669
109. Pringa E, Martinez-Noel G, Muller U, Harbers K (2001) Interaction of the ring finger-related U-box motif of a nuclear dot protein with ubiquitin-conjugating enzymes. J Biol Chem 276(22):19617–19623
110. Ballinger CA, Connell P, Wu Y, Hu Z, Thompson LJ, Yin LY et al (1999) Identification of CHIP, a novel tetratricopeptide repeat-containing protein that interacts with heat shock proteins and negatively regulates chaperone functions. Mol Cell Biol 19(6):4535–4545
111. Imai Y, Soda M, Hatakeyama S, Akagi T, Hashikawa T, Nakayama KI et al (2002) CHIP is associated with Parkin, a gene responsible for familial Parkinson's disease, and enhances its ubiquitin ligase activity. Mol Cell 10(1):55–67
112. Wang BB, Hayenga KJ, Payan DG, Fisher JM (1996) Identification of a nuclear-specific cyclophilin which interacts with the proteinase inhibitor eglin c. Biochem J 314(Pt 1):313–319
113. Spence J, Sadis S, Haas AL, Finley D (1995) A ubiquitin mutant with specific defects in DNA repair and multiubiquitination. Mol Cell Biol 15(3):1265–1273
114. Arnason T, Ellison MJ (1994) Stress resistance in Saccharomyces cerevisiae is strongly correlated with assembly of a novel type of multiubiquitin chain. Mol Cell Biol 14(12):7876–7883
115. Spence J, Gali RR, Dittmar G, Sherman F, Karin M, Finley D (2000) Cell cycle-regulated modification of the ribosome by a variant multiubiquitin chain. Cell 102(1):67–76
116. Freemont PS, Hanson IM, Trowsdale J (1991) A novel cysteine-rich sequence motif. Cell 64(3):483–484
117. Barlow PN, Luisi B, Milner A, Elliott M, Everett R (1994) Structure of the C3HC4 domain by 1H-nuclear magnetic resonance spectroscopy. A new structural class of zinc-finger. J Mol Biol 237(2):201–211
118. Borden KL, Boddy MN, Lally J, O'Reilly NJ, Martin S, Howe K et al (1995) The solution structure of the RING finger domain from the acute promyelocytic leukaemia proto-oncoprotein PML. EMBO J 14(7):1532–1541
119. Petroski MD, Deshaies RJ (2005) Function and regulation of cullin-RING ubiquitin ligases. Nat Rev Mol Cell Biol 6(1):9–20
120. Meroni G, Diez-Roux G (2005) TRIM/RBCC, a novel class of 'single protein RING finger' E3 ubiquitin ligases. BioEssays 27(11):1147–1157
121. Sardiello M, Cairo S, Fontanella B, Ballabio A, Meroni G (2008) Genomic analysis of the TRIM family reveals two groups of genes with distinct evolutionary properties. BMC Evol Biol 8:225
122. Eisenhaber B, Chumak N, Eisenhaber F, Hauser MT (2007) The ring between ring fingers (RBR) protein family. Genome Biol 8(3):209
123. Marin I, Ferrus A (2002) Comparative genomics of the RBR family, including the Parkinson's disease-related gene parkin and the genes of the ariadne subfamily. Mol Biol Evol 19(12):2039–2050

124. Joazeiro CA, Wing SS, Huang H, Leverson JD, Hunter T, Liu YC (1999) The tyrosine kinase negative regulator c-Cbl as a RING-type, E2-dependent ubiquitin-protein ligase. Science 286(5438):309–312
125. Albert TK, Hanzawa H, Legtenberg YI, de Ruwe MJ, van den Heuvel FA, Collart MA et al (2002) Identification of a ubiquitin-protein ligase subunit within the CCR4-NOT transcription repressor complex. EMBO J 21(3):355–364
126. Seol JH, Feldman RM, Zachariae W, Shevchenko A, Correll CC, Lyapina S et al (1999) Cdc53/cullin and the essential Hrt1 RING-H2 subunit of SCF define a ubiquitin ligase module that activates the E2 enzyme Cdc34. Genes Dev 13(12):1614–1626
127. Lahav-Baratz S, Sudakin V, Ruderman JV, Hershko A (1995) Reversible phosphorylation controls the activity of cyclosome-associated cyclin-ubiquitin ligase. Proc Natl Acad Sci U S A 92(20):9303–9307
128. Zhou P, Howley PM (1998) Ubiquitination and degradation of the substrate recognition subunits of SCF ubiquitin-protein ligases. Mol Cell 2(5):571–580
129. Fang S, Jensen JP, Ludwig RL, Vousden KH, Weissman AM (2000) Mdm2 is a RING finger-dependent ubiquitin protein ligase for itself and p53. J Biol Chem 275(12):8945–8951
130. Craig KL, Tyers M (1999) The F-box: a new motif for ubiquitin dependent proteolysis in cell cycle regulation and signal transduction. Prog Biophys Mol Biol 72(3):299–328
131. Nelson DE, Randle SJ, Laman H (2013) Beyond ubiquitination: the atypical functions of Fbxo7 and other F-box proteins. Open Biol 3(10):130131
132. Mason B, Laman H (2020) The FBXL family of F-box proteins: variations on a theme. Open Biol 10(11):200319
133. Nash P, Tang X, Orlicky S, Chen Q, Gertler FB, Mendenhall MD et al (2001) Multisite phosphorylation of a CDK inhibitor sets a threshold for the onset of DNA replication. Nature 414(6863):514–521
134. Orlicky S, Tang X, Willems A, Tyers M, Sicheri F (2003) Structural basis for phosphodependent substrate selection and orientation by the SCFCdc4 ubiquitin ligase. Cell 112(2):243–256
135. Yoshida Y, Tokunaga F, Chiba T, Iwai K, Tanaka K, Tai T (2003) Fbs2 is a new member of the E3 ubiquitin ligase family that recognizes sugar chains. J Biol Chem 278(44):43877–43884
136. Yoshida Y, Chiba T, Tokunaga F, Kawasaki H, Iwai K, Suzuki T et al (2002) E3 ubiquitin ligase that recognizes sugar chains. Nature 418(6896):438–442
137. Kuchay S, Duan S, Schenkein E, Peschiaroli A, Saraf A, Florens L et al (2013) FBXL2- and PTPL1-mediated degradation of p110-free p85beta regulatory subunit controls the PI(3)K signalling cascade. Nat Cell Biol 15(5):472–480
138. Du Y, Song W, Chen J, Chen H, Xuan Z, Zhao L et al (2019) The potassium channel KCa3.1 promotes cell proliferation by activating SKP2 and metastasis through the EMT pathway in hepatocellular carcinoma. Int J Cancer 145(2):503–516
139. Wei X, Li X, Yan W, Zhang X, Sun Y, Zhang F (2018) SKP2 promotes hepatocellular carcinoma progression through nuclear AMPK-SKP2-CARM1 signaling transcriptionally regulating nutrient-deprived autophagy induction. Cell Physiol Biochem 47(6):2484–2497
140. Shapira M, Kakiashvili E, Rosenberg T, Hershko DD (2006) The mTOR inhibitor rapamycin down-regulates the expression of the ubiquitin ligase subunit Skp2 in breast cancer cells. Breast Cancer Res 8(4):R46
141. Liu K, Zhang L, Zhao Q, Zhao Z, Zhi F, Qin Y et al (2018) SKP2 attenuates NF-kappaB signaling by mediating IKKbeta degradation through autophagy. J Mol Cell Biol 10(3):205–215
142. Bornstein G, Bloom J, Sitry-Shevah D, Nakayama K, Pagano M, Hershko A (2003) Role of the SCFSkp2 ubiquitin ligase in the degradation of p21Cip1 in S phase. J Biol Chem 278(28):25752–25757
143. Carrano AC, Eytan E, Hershko A, Pagano M (1999) SKP2 is required for ubiquitin-mediated degradation of the CDK inhibitor p27. Nat Cell Biol 1(4):193–199
144. Kamura T, Hara T, Kotoshiba S, Yada M, Ishida N, Imaki H et al (2003) Degradation of p57Kip2 mediated by SCFSkp2-dependent ubiquitylation. Proc Natl Acad Sci U S A 100(18):10231–10236

145. Huang H, Regan KM, Wang F, Wang D, Smith DI, van Deursen JM et al (2005) Skp2 inhibits FOXO1 in tumor suppression through ubiquitin-mediated degradation. Proc Natl Acad Sci U S A 102(5):1649–1654
146. Li C, Du L, Ren Y, Liu X, Jiao Q, Cui D et al (2019) SKP2 promotes breast cancer tumorigenesis and radiation tolerance through PDCD4 ubiquitination. J Exp Clin Cancer Res 38(1):76
147. Chan CH, Li CF, Yang WL, Gao Y, Lee SW, Feng Z et al (2012) The Skp2-SCF E3 ligase regulates Akt ubiquitination, glycolysis, herceptin sensitivity, and tumorigenesis. Cell 151(4):913–914
148. Chen BB, Glasser JR, Coon TA, Zou C, Miller HL, Fenton M et al (2012) F-box protein FBXL2 targets cyclin D2 for ubiquitination and degradation to inhibit leukemic cell proliferation. Blood 119(13):3132–3141
149. Chen BB, Glasser JR, Coon TA, Mallampalli RK (2012) F-box protein FBXL2 exerts human lung tumor suppressor-like activity by ubiquitin-mediated degradation of cyclin D3 resulting in cell cycle arrest. Oncogene 31(20):2566–2579
150. Li LQ, Pan D, Chen H, Zhang L, Xie WJ (2016) F-box protein FBXL2 inhibits gastric cancer proliferation by ubiquitin-mediated degradation of forkhead box M1. FEBS Lett 590(4):445–452
151. Kuchay S, Giorgi C, Simoneschi D, Pagan J, Missiroli S, Saraf A et al (2017) PTEN counteracts FBXL2 to promote IP3R3- and Ca(2+)-mediated apoptosis limiting tumour growth. Nature 546(7659):554–558
152. Yaron A, Hatzubai A, Davis M, Lavon I, Amit S, Manning AM et al (1998) Identification of the receptor component of the IkappaBalpha-ubiquitin ligase. Nature 396(6711):590–594
153. Busino L, Donzelli M, Chiesa M, Guardavaccaro D, Ganoth D, Dorrello NV et al (2003) Degradation of Cdc25A by beta-TrCP during S phase and in response to DNA damage. Nature 426(6962):87–91
154. Akhoondi S, Sun D, von der Lehr N, Apostolidou S, Klotz K, Maljukova A et al (2007) FBXW7/hCDC4 is a general tumor suppressor in human cancer. Cancer Res 67(19):9006–9012
155. Close V, Close W, Kugler SJ, Reichenzeller M, Yosifov DY, Bloehdorn J et al (2019) FBXW7 mutations reduce binding of NOTCH1, leading to cleaved NOTCH1 accumulation and target gene activation in CLL. Blood 133(8):830–839
156. Yoshida K (2005) Characterization of estrogen-induced F-box protein FBXO45. Oncol Rep 14(2):531–535
157. Saiga T, Fukuda T, Matsumoto M, Tada H, Okano HJ, Okano H et al (2009) Fbxo45 forms a novel ubiquitin ligase complex and is required for neuronal development. Mol Cell Biol 29(13):3529–3543
158. Chung FZ, Sahasrabuddhe AA, Ma K, Chen X, Basrur V, Lim MS et al (2014) Fbxo45 inhibits calcium-sensitive proteolysis of N-cadherin and promotes neuronal differentiation. J Biol Chem 289(41):28448–28459
159. Shimanoe C, Hachiya T, Hara M, Nishida Y, Tanaka K, Sutoh Y et al (2019) A genome-wide association study of coping behaviors suggests FBXO45 is associated with emotional expression. Genes Brain Behav 18(2):e12481
160. Wang K, Qu X, Liu S, Yang X, Bie F, Wang Y et al (2018) Identification of aberrantly expressed F-box proteins in squamous-cell lung carcinoma. J Cancer Res Clin Oncol 144(8):1509–1521
161. Chen X, Sahasrabuddhe AA, Szankasi P, Chung F, Basrur V, Rangnekar VM et al (2014) Fbxo45-mediated degradation of the tumor-suppressor Par-4 regulates cancer cell survival. Cell Death Differ 21(10):1535–1545
162. Kogure N, Yokobori T, Ogata K, Altan B, Mochiki E, Ohno T et al (2017) Low expression of FBXO45 is associated with gastric cancer progression and poor prognosis. Anticancer Res 37(1):191–196
163. Abshire CF, Carroll JL, Dragoi AM (2016) FLASH protects ZEB1 from degradation and supports cancer cells' epithelial-to-mesenchymal transition. Oncogenesis 5(8):e254

164. Peschiaroli A, Scialpi F, Bernassola F, Pagano M, Melino G (2009) The F-box protein FBXO45 promotes the proteasome-dependent degradation of p73. Oncogene 28(35):3157–3166
165. Wang X, Trotman LC, Koppie T, Alimonti A, Chen Z, Gao Z et al (2007) NEDD4-1 is a proto-oncogenic ubiquitin ligase for PTEN. Cell 128(1):129–139
166. Oliner JD, Kinzler KW, Meltzer PS, George DL, Vogelstein B (1992) Amplification of a gene encoding a p53-associated protein in human sarcomas. Nature 358(6381):80–83
167. Uchida C, Miwa S, Kitagawa K, Hattori T, Isobe T, Otani S et al (2005) Enhanced Mdm2 activity inhibits pRB function via ubiquitin-dependent degradation. EMBO J 24(1):160–169
168. Tedesco D, Lukas J, Reed SI (2002) The pRb-related protein p130 is regulated by phosphorylation-dependent proteolysis via the protein-ubiquitin ligase SCF(Skp2). Genes Dev 16(22):2946–2957
169. von der Lehr N, Johansson S, Wu S, Bahram F, Castell A, Cetinkaya C et al (2003) The F-box protein Skp2 participates in c-Myc proteosomal degradation and acts as a cofactor for c-Myc-regulated transcription. Mol Cell 11(5):1189–1200
170. Hiramatsu Y, Kitagawa K, Suzuki T, Uchida C, Hattori T, Kikuchi H et al (2006) Degradation of Tob1 mediated by SCFSkp2-dependent ubiquitination. Cancer Res 66(17):8477–8483
171. Sells SF, Wood DP Jr, Joshi-Barve SS, Muthukumar S, Jacob RJ, Crist SA et al (1994) Commonality of the gene programs induced by effectors of apoptosis in androgen-dependent and -independent prostate cells. Cell Growth Differ 5(4):457–466
172. El-Guendy N, Rangnekar VM (2003) Apoptosis by Par-4 in cancer and neurodegenerative diseases. Exp Cell Res 283(1):51–66
173. Moreno-Bueno G, Fernandez-Marcos PJ, Collado M, Tendero MJ, Rodriguez-Pinilla SM, Garcia-Cao I et al (2007) Inactivation of the candidate tumor suppressor par-4 in endometrial cancer. Cancer Res 67(5):1927–1934
174. Masters SL, Yao S, Willson TA, Zhang JG, Palmer KR, Smith BJ et al (2006) The SPRY domain of SSB-2 adopts a novel fold that presents conserved Par-4-binding residues. Nat Struct Mol Biol 13(1):77–84
175. Johnstone RW, See RH, Sells SF, Wang J, Muthukkumar S, Englert C et al (1996) A novel repressor, par-4, modulates transcription and growth suppression functions of the Wilms' tumor suppressor WT1. Mol Cell Biol 16(12):6945–6956
176. Diaz-Meco MT, Municio MM, Frutos S, Sanchez P, Lozano J, Sanz L et al (1996) The product of par-4, a gene induced during apoptosis, interacts selectively with the atypical isoforms of protein kinase C. Cell 86(5):777–786
177. Goswami A, Burikhanov R, de Thonel A, Fujita N, Goswami M, Zhao Y et al (2005) Binding and phosphorylation of par-4 by akt is essential for cancer cell survival. Mol Cell 20(1):33–44
178. Gurumurthy S, Goswami A, Vasudevan KM, Rangnekar VM (2005) Phosphorylation of Par-4 by protein kinase A is critical for apoptosis. Mol Cell Biol 25(3):1146–1161
179. Hebbar N, Burikhanov R, Shukla N, Qiu S, Zhao Y, Elenitoba-Johnson KSJ et al (2017) A Naturally Generated Decoy of the Prostate Apoptosis Response-4 Protein Overcomes Therapy Resistance in Tumors. Cancer Res 77(15):4039–4050
180. Woo JS, Suh HY, Park SY, Oh BH (2006) Structural basis for protein recognition by B30.2/SPRY domains. Mol Cell 24(6):967–976
181. Uljon S, Xu X, Durzynska I, Stein S, Adelmant G, Marto JA et al (2016) Structural Basis for Substrate Selectivity of the E3 Ligase COP1. Structure 24(5):687–696
182. Ikeda K, Inoue S, Muramatsu M (2005) RING finger-B box-coiled coil (RBCC) proteins as ubiquitin ligase in the control of protein degradation and gene regulation. In: Iuchi S, Kuldell N (eds) Zinc finger proteins: from atomic contact to cellular function. Springer US, Boston, pp 106–113
183. Styhler S, Nakamura A, Lasko P (2002) VASA localization requires the SPRY-domain and SOCS-box containing protein. GUSTAVUS Dev Cell 3(6):865–876
184. Ozato K, Shin DM, Chang TH, Morse HC 3rd. (2008) TRIM family proteins and their emerging roles in innate immunity. Nat Rev Immunol 8(11):849–860

185. Freemont PS (2000) RING for destruction? Curr Biol 10(2):R84–R87
186. Zhang Z, Bao M, Lu N, Weng L, Yuan B, Liu YJ (2013) The E3 ubiquitin ligase TRIM21 negatively regulates the innate immune response to intracellular double-stranded DNA. Nat Immunol 14(2):172–178
187. Lazzari E, Korczeniewska J, Ni Gabhann J, Smith S, Barnes BJ, Jefferies CA (2014) TRIpartite motif 21 (TRIM21) differentially regulates the stability of interferon regulatory factor 5 (IRF5) isoforms. PLoS One 9(8):e103609
188. Reddy BA, van der Knaap JA, Bot AG, Mohd-Sarip A, Dekkers DH, Timmermans MA et al (2014) Nucleotide biosynthetic enzyme GMP synthase is a TRIM21-controlled relay of p53 stabilization. Mol Cell 53(3):458–470
189. Nguyen JQ, Irby RB (2017) TRIM21 is a novel regulator of Par-4 in colon and pancreatic cancer cells. Cancer Biol Ther 18(1):16–25
190. Dornan D, Bheddah S, Newton K, Ince W, Frantz GD, Dowd P et al (2004) COP1, the negative regulator of p53, is overexpressed in breast and ovarian adenocarcinomas. Cancer Res 64(20):7226–7230
191. Su CH, Zhao R, Zhang F, Qu C, Chen B, Feng YH et al (2011) 14-3-3sigma exerts tumor-suppressor activity mediated by regulation of COP1 stability. Cancer Res 71(3):884–894
192. Marine JC (2012) Spotlight on the role of COP1 in tumorigenesis. Nat Rev Cancer 12(7):455–464
193. Zhang Y, Zhou L, Rouge L, Phillips AH, Lam C, Liu P et al (2013) Conformational stabilization of ubiquitin yields potent and selective inhibitors of USP7. Nat Chem Biol 9(1):51–58
194. Meng X, Liu X, Guo X, Jiang S, Chen T, Hu Z et al (2018) FBXO38 mediates PD-1 ubiquitination and regulates anti-tumour immunity of T cells. Nature 564(7734):130–135
195. Lim SO, Li CW, Xia W, Cha JH, Chan LC, Wu Y et al (2016) Deubiquitination and Stabilization of PD-L1 by CSN5. Cancer Cell 30(6):925–939
196. Toure M, Crews CM (2016) Small-molecule PROTACS: new approaches to protein degradation. Angew Chem Int Ed Engl 55(6):1966–1973
197. Sun X, Gao H, Yang Y, He M, Wu Y, Song Y et al (2019) PROTACs: great opportunities for academia and industry. Signal Transduct Target Ther 4(1):64

Regulation of Par-4 Function by Phosphorylation

Margaux Sevin, John E. Eriksson, Valérie Mezger, Carmen Garrido, and Aurélie de Thonel

Abstract This chapter discusses the regulatory role of the phosphorylation of the tumor suppressor protein, Par-4 (PAWR, PRKC, apoptosis Wilm's tumor 1 (WT1)), which is highly expressed in various tissues. One of the most appealing characteristics of Par-4 is its ability to selectively induce apoptosis in cancer cells but not in normal or immortalized cells. This feature can mainly be explained by the fact that Par-4 is regulated by various posttranslational modifications and primarily by phosphorylation. Indeed, Par-4 contains multiple phosphorylation sites, located in different domains of the protein, which modulate its pro-apoptotic functions, localization, cleavage by caspases, as well as its dimerization with partners of biological consequence. Notably, in mammals Par-4 phosphorylation is tightly regulated by kinases in a species-specific manner that either favors (PKA) or inhibits its pro-apoptotic activity (AKT and CK2). A complex interplay exists between Par-4, its cleaved form, and the kinase, AKT1, whose mechanisms entail an additional level of regulation, involving the casein kinase 2 (CK2). The phosphorylation of Par-4, which is induced by different kinases in a coordinated way and underlie Par-4 multifaceted

M. Sevin
INSERM, UMR1287, Gustave Roussy Cancer Campus, Villejuif, France

J. E. Eriksson
Faculty of Science and Engineering, Åbo Akademi University and Turku Bioscience Centre, Turku, Finland

V. Mezger · A. de Thonel (✉)
CNRS, UMR7216 Épigénétique et Destin Cellulaire, Paris Cedex 13, France

Univ Paris Diderot, Sorbonne Paris Cité, Paris Cedex 13, France

Département Hospitalo-Universitaire DHU PROTECT, Paris, France
e-mail: aurelie.dethonel@univ-paris-diderot.fr

C. Garrido
INSERM, UMR1231, Laboratoire d'Excellence LipSTIC, Dijon, France

University of Bourgogne Franche-Comté, Dijon, France

Département d'Oncologie Médicale, Centre Georges-François Leclerc, Dijon, France

V. M. Rangnekar (ed.), *Tumor Suppressor Par-4*,
https://doi.org/10.1007/978-3-030-73572-2_6

impacts, suggests that Par-4 could be an "Achilles heel" of cancer cells that could be exploited in the framework of anticancer research.

Keywords Par-4 · CK2 · AKT · PKA · ZIP kinase · Phosphorylation · Interaction · Caspase cleavage · Apoptosis · Cancers

1 Introduction

The fine-tuned regulation of protein functions is central to the organization and dispensation of biological systems. Among these several regulatory processes, posttranslational modifications (PTMs) represent a multifaceted mechanism to govern protein functions by controlling protein stability, catalytic or signaling activity, cellular localization, and interaction with other molecules, thereby calibrating cellular responses to the slightest change in the microenvironment. There are various types of PTMs (acting alone or in combinatorial manner) contributing to this regulation and are often readily reversible, such as phosphorylation, acetylation, ubiquitination, and methylation.

Par-4 (PAWR, PRKC, apoptosis Wilm's tumor 1 (WT1) regulator) is a tumor suppressor protein highly expressed in several tissues such as prostate, lung, and endometrium [31, 40, 42]. As many tumor suppressors, Par-4 is downregulated by approximately 40% of human endometrial carcinomas, human prostate carcinomas, and human lung adenocarcinomas [28, 42, 63]. One of the most appealing characteristics of Par-4 is its ability to selectively induce apoptosis in cancer cells but not in normal or immortalized cells. Depending on the cellular context, exogenous Par-4 overexpression in cancer cells has been shown to be either sufficient to induce apoptosis [4, 13, 22, 24, 65] or to increase the sensitivity of the cells to apoptotic agents, like TNF-related apoptosis-inducing ligand (TRAIL), vincristine, doxorubicin, and radiation [6, 44, 69, 81]. However, and importantly, endogenous Par-4, despite its constitutive expression in normal and cancer cells, does not induce apoptosis by itself but is essential for the response to the apoptotic properties of various chemotherapeutic agents [37]. Par-4 is primarily localized in the cytoplasm in normal cells, [24, 37, 81], whereas it localizes both in the cytoplasm and the nucleus in many cancer cells and tumor specimens [18, 24, 37].

The involvement of Par-4 in apoptosis is multifaceted. It initiates cell death program by acting at two major levels: (i) the activation of the Fas apoptotic pathway, facilitating the interaction of Fas-FasL (Fas Ligand), which sequentially recruits the Fas-associated death domain adapter protein (FADD) and the procaspase-8 to form the death-inducing signaling complex (DISC) [92], and (ii) the inhibition of the pro-survival NF-κB pathway. In addition, it also represses anti-apoptotic factors like Bcl2 (B cell lymphoma-2) *via* its interaction with WT1 [16].

The mechanisms that tightly regulate Par-4 functions and/or localization operate at transcriptional levels and posttranslational levels, which may explain why endogenous Par-4, despite its constitutive expression, does not induce apoptosis on its own. Upregulation of Par-4 at the transcriptional level has been detected in response to diverse apoptotic stimuli including trophic factor withdrawal, exposure to beta-amyloid peptides, ischemia, staurosporine, and increased Ca^{2+} levels [36, 59]. Conversely, Mabe et al. have shown that epithelial-to-mesenchymal transition promotes epigenetic silencing of Par-4 in recurrent breast tumors [56]. In that respect, the promoter of Par-4 contains a CpG island that has methylated in endometrial tumors and in Ras-transformed fibroblasts [63, 73].

In this chapter, we will focus on the posttranslational regulation of Par-4 through phosphorylation. Par-4 is regulated by a number of posttranslational modifications (PTMs; www.Phosphosite.org) including ubiquitination (See Chap. 5, vol 1) and phosphorylation. Given the similarity between humans and other species including rodents, PTMs are generally well preserved. Mouse Par-4 shows 83% and 91% identity to human and rat Par-4, respectively [41]. Furthermore, Par-4 protein, in specific regions, e.g., leucine zipper domain (LZ), C-terminal region, and the two nuclear localization sequences (NLS), exhibits 100% conservation across species [24]. Of note, the LZ domain (spanning amino acids 293–333 in mouse and 300–340 in human) enables Par-4 to interact with other proteins as either homo- or heterodimers (Fig. 1).

Par-4 contains multiple phosphorylation sites located in different domains of the protein, which modulate its functions, localization, and dimerization with partners

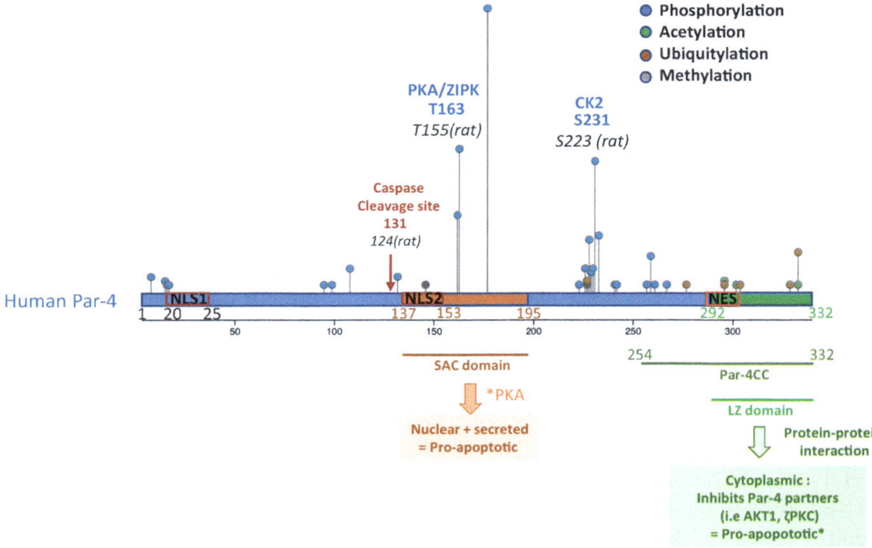

Fig. 1 Structure and post-translational modifications (PTMs) in human Par-4. Par-4 contains 2 nuclear localization sequences (NLS), a SAC (selective for apoptosis induction in cancer cell) domain and a Leucine Zipper domain (LZ) that allows the interaction of Par-4 with its partners such as AKT, PKC and Wilm's tumor suppressor. PTMs are indicated by circles and colours correspond to a specific PTM. Orthologue sites in rat Par-4 are indicated in italic. *www.Phosphosite.org*

of biological consequence. Phosphorylation modifications alter the properties of the substrate Par-4 protein, since phosphate is bulky and displays negative charges. This chemical modification, which is stable unless removed by a protein phosphatase, is performed by protein kinases that catalyze the transfer of a phosphoryl group (PO32−) from ATP to an amino acid with a hydroxyl group (e.g., serine, threonine, or tyrosine). Studies have demonstrated that phosphorylation of Par-4 is mainly driven by three serine/threonine (S/T) kinases at consensus sites, namely, protein kinase A (PKA), AKT kinase, and casein kinase II (CK2), all being essential enzymes involved in a wide range of cellular functions that are often highjacked by tumor cells to promote survival and proliferation.

2 Positive Regulation of Par-4 Pro-apoptotic Activity

2.1 Phosphorylation by PKA

PKA is a universal serine/threonine (S/T) kinase that phosphorylates a wide variety of substrates, which are involved in the regulation of metabolism, cell growth, and differentiation. PKA holoenzymes are inactive hetero-tetramers composed of two regulatory (R) subunits and two catalytic subunits (C) [51]. PKA becomes active by cooperatively binding two cAMP molecules to each R subunit, triggering the release of the two C subunits from the holoenzyme, which then phosphorylate specific substrate proteins on serine and threonine residues, both in the cytoplasm and in the nucleus [83, 93]. Depending on the regulatory subunit that is associated with the kinase, PKA that contain either RI or RII are termed PKA type I or type II, respectively. The RI and RII isoforms differ mainly in their localization and expression level [10, 83]. Although PKA is not considered as an oncogene *per se*, the ubiquitously expressed isoform RI is overexpressed in human cancer cells and primary tumors and may contribute to the progression of cancer. Accordingly, the PKA enzymatic activity is higher in transformed cells than in normal or immortalized cells [37]. In this context, the enhanced PKA activity was shown to participate to the cancer-specific activation of Par-4. Gurumurthy et al. demonstrated that two distinct events were required for Par-4 activation and, thus, for the induction of its apoptotic function: Par-4 entry into the nucleus and its phosphorylation on threonine T155 by PKA [37], as described below.

Studies have highlighted the key role of the bipartite nuclear localization signal-2 of Par-4 (NLS2; [137–153aa]) on its activity (Fig. 1). Indeed, the deletion of NLS2 results in the loss of Par-4 ability to induce apoptosis [24]. Interestingly, the flanking phosphorylation sites of the NLS2 sequence do not seem to be involved in Par-4 nuclear localization. In this regard, it was observed that Par-4 was mainly cytoplasmic in androgen-dependent cancer cells (e.g., LNCaP) as well as in normal/immortalized cells (which are resistant to Par-4-induced apoptosis). In contrast, in sensitive cancer cell lines (androgen-independent cancer cells, as well as many other cancer cells such as breast, lung, head and neck cancers) into which Par-4 overexpression induces apoptosis, the localization of Par-4 is both nuclear and cytoplasmic. In sensitive cancer cells, it was also reported that a truncated form of Par-4, containing amino acids 137–195 in rodent or amino acids 146–204 in human, therefore carrying a deletion of 58 aa but retaining an intact NLS2 domain (Fig. 1), constitutively enters the nucleus and displays an increased potential to trigger apoptosis [24]. This region containing the NLS2 represents a unique core domain of Par-4, which is 100% conserved in rodent and human, and orchestrates the Par-4 pro-apoptotic functions (e.g., inhibition of NF-κB activity and Fas activation) [13]. However, even though this core domain also localizes to the nucleus in normal or immortalized cells (in which full-length Par-4 is strictly cytosolic), it does not trigger apoptosis by itself and was therefore named the SAC (selective for apoptosis induction in cancer cells) domain [24]. Nuclear entry is thus essential but not sufficient for the triggering of apoptosis by Par-4, and additional activation events must selectively occur in cancer cells, and other Par-4 domains must be involved, to lead to the induction of apoptosis.

The second prerequisite, for inducing apoptosis *via* Fas/FasL trafficking and NF-κB inhibition, is the phosphorylation of Par-4 by PKA within the SAC domain, at the threonine residue T155 in rodent (T163 is the human orthologue residue) [37]. Of note, PKA also phosphorylates the residue T154, but it does not seem crucial, since the Par-4-154A mutant retains its apoptotic ability. The phosphorylation by PKA of ectopically expressed wild-type Par-4 has been confirmed by metabolic labeling, as well as the phosphorylation of endogenous Par-4, by using an antibody recognizing the phosphorylated residue T155 [75]. In addition, chemotherapeutic agents, such as vincristine, were shown to induce PKA-dependent phosphorylation of Par-4 and its subsequent nuclear translocation, leading to apoptosis both in lung cancer cells and primary human lung fibroblasts [37]. Accordingly, PKA activity in cancer and normal cells correlates with the respective phosphorylation status of Par-4 in these cells, as well as with cell susceptibility to Par-4-induced apoptosis. The impact of T155 phosphorylation event on the nuclear entry of Par-4 is not yet resolved. It is possible that this phosphorylation event prevents the potential masking of the NLS2 by regulating the intramolecular folding of the LZ domain, thus facilitating nuclear translocation. However, the mutation of T155 in the SAC does not impact *per se* its nuclear localization, whereas it hampers its capacity to kill transformed and androgen-independent cell types such as NIH3T3/Ras, as well as various lung and breast cancer cells lines. These data thus support the fact that Par-4 is phosphorylated in its active state and that its phosphorylation at T155 represents a specific feature of cancer cells when compared with normal cells.

Therefore, Par-4-inducible apoptosis can be explained by this dual requirement of PKA phosphorylation and NLS2, which are present and fully active in cancer cells that are susceptible to Par-4. However, in cells that are resistant to Par-4, such as hormone-dependent cancer cells and normal or immortalized cells, one or both of these activating events are not enabled, and, therefore, cells remain resistant to apoptosis [37]. This activation scheme is illustrated in Fig. 2.

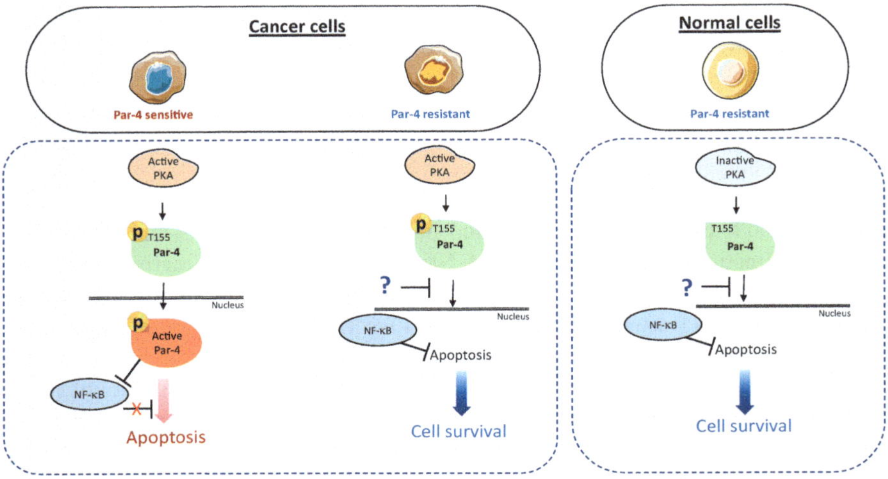

Fig. 2 Model for cancer-selective apoptosis by Par-4. Par-4-induced apoptosis in cancer cells depends on two independent events, the nuclear entry of Par-4 and its phosphorylation by PKA. In cells showing low levels of active PKA (*e.g.*, normal or immortalized cells), Par-4 is not phosphorylated at the T155 residue by PKA and is sequestrated in the cytoplasm, thus Par-4 remains inactive. In cancer cells characterized by high levels of PKA activity and resistant to Par-4 induced apoptosis, the phosphorylation of Par-4 by PKA is not sufficient, as the entry of Par-4 into the nucleus is blocked by pro-survival factors (*e.g.*, NFκB). In cancer cells sensitive to Par-4 induced apoptosis, the nuclear entry is not hampered by pro-survival factors and the high levels of PKA activity causes the phosphorylation of Par-4 at T155. Thus, Par-4 which is activated by phosphorylation and nuclear entry, triggers apoptosis of these cancer cells. (From Ref. [37])

2.2 *Phosphorylation by ZIP Kinase*

Zipper-interacting protein kinase (ZIPK) is a serine/threonine (Ser/Thr) protein kinase which belongs to the large family of death-associated protein kinases (DAP), (for review, see [46]) known for their role in the mitochondria (intrinsic) and in the death receptor (extrinsic) apoptotic pathways. The homology between ZIPK and DAP kinase is restricted to their N-terminal domains, and they differ in terms of subcellular localization and mechanism of activation. While DAP kinase is associated with the actin filaments of the cytoskeleton [17], ZIPK colocalizes with promyelocytic leukemia (PML) nuclear bodies [46]. ZIPK is also characterized by a functional leucine zipper domain that serves for homodimerization and heterodimerization with partners, such as the transcription factors ATF4 (activating transcription factor 4) and AATF (apoptosis-antagonizing transcription factor) [43]. The kinase activity of ZIPK is critical for its death-promoting effects and is abolished by mutations of the LZ domain indicating that this domain, which mediates homodimerization, is necessary for activation of the kinase. ZIPK has also been associated with the regulation of other cellular processes such as cell motility [49] and mitotic processes [26], as well as smooth muscle contraction [60, 67].

Cytoplasmic Par-4, which was shown to directly associate with actin cytoskeleton in rat fibroblasts, has been identified as a binding partner of ZIPK in these cells. This interaction involves the LZ domain of Par-4 and the arginine-rich domain of ZIPK [69]. The binding of Par-4 to ZIPK activates the apoptotic function of ZIPK, which is impaired by the transcription factor AATF, an endogenous antagonist of both Par-4 and ZIPK [35, 69, 70]. This proapoptotic function of ZIPK may imply a switch in its subcellular localization. Indeed, the coexpression of ZIPK and Par-4 leads to the relocation of ZIPK from nuclear speckles to the microfilament system, where it enhances the phosphorylation of the regulatory light chain of myosin II (MLC) leading to dramatic cytoskeletal rearrangements associated with morphological signs of apoptosis [7, 17, 69]. This colocalization with actin filaments appears to be a general prerequisite for induction of apoptosis by DAP kinase family members [17]. Boosen et al. have demonstrated that ZIPK also enhanced Par-4 phosphorylation at T155 as an important step to induce apoptosis in rat fibroblasts. Interestingly, phosphorylated Par-4 at T155 has not always been associated to a pro-apoptotic function, since it was also shown to support contractility in smooth muscle cells by acting as a cytoskeletal scaffold for ZIPK [61, 88]. Of note, the formation of the complex ZIPK/Par-4 does not seem to occur in all cell types and species [7, 82]. Therefore, phosphorylated Par-4 at T155, which displays both pro- and non-apoptotic functions, might be regulated by different kinases, depending on their level of activity in a given cell type (e.g., PKA in prostate cancer cells, ZIPK in rat fibroblasts, and potentially other kinases in smooth muscle cells) [61].

3 Negative Regulation of Par-4 Pro-apoptotic Activity

3.1 Par-4 and AKT Interplay

Par-4 both regulates and is regulated by the AKT survival kinase. The PI3K/AKT (phosphatidylinositol 3-kinase/protein kinase-B) signaling pathway is considered to be one of the most frequently altered pathways in human cancer. This pathway, which is initiated by assorted oncogenes and growth factor receptors, regulates various cellular functions, such as cell proliferation, cell growth, cell migration, vesicular trafficking, and cell survival [94]. AKT is a proto-oncogene, consisting three highly related isoforms AKT1, AKT2, and AKT3 which possess both overlapping and distinct cellular functions to regulate a broad range of cellular processes in pathophysiology [29, 52, 72]. They all possess similar phosphorylation sites. AKT1 and AKT2 are expressed ubiquitously, AKT2 being mainly expressed in insulin-responsive cells, and AKT3 in neurons and testes in mice [50]. AKT is one of the major downstream effectors of PI3K, and genetic alterations driving aberrant activation of this S/T kinase are commonly detected in malignant tumors: these include oncogenic activation of the receptor tyrosine kinases RAS or PI3K, loss of tumor suppressor PTEN (phosphatase and tensin homolog deleted on chromosome 10), or gain of function (mutations/amplification) in AKT isoforms [12, 84, 87, 89]. Most of the studies described a tumor-initiating role of AKT1, whereas AKT2 seems mainly responsible for tumor progression and metastasis.

Molecularly, following PI3K activation, a second messenger phosphatidylinositol (3-5)-trisphosphate (PIP3) is generated from phosphatidylinositol 4,5-bisphosphate (PIP2). PIP3 recruits pleckstrin homology (PH) domain-containing proteins, like AKT kinase, to the cell membrane, driving AKT conformational change and phosphorylation on different residues. Threonine T308 [27], located in the catalytic domain of human AKT, is mainly phosphorylated by 3-phosphoinositide-dependent protein kinase I (PDK1) and the serine S473 in the C-terminal region by PDK2 (mammalian target of rapamycin complex 2 (mTORC2)) [96], which contributes to the full activation of AKT. In parallel, other kinases are known to regulate the level of amplitude of AKT activation. Phosphorylation of AKT at Ser129 by CK2 (casein kinase 2) is responsible for the hyperactivation of AKT [23]. Similarly, protein kinase C zeta (PKCζ) phosphorylates AKT at S124, which helps to control the full activation of AKT (e.g., Ph-T308/S473) [28, 42]. Upon phosphorylation, AKT detaches from the membrane and translocates into the cytoplasm and the nucleus, where it phosphorylates downstream targets like mTOR complex 1 (mTORC1), an S/T kinase that regulates different processes including protein translation and synthesis, survival, and cell cycle progression. Inactivation of AKT signaling is mainly mediated by phosphatases such as phosphatase 2A (PP2A). AKT activity is also indirectly inactivated by the lipid phosphatase PTEN that dephosphorylates PIP3. PTEN is a tumor suppressor often mutated in the advanced stages of various human tumors, including prostate, glioblastoma, melanoma, and endometrial carcinoma [11, 14]. As a consequence of the PTEN loss of function, AKT becomes constitutively active, bearing an aberrant and constitutive T308 phosphorylation and promoting an accelerated proliferation of cancer cells [79].

The tumor suppressor Par-4 is an integral player in the PTEN network including in human prostate carcinomas, where its downregulation is concomitantly associated with PTEN loss. Consistent with this, while Par-4 knockout mice, similar to PTEN-heterozygous mice, only develop benign prostate neoplasia, the genetic ablation of Par-4 in PTEN-heterozygous mice induces fully invasive prostate carcinomas in a cell-autonomous manner [28]. Thus, Par-4 inactivation cooperates with PTEN haploinsufficiency affecting the initiation, progression, and severity of lesions from benign prostate neoplasia to aggressive carcinomas. This could be explained by the fact that the genetic ablation of Par-4 leads to an increase in AKT activation, which is further enhanced in the context of PTEN heterozygosity. Par-4 was actually described to act as an inhibitor of AKT, both *in vitro* and *in vivo*, by controlling the phosphorylation of two key residues to the AKT's activity, S473 and T308. This regulatory role of Par-4 on AKT activation is cell autonomous and of physiological relevance since it is observed in mouse tissues, in embryonic fibroblasts, and in human lung cancer cells. The higher levels of activated AKT detected in Par-4-deficient mice are partly driven by the complex Par-4/PKCζ, since it was shown to modulate the extent of AKT activation by directly targeting S124 [42]. Hence, Par-4 impairs the PKCζ signaling pathway leading not only to the transcriptional activation of NF-κB, which promotes cell survival and proliferation [30, 54, 85], but also to enhanced AKT activity [28, 42]. Therefore, Par-4 deficiency cooperates with PTEN haploinsufficiency in prostate cancer initiation and progression, and their concurrent inactivation provides a unique mechanism that targets simultaneous activation of AKT and NF-κB signaling pathways [28].

Conversely, the isoform AKT1, which acts *via* the activation of survival pathways (e.g., mTOR complex 1, NF-κB) and the inactivation of pro-apoptotic actors (e.g., caspase-9, Bad), can also directly bind to and inhibit the pro-apoptotic protein Par-4. This is highlighted in prostate cancer where Par-4 is not downregulated, silenced, or mutated but inactivated due to phosphorylation by AKT1 [32]. Interestingly, apart from Par-4, none of the other AKT1 targets has been directly shown to be involved in either apoptosis in the rodent ventral prostate upon androgen ablation or in preneoplastic and neoplastic lesions. The residue phosphorylated by AKT1 has been identified in rat Par-4 as S249 but is not conserved between species. Nevertheless, AKT1 is able to associate with human Par-4 in various cell lines from different species. Proteomic analyses using phospho-AKT substrate motif (RXXS*/T*) immunoaffinity beads suggest that residue S228, which belongs to a consensus site for 14-3-3 binding, might be a potential AKT1 site in human cells. Studies have reported that Par-4 mutant S228A slightly increased the sensitivity to TRAIL-mediated apoptosis in Caki cells, but its localization has not been analyzed yet. Given that this serine residue is localized in the conserved sequence 226-Y**K**S**TT**S**V**S-233, which is highly targeted by several PTMs (including phosphorylation and ubiquitination), S228 might be regulated by these surrounding PTMs, a process which still remains to be investigated.

The C-terminal region of Par-4 (Par-4CC, mouse residues 254–332, Fig. 1), predicted as a coiled-coil domain [55], was shown to sensitize cells to various apoptotic stimuli [81]. This region contains the LZ domain of Par-4 and a nuclear export sequence [55], which has an undeniable effect on Par-4 cytoplasmic localization, depending on the cellular context (Fig. 1). The C-terminal truncated form of Par-4 (Par-4CC) has indeed the capacity to interact with different cytoplasmic partners (e.g., Wilm's tumor protein 1, PKCζ) including AKT1 and to inhibit their activity, thereby promoting apoptosis. These interacting partners, as demonstrated with AKT1 in prostate cancer cells, have the ability to affect, in turn, the nuclear translocation of full-length Par-4 and its pro-apoptotic functions. In this regard, Par-4 mutants deleted for the LZ domain or the AKT1 phosphorylation site are able to induce apoptosis in cancer cells, regardless of the basal activity of AKT1 [32]. Similarly, the SAC domain of Par-4, which is also devoid of LZ domain, translocates to the nucleus in all cell types where it triggers apoptosis in cancer cells provided of an elevated level of PKA activity.

Goswani et al. [32] described the mechanism of action of the phosphorylation site S249, by showing that it allows the recruitment of the chaperone proteins 14-3-3σ, which promotes the sequestration of Par-4 in the cytoplasm, thereby preventing Par-4 nuclear translocation and induction of apoptosis. In androgen-dependent human prostate cancer cells, LNCaP, which are resistant to Par-4, AKT1 depletion by siRNA allowed the translocation of human Par-4 to the nucleus and apoptosis [13]. This inactivation of human Par-4 by AKT1 in LNCaP cells is essential for cell survival since apoptosis upon AKT1 inhibition is primarily driven by Par-4 and not by other AKT1 substrates, such as the Forkhead family proteins, Bad, p21, or p27 [80] (Fig. 3a).

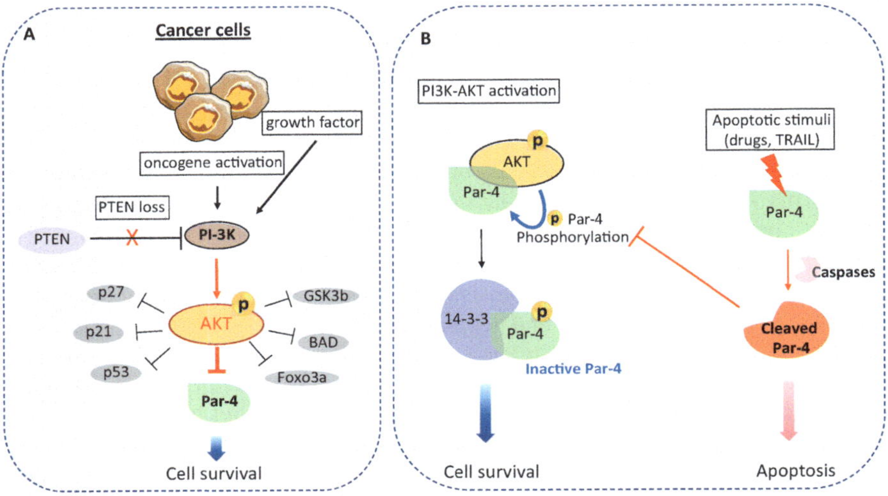

Fig. 3 PI-3K/AKT1 signaling induces Par-4 inactivation is key to induce survival of prostate cancer cells. **(a)** In most cancers, AKT1 activity is often elevated and can be due to (i) the loss of function of an inhibitor of the PI-3K pathway, the tumor suppressor PTEN (phosphatase and tensin homologue) (ii) the stimulation, by various growth and survival factors or oncogenes, of the signaling PI-3K pathway. AKT1 is key actor in cell survival by inactivating several players in apoptotic signaling (from [32]). **(b)** Phosphorylation of Par-4 by AKT1 promotes its binding to the 14-3-3 protein. As a result, Par-4 is sequestered in the cytoplasm, which blocks its pro-apoptotic activity. Conversely, inactivation of AKT1 by the caspase-dependent cleaved form of Par-4 triggers apoptosis [21]

Altogether, these data reveal a bidirectional interplay between Par-4 and AKT whose regulation still deserves further consideration. Regarding the marked inhibitory effect of Par-4 on AKT1, this was partly attributed to an indirect effect requiring the inhibition of PKCζ by Par-4 [28]. However, a direct effect can also be considered. It could be assumed that under normal conditions, Par-4 is inhibited by AKT1 and only upon apoptotic stimuli, at its turn, gains the ability to inhibit AKT1. Interestingly, Koh et al. have demonstrated that the protein kinase C-related kinase 2, once cleaved by caspases, acts as an AKT inhibitor [48]. Following apoptotic stimuli, cleavage of Par-4 by caspases is a crucial step that leads to the production of a cleaved Par-4 form, comprising both the SAC domain and the LZ domain (amino acids 132–340 in humans and 124–332 in rats). Thus, it is possible that the caspase-cleaved form of Par-4 loses its ability to be sequestrated by 14-3-3σ, thereby enabling the inhibition of AKT [32]. Accordingly, immunoprecipitation experiments from COS cells co-transfected with 14-3-3σ (HA-tag) together with either Par-4 wild-type or truncated Par-4 (amino acids 124–332) show no evidence of interaction between 14-3-3σ and truncated Par-4 (aa 124–332), whereas 14-3-3σ perfectly interacted with wild-type Par-4 (Fig. 4a). Furthermore, treatment with the death ligand TRAIL induces a strong inhibition of AKT1 phosphorylation in PC3 prostate cells expressing truncated Par-4 (aa 124–332) compared to the control

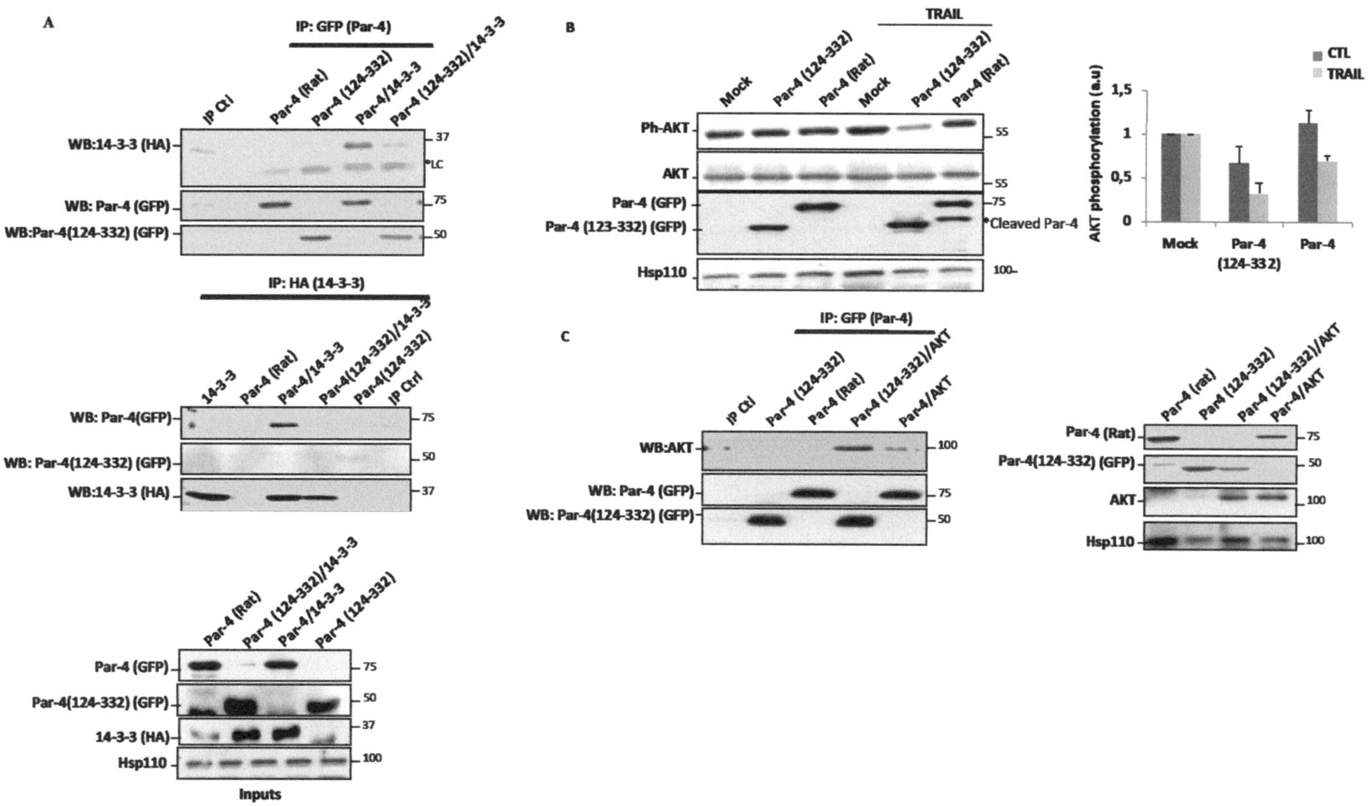

Fig. 4 Cleaved Par-4 escapes 14-3-3 sequestration and inhibits AKT activation. (**a**) Cos cells were transfected with GFP tagged Par4 or Par-4(124-332) in presence or absence of HA tagged 14-3-3 and constitutively active AKT kinase to favour Par-4/14-3-3 complex formation. Immunoprecipitation using either (continued)

Fig 4 (continued) HA antibody (upper panel) or GFP antibody (lower panel) was performed on cell extracts, was followed by immunoblotting with Par-4 or HA antibody. Hsp110 is the loading control. Inputs: protein expression in total cell lysate. **(b)** PC3 cells transfected with the indicated constructs were treated or not with TRAIL (3h, 500ng/ml) and then AKT activity was assessed by western-blotting, using an anti-phospho AKT (S473) antibody. Graph, densitometric analysis of phosphorylated AKT ratio in the different conditions**.** **(C)** As in A. Immunoprecipitation using GFP antibody, followed by immunoblotting with GFP or AKT antibody

(mock) and wild-type Par-4 (Fig. 4b). Accordingly, we noticed that the interaction of the cleaved form of Par-4 with AKT1 was favored in comparison with the wild-type Par-4 (Fig. 4c). These results suggest that Par-4, once cleaved by caspases during apoptosis, binds to and impairs AKT activation and escapes to 14-3-3σ sequestration to enter the nucleus (Fig. 3b).

Altogether, these data reveal a complex interplay between Par-4 and AKT1 whose mechanisms entail an additional level of regulation involving the casein kinase 2 (CK2) that we are developing hereafter.

3.2 Regulation of Par-4 by the Protein Kinase CK2

The protein kinase CK2 (formerly known as casein kinase 2) is a constitutive, highly conserved serine/threonine kinase ubiquitously distributed in a variety of eukaryotic cells which control almost all kinds of key processes such as cell cycle, proliferation, apoptosis, and transformation. CK2 kinase may function as a monomeric kinase or within a tetrameric complex consisting of two catalytic subunits (CK2α and/or CK2α') and two regulatory subunits CK2β. The regulatory subunit CK2β regulates both the substrate specificity of CK2 and enhances the catalytic subunit stability of CK2 kinase [5].

Tumor cells tend to depend on CK2 activity for their survival, in the paradigm of "non-oncogene addiction." Elevated levels of CK2α, CK2α', and/or CK2β are observed in various types of carcinomas in human (including prostate, colon, and lung [20, 64, 97] as well as in experimental rodent tumors, and correlates often with poor prognosis. Expression of nuclear CK2α protein was shown to be higher in colon and prostate tumor cells than in normal cells and has the potential to be used as a diagnostic and prognostic marker [68]. CK2 favors tumorigenesis by promoting pro-survival pathways such as the activation of NF-κB [77], inhibition of PTEN [86], as well as apoptotic pathways. Indeed, CK2 has been reported to (i) prevent caspase-mediated degradation of some pro-apoptotic proteins, (ii) directly modulate the activities of members of the endoplasmic reticulum (ER) stress signaling pathway, including the ER chaperone GRP78 (glucose-regulated protein-78, also known as BiP (binding immunoglobulin protein), and (iii) inhibit the mitochondrial pathway [2, 33, 78, 91]. Supporting the anti-apoptotic role of CK2, inhibitors of CK2 sensitize cancer cells to apoptosis induced by death ligands (such as TRAIL, Fas ligand) or chemotherapeutic drugs (such as etoposide), whereas the increased activity and/or expression of CK2, observed in cancer cells (such as prostate cancer cells), protects from cell death [38, 76, 90]. Given that CK2 expression is invariantly elevated in cancer when compared to normal cells and tissues [78], it further supports the notion that CK2 can be considered as a biomarker and a promising drug target for cancer studies, despite its ubiquitous expression [34, 68]. In this respect, Silmitasertib (codenamed CX-4945), the first CK2 inhibitor discovered by *Cylene Pharmaceutical*, is in clinical trials for the treatment of bile duct cancer (Phase II)

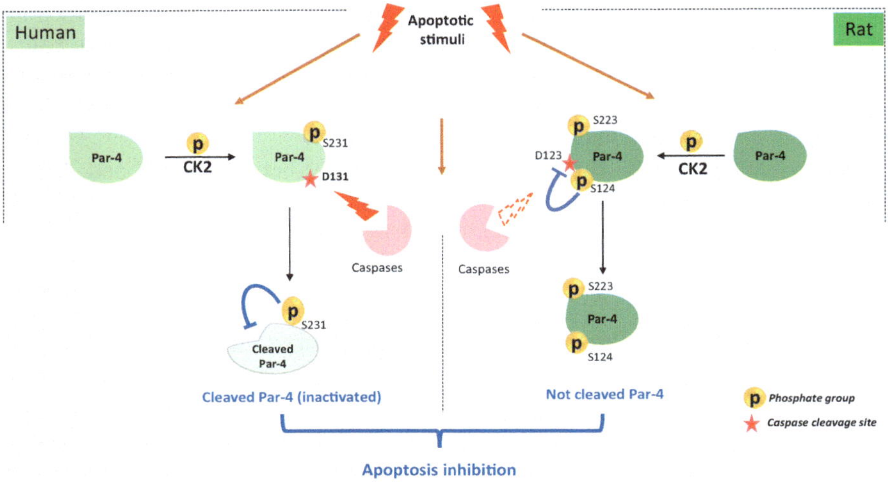

Fig. 5 Proposed model of negative regulation of Par-4 by CK2. Resistant prostate cancer cells – characterized by a elevated CK2 activity – apoptotic stimuli favours Par-4 phosphorylation by CK2 at residue S231 and S124/S223, in human and rodent Par-4, respectively. As the consequence, the apoptotic functions of Par-4 are dramatically altered. Interestingly, the inhibitory effect of CK2 among species slightly differs since **(i)** in rodent, CK2 phosphorylates Par-4 mainly at S124 thereby inhibiting caspase-mediated Par-4 cleavage (D123) and apoptosis (right panel) whereas the cleavage of human Par-4 (D131) is not affected by CK2 phosphorylation (left panel); **(ii)** the phosphorylation at residue S231 of human Par-4 – in contrast to the rat ortholog residue S223 – is required to inhibit the pro-apoptotic functions of Par-4 (both full length and cleaved form). [21]

and in preclinical development for solid cancers and hematological malignancies [34, 74].

CK2 was also shown to directly associate with and phosphorylate the pro-apoptotic protein Par-4, leading to its inactivation both in human and rodents (Fig. 5). Par-4 directly interacts with the catalytic subunit alpha of CK2, likely independently of its LZ domain (unpublished data). Notably, the heterotetrametric holoenzyme CK2 was reported to phosphorylate Par-4 on different serine residues specifically, S124/S223 in rat Par-4 and S231 in human (the orthologue of rat S223). Therefore, this difference between species unravels different levels of regulation of Par-4 apoptotic functions by CK2 [21].

The proapoptotic properties of Par-4 depend on the inhibition of survival factors but also on caspase activation. Upon treatment with chemotherapeutic agents or death ligand, Par-4 is cleaved by caspases on residue D131 in human (D123 in rodents) ([15, 21]. As previously commented, cleavage of Par-4 by caspases is a crucial step that leads to the production of a Par-4 cleaved form comprising both the SAC domain and the LZ domain. This caspase-cleaved product of Par-4 may exert its apoptotic activity *via* its ability (i) to escape the sequestration by 14-3-3 (rodent Par-4, Fig. 4a), (ii) to interact with and inhibit AKT (Fig. 4b), (iii) to be activated by PKA and to translocate to the nucleus (Fig. 2), (iiii) and to be secreted. Indeed, thanks to its SAC domain, the cleaved fragment of Par-4 is secreted, hence inducing a "bystander" effect that results in the killing of the surrounding cells through the activation of the caspase signaling pathway [9, 21]. CK2 was shown to prevent caspase-mediated cleavage of rat Par-4 at residue D123 by phosphorylating the juxtaposed residue S124, thus dramatically reducing Par-4 apoptotic functions (Fig. 5). As stated above, this inhibitory effect of CK2 on Par-4 cleavage occurs in rat but not in human, implying species-specific effects of CK2 phosphorylation. Similarly, the proteolytic cleavage of caspase-9 is specifically blocked, in rodent cells, by CK2 phosphorylation, suggesting an evolutionary role of CK2 [62]. In contrast, mutation of rat Par-4 S223 does not by itself affect caspase cleavage, but this phosphorylation site, unlike the phosphorylation site S124, is well conserved among species. The human CK2 phosphorylation site S231, the ortholog of rodent S223, has an essential regulatory function in human, *via* its ability to impair the Par-4 proapoptotic function ([21]; Fig. 5). Consistent with this, expression of the S231A Par-4 mutant enhanced sensitivity to TRAIL and CK2 knockdown impairs the phosphorylation of S231 leading also to the sensitization of prostate cancer cells to apoptosis. Furthermore, a higher phosphorylation level of the human Par-4 residue S231 was observed in prostate cancer cells – which display elevated CK2 activity – compared with their normal counterparts [21]. This suggests that this phosphorylation site might represent a potential biomarker in prostate cancer.

The mechanism by which phosphorylated S231 prevents Par-4-induced apoptosis yet remains to be determined. Of note, based on proteomic analysis (https://www.phosphosite.org), the highly conserved sequence 226-YK**STTS**V**S**-233 seems to be a "hot spot" of regulation by S/T kinases including CK2 on S231 (and potentially AKT on S228), which suggest additional layers of complexity that deserve further consideration. While it is not clear whether the human phosphorylation site S231 of Par-4 (as its orthologue residue S223, in rat) affects its nuclear localization, it is possible that this phosphosite regulates the interaction of Par-4 *via* its LZ domain with its partners (e.g., NF-κB, PKCζ, and also AKT). The residue S231 is indeed located in the C-terminal part of Par-4 and CK2 is a partner of both PKCζ and p62/SQSTM1 [8, 58]. The protein p62, which is also overexpressed in prostate cancer [45], is known to counteract the inhibitory effect of Par-4 on PKCζ activity during apoptosis [1]. Therefore, when part of the interacting complex p62-CK2-PKCζ-Par-4, CK2 may further enhance the phosphorylation of Par-4 upon apoptotic stimuli, thereby strongly inhibiting the Par-4 proapoptotic functions. Consistent with this hypothesis, endogenous Par-4 phosphorylation on the residue S231 is increased during apoptosis induced by TRAIL in the human prostate cancer PC-3 cells, which are resistant to TRAIL-induced apoptosis [21].

CK2 might also affect the extracellular function of Par-4 since a key role of this kinase is represented by its involvement in the ER stress, particularly by regulating the induction of the chaperone GRP78 in glial and multiple myeloma cells [39, 57]. GRP78, which is highly conserved across species, is mainly localized in the ER where it regulates protein folding and induces the unfolded protein response pathway (UPR) upon stress conditions [53]. GRP78 is also highly expressed on the cell surface of several types of cancer, including prostate cancer cells, owing to their inherently elevated ER stress levels, whereas it is low expressed on normal cells [3, 66]. GRP78 (intra- and extracellular) generates pro-oncogenic signals promoting survival. However, in some situations, GRP78 has pro-apoptotic functions; this is observed with Par-4 but also with Kringle 5, an angiogenesis inhibitor [19].

Previous studies indicate that intracellular Par-4 colocalizes with GRP78 in the ER and is secreted spontaneously via the classical pathway that involves the ER/Golgi network. This secretion of Par-4 has been detected both in normal and cancer cell cultures and was further enhanced following ER stress induced by treatment with drugs such as thapsigargin or TRAIL. Extracellular Par-4 has the ability to bind to the cell surface GRP78 receptor via its SAC domain, thereby forming a complex, which is functionally active. This Par-4/GRP78 complex displays pro-apoptotic effects that requires both (i) intracellular Par-4 function for the trafficking of GRP78 to the cell membrane and (ii) ER stress to trigger the caspases pathway in a FADD-dependent manner [9]. Therefore, CK2, which is also present in the ER, could, one the hand, promote the induction of GRP78 upon ER stress to allow the cell to cope with stress and, on the other hand, inhibit the pro-apoptotic function of the Par-4/GRP78 complex by phosphorylating Par-4. Therefore, CK2 could be able to modulate both the intra- and extracellular functions of Par-4 [39].

Altogether, these data suggest that both AKT and CK2 phosphorylate and control Par-4 activity. One could hypothesize that these two kinases act synergistically on Par-4 to hamper its activity given the crosstalk existing between CK2 and AKT. AKT is a known substrate of CK2, which further increases its catalytic activity, and constitutive phosphorylation of AKT by CK2 seems required for maximal activation of AKT. This is supported by the fact that CK2 inhibition impairs the activation of AKT, even in PTEN-deficient mice [23]. With respect to their coordinated role on Par-4 regulation, one might assume that CK2 prevents – by phosphorylation of the residue S231 – the cleaved form of Par-4 to inhibit AKT. The phosphorylation site S231 might indeed affect the pro-apoptotic activity of Par-4 by negatively regulating its interaction with AKT and/or PKCζ (via its LZ domain) and also its phosphorylation by PKA. These data suggest different levels of regulation of Par-4 by different survival kinases in a species-specific manner that might also counteract PKA effects.

Besides the complex interplay between Par-4 and kinases, it is noteworthy that these phosphorylated events are counteracted by phosphatases, especially phosphatase 1 (PP1) [21]. In this regard, treatment with the phosphatase inhibitor calyculin, at doses that selectively inhibit PP1, promotes Par-4 nuclear translocation and seems to be associated to a strong increase of its phosphorylation levels, in human prostate cancer cells PC3 (high kinase activities) but not in LNCaP (low kinase activities) (unpublished data). These results support the fact that the balance phosphorylation/dephosphorylation tightly controls the functions of Par-4 in a cellular context-dependent manner.

4 Conclusion

Par-4 phosphorylation is tightly regulated by kinases to either favor (PKA) or inhibit (AKT and CK2) its pro-apoptotic activity. Several phosphorylation sites, in particular those localizing in the "hotspot" at the C terminus of Par-4 including potential PKC phosphosites, remain to be characterized (Fig. 1). Par-4 is also subjected to other PTM like ubiquitination, which regulates its stability, as described in Chap. 5 (vol 1), but also acetylation, both PTMs being mainly located and on common residues in the LZ domain. Interestingly, acetylation is known to regulate positively the interaction with partners as well as the protein stability [71, 95] and seems to target in Par-4 the same residues as ubiquitin, thus providing a further level of Par-4 regulation (phosphosite.org). On the whole, the presence of multiple Par-4 phosphorylation sites, which are induced in a coordinated manner by different survival kinases, suggest that Par-4 could be an "Achilles heel" of cancer cells. Thus, Par-4 represents a new promising therapeutic target in cancer research, and currently, clinical approaches are being developed as discussed in volume 2 on Tumor Suppressor Par-4.

Acknowledgments VM teams has been awarded "Équipe labellisée" by the "Fondation pour la Recherche Medicale" (EQU201903007924) and is supported by the Agence Nationale de la Recherche (ANR-19-CE16-0030) and "Fondation pour la recherche sur le Cerveau" (FRC), CG team has the label of excellence from "la is Ligue National Contre le Cancer" and is supported by the ANR-11-LABX-0021, Conseil Regional de Bourgogne-Franche Comté and the European Union program FEDER.

Bibliography

1. Ahmad KA, Wang G, Slaton J, Unger G, Ahmed K (2005) Targeting CK2 for cancer therapy. Anticancer Drugs 16(10):1037–1043. https://doi.org/10.1097/00001813-200511000-00001. PMID: 16222144
2. Ahmed K, Gerber DA, Cochet C (2002) Joining the cell survival squad: an emerging role for protein kinase CK2. Trends Cell Biol. 12(5):226–230. https://doi.org/10.1016/s0962-8924(02)02279-1. PMID: 12062170
3. Arap MA, Lahdenranta J, Mintz PJ, Hajitou A, Sarkis AS, Arap W, Pasqualini R (2004) Cell surface expression of the stress response chaperone GRP78 enables tumor targeting by circulating ligands. Cancer Cell 6(3):275–284. https://doi.org/10.1016/j.ccr.2004.08.018. PMID: 15380518
4. Berra E, Municio MM, Sanz L, Frutos S, Diaz-Meco MT, Moscat J (1997) Positioning atypical protein kinase C isoforms in the UV-induced apoptotic signaling cascade. Mol Cell Biol 17(8):4346–4354. https://doi.org/10.1128/mcb.17.8.4346. PMID: 9234692; PMCID: PMC232288
5. Bibby AC, Litchfield DW (2005) The multiple personalities of the regulatory subunit of protein kinase CK2: CK2 dependent and CK2 independent roles reveal a secret identity for CK2beta. Int J Biol Sci. 1(2):67–79. https://doi.org/10.7150/ijbs.1.67. Epub 2005 Apr 1. PMID: 15951851; PMCID: PMC1142214
6. Boehrer S, Chow KU, Beske F, Kukoc-Zivojnov N, Puccetti E, Ruthardt M, Baum C, Rangnekar VM, Hoelzer D, Mitrou PS, Weidmann E (2002) In lymphatic cells par-4 sensitizes

to apoptosis by down-regulating bcl-2 and promoting disruption of mitochondrial membrane potential and caspase activation. Cancer Res. 62(6):1768–1775. PMID: 11912153
7. Boosen M, Vetterkind S, Kubicek J, Scheidtmann KH, Illenberger S, Preuss U (2009) Par-4 is an essential downstream target of DAP-like kinase (Dlk) in Dlk/Par-4-mediated apoptosis. Mol Biol Cell 20(18):4010–4020. https://doi.org/10.1091/mbc.e09-02-0173. Epub 2009 Jul 22. PMID: 19625447; PMCID: PMC2743620
8. Bren GD, Pennington KN, Paya CV (2000) PKC-zeta-associated CK2 participates in the turnover of free IkappaBalpha. J Mol Biol 297(5):1245–1258. https://doi.org/10.1006/jmbi.2000.3630. PMID: 10764587
9. Burikhanov R, Zhao Y, Goswami A, Qiu S, Schwarze SR, Rangnekar VM (2009) The tumor suppressor Par-4 activates an extrinsic pathway for apoptosis. Cell 138(2):377–388. https://doi.org/10.1016/j.cell.2009.05.022. Erratum in: Cell. 2009 Sep 4;138(5):1032. PMID: 19632185; PMCID: PMC2774252
10. Cadd G, McKnight GS (1989) Distinct patterns of cAMP-dependent protein kinase gene expression in mouse brain. Neuron. 3(1):71–79. https://doi.org/10.1016/0896-6273(89)90116-5. PMID: 2619996
11. Carnero A, Paramio JM (2014) The PTEN/PI3K/AKT Pathway in vivo, Cancer Mouse Models. Front Oncol 4:252. https://doi.org/10.3389/fonc.2014.00252. PMID: 25295225; PMCID: PMC4172058
12. Carpten JD, Faber AL, Horn C, Donoho GP, Briggs SL, Robbins CM, Hostetter G, Boguslawski S, Moses TY, Savage S, Uhlik M, Lin A, Du J, Qian YW, Zeckner DJ, Tucker-Kellogg G, Touchman J, Patel K, Mousses S, Bittner M, Schevitz R, Lai MH, Blanchard KL, Thomas JE (2007) A transforming mutation in the pleckstrin homology domain of AKT1 in cancer. Nature. 448(7152):439–444. https://doi.org/10.1038/nature05933. Epub 2007 Jul 4. PMID: 17611497
13. Chakraborty M, Qiu SG, Vasudevan KM, Rangnekar VM (2001) Par-4 drives trafficking and activation of Fas and Fasl to induce prostate cancer cell apoptosis and tumor regression. Cancer Res. 61(19):7255–7263. PMID: 11585763
14. Chalhoub N, Baker SJ (2009) PTEN and the PI3-kinase pathway in cancer. Annu Rev Pathol 4:127–150. https://doi.org/10.1146/annurev.pathol.4.110807.092311. PMID: 18767981; PMCID: PMC2710138
15. Chaudhry P, Singh M, Parent S, Asselin E (2012) Prostate apoptosis response 4 (Par-4), a novel substrate of caspase-3 during apoptosis activation. Mol Cell Biol 32(4):826–839. https://doi.org/10.1128/MCB.06321-11. Epub 2011 Dec 19. PMID: 22184067; PMCID: PMC3272980
16. Cheema SK, Mishra SK, Rangnekar VM, Tari AM, Kumar R, Lopez-Berestein G (2003) Par-4 transcriptionally regulates Bcl-2 through a WT1-binding site on the bcl-2 promoter. J Biol Chem. 278(22):19995–20005
17. Cohen O, Feinstein E, Kimchi A (1997) DAP-kinase is a Ca2+/calmodulin-dependent, cytoskeletal-associated protein kinase, with cell death-inducing functions that depend on its catalytic activity. EMBO J 16(5):998–1008. https://doi.org/10.1093/emboj/16.5.998. PMID: 9118961; PMCID: PMC1169700
18. Cook J, Krishnan S, Ananth S, Sells SF, Shi Y, Walther MM, Linehan WM, Sukhatme VP, Weinstein MH, Rangnekar VM (1999) Decreased expression of the pro-apoptotic protein Par-4 in renal cell carcinoma. Oncogene. 18(5):1205–1208. https://doi.org/10.1038/sj.onc.1202416. PMID: 10022126
19. Davidson DJ, Haskell C, Majest S, Kherzai A, Egan DA, Walter KA, Schneider A, Gubbins EF, Solomon L, Chen Z, Lesniewski R, Henkin J (2005) Kringle 5 of human plasminogen induces apoptosis of endothelial and tumor cells through surface-expressed glucose-regulated protein 78. Cancer Res. 65(11):4663–4672. https://doi.org/10.1158/0008-5472.CAN-04-3426. PMID: 15930284
20. Daya-Makin M, Sanghera JS, Mogentale TL, Lipp M, Parchomchuk J, Hogg JC, Pelech SL (1994) Activation of a tumor-associated protein kinase (p40TAK) and casein kinase 2 in human

squamous cell carcinomas and adenocarcinomas of the lung. Cancer Res. 54(8):2262–2268. PMID: 7513612.

21. de Thonel A, Hazoumé A, Kochin V, Isoniemi K, Jego G, Fourmaux E, Hammann A, Mjahed H, Filhol O, Micheau O, Rocchi P, Mezger V, Eriksson JE, Rangnekar VM, Garrido C (2014) Regulation of the proapoptotic functions of prostate apoptosis response-4 (Par-4) by casein kinase 2 in prostate cancer cells. Cell Death Dis. 5:e1016.
22. Díaz-Meco MT, Municio MM, Frutos S, Sanchez P, Lozano J, Sanz L, Moscat J (1996) The product of par-4, a gene induced during apoptosis, interacts selectively with the atypical isoforms of protein kinase C. Cell. 86(5):777–786. https://doi.org/10.1016/s0092-8674(00)80152-x. PMID: 8797824
23. Di Maira G, Salvi M, Arrigoni G, Marin O, Sarno S, Brustolon F, Pinna LA, Ruzzene M (2005) Protein kinase CK2 phosphorylates and upregulates Akt/PKB. Cell Death Differ. 12(6):668–677. https://doi.org/10.1038/sj.cdd.4401604
24. El-Guendy N, Zhao Y, Gurumurthy S, Burikhanov R, Rangnekar VM (2003) Identification of a unique core domain of par-4 sufficient for selective apoptosis induction in cancer cells. Mol Cell Biol 23(16):5516–5525. https://doi.org/10.1128/mcb.23.16.5516-5525.2003. PMID: 12897127; PMCID: PMC166354
25. El-Guendy N, Rangnekar VM (2003) Apoptosis by Par-4 in cancer and neurodegenerative diseases. Exp Cell Res. 283(1):51–66. https://doi.org/10.1016/s0014-4827(02)00016-2. PMID: 12565819
26. Engemann H, Heinzel V, Page G, Preuss U, Scheidtmann KH (2002) DAP-like kinase interacts with the rat homolog of Schizosaccharomyces pombe CDC5 protein, a factor involved in pre-mRNA splicing and required for G2/M phase transition. Nucleic Acids Res 30(6):1408–1417. https://doi.org/10.1093/nar/30.6.1408. PMID: 11884640; PMCID: PMC101352
27. Engelman JA, Luo J, Cantley LC (2006) The evolution of phosphatidylinositol 3-kinases as regulators of growth and metabolism. Nat Rev Genet. 7(8):606–619. https://doi.org/10.1038/nrg1879. PMID: 16847462
28. Fernandez-Marcos PJ, Abu-Baker S, Joshi J, Galvez A, Castilla EA, Cañamero M, Collado M, Saez C, Moreno-Bueno G, Palacios J, Leitges M, Serrano M, Moscat J, Diaz-Meco MT (2009) Simultaneous inactivation of Par-4 and PTEN in vivo leads to synergistic NF-kappaB activation and invasive prostate carcinoma. Proc Natl Acad Sci U S A. 106(31):12962–12967. https://doi.org/10.1073/pnas.0813055106. PMID: 19470463; PMCID: PMC2722271
29. Follo MY, Manzoli L, Poli A, McCubrey JA, Cocco L (2015) PLC and PI3K/Akt/mTOR signalling in disease and cancer. Adv Biol Regul. 57:10–16. https://doi.org/10.1016/j.jbior.2014.10.004. Epub 2014 Nov 26. PMID: 25482988
30. Garcia-Cao I, Lafuente MJ, Criado LM, Diaz-Meco MT, Serrano M, Moscat J (2003) Genetic inactivation of Par4 results in hyperactivation of NF-kappaB and impairment of JNK and p38. EMBO Rep 4(3):307–312. https://doi.org/10.1038/sj.embor.embor769. PMID: 12634851; PMCID: PMC1315897
31. García-Cao I, Duran A, Collado M, Carrascosa MJ, Martín-Caballero J, Flores JM, Diaz-Meco MT, Moscat J, Serrano M (2005) Tumour-suppression activity of the proapoptotic regulator Par4. EMBO Rep 6(6):577–583. https://doi.org/10.1038/sj.embor.7400421. PMID: 15877079; PMCID: PMC1369092
32. Goswami A, Burikhanov R, de Thonel A, Fujita N, Goswami M, Zhao Y, Eriksson JE, Tsuruo T, Rangnekar VM (2005) Binding and phosphorylation of Par-4 by AKT is essential for cancer cell survival. Mol Cell. 20:33–44
33. Götz C, Montenarh M (2013) Protein kinase CK2 in the ER stress response. Adv Biol Chem 2013:1–5. https://doi.org/10.4236/abc.2013.33A001
34. Gowda C, Sachdev M, Muthusami S, Kapadia M, Petrovic-Dovat L, Hartman M, Ding Y, Song C, Payne JL, Tan BH, Dovat S (2017) Casein Kinase II (CK2) as a Therapeutic Target for Hematological Malignancies. Curr Pharm Des. 23(1):95–107. https://doi.org/10.2174/1381612822666161006154311. PMID: 27719640

35. Guo Q, Xie J (2004) AATF inhibits aberrant production of amyloid beta peptide 1-42 by interacting directly with Par-4. J Biol Chem. 279(6):4596–4603. https://doi.org/10.1074/jbc.M309811200. Epub 2003 Nov 18. PMID: 14627703
36. Guo Q, Fu W, Xie J, Luo H, Sells SF, Geddes JW, Bondada V, Rangnekar VM, Mattson MP (1998) Par-4 is a mediator of neuronal degeneration associated with the pathogenesis of Alzheimer disease. Nat Med. 4(8):957–962. https://doi.org/10.1038/nm0898-957. PMID: 9701251
37. Gurumurthy S, Goswami A, Vasudevan KM, Rangnekar VM (2005) Phosphorylation of Par-4 by protein kinase A is critical for apoptosis. Mol Cell Biol. 25(3):1146–1161. https://doi.org/10.1128/MCB.25.3.1146-1161.2005
38. Izeradjene K, Douglas L, Delaney A, Houghton JA (2005) Casein kinase II (CK2) enhances death-inducing signaling complex (DISC) activity in TRAIL-induced apoptosis in human colon carcinoma cell lines. Oncogene. 24(12):2050–2058. https://doi.org/10.1038/sj.onc.1208397. PMID: 15688023
39. Hosoi T, Korematsu K, Horie N, Suezawa T, Okuma Y, Nomura Y, Ozawa K (2012) Inhibition of casein kinase 2 modulates XBP1-GRP78 arm of unfolded protein responses in cultured glial cells. PLoS One. 7(6):e40144. https://doi.org/10.1371/journal.pone.0040144. Epub 2012 Jun 29. PMID: 22768244; PMCID: PMC3387139
40. Johnstone RW, See RH, Sells SF, Wang J, Muthukkumar S, Englert C, Haber DA, Licht JD, Sugrue SP, Roberts T, Rangnekar VM, Shi Y (1996) A novel repressor, par-4, modulates transcription and growth suppression functions of the Wilms' tumor suppressor WT1. Mol Cell Biol. 16(12):6945–6956. https://doi.org/10.1128/mcb.16.12.6945. PMID: 8943350; PMCID: PMC231698
41. Johnstone RW, Tommerup N, Hansen C, Vissing H, Shi Y (1998) Mapping of the human PAWR (par-4) gene to chromosome 12q21. Genomics. 53(2):241–243. https://doi.org/10.1006/geno.1998.5494. PMID: 9790775
42. Joshi J, Fernandez-Marcos PJ, Galvez A, Amanchy R, Linares JF, Duran A, Pathrose P, Leitges M, Cañamero M, Collado M, Salas C, Serrano M, Moscat J, Diaz-Meco MT (2008) Par-4 inhibits Akt and suppresses Ras-induced lung tumorigenesis. EMBO J. 27(16):2181–2193. https://doi.org/10.1038/emboj.2008.149. Epub 2008 Jul 24. PMID: 18650932; PMCID: PMC2519103
43. Kawai T, Matsumoto M, Takeda K, Sanjo H, Akira S (1998) ZIP kinase, a novel serine/threonine kinase which mediates apoptosis. Mol Cell Biol. 18(3):1642–1651. https://doi.org/10.1128/mcb.18.3.1642. PMID: 9488481; PMCID: PMC108879
44. Kawai T, Akira S, Reed JC (2003) ZIP kinase triggers apoptosis from nuclear PML oncogenic domains. Mol Cell Biol. 23(17):6174–6186. https://doi.org/10.1128/mcb.23.17.6174-6186.2003. PMID: 12917339; PMCID: PMC180930
45. Kitamura H, Torigoe T, Asanuma H, Hisasue SI, Suzuki K, Tsukamoto T, Satoh M, Sato N (2006) Cytosolic overexpression of p62 sequestosome 1 in neoplastic prostate tissue. Histopathology. 48(2):157–161. https://doi.org/10.1111/j.1365-2559.2005.02313.x. PMID: 16405664
46. Kögel D, Bierbaum H, Preuss U, Scheidtmann KH (1999) C-terminal truncation of Dlk/ZIP kinase leads to abrogation of nuclear transport and high apoptotic activity. Oncogene. 18(51):7212–7218. https://doi.org/10.1038/sj.onc.1203169. PMID: 10602474
47. Kögel D, Prehn JH, Scheidtmann KH (2001) The DAP kinase family of pro-apoptotic proteins: novel players in the apoptotic game. Bioessays. 23(4):352–358. https://doi.org/10.1002/bies.1050. PMID: 11268041
48. Koh H, Lee KH, Kim D, Kim S, Kim JW, Chung J (2000) Inhibition of Akt and its anti-apoptotic activities by tumor necrosis factor-induced protein kinase C-related kinase 2 (PRK2) cleavage. J Biol Chem. 275(44):34451–34458. https://doi.org/10.1074/jbc.M001753200. PMID: 10926925

49. Komatsu S, Ikebe M (2004) ZIP kinase is responsible for the phosphorylation of myosin II and necessary for cell motility in mammalian fibroblasts. J Cell Biol. 165(2):243–254. https://doi.org/10.1083/jcb.200309056. Epub 2004 Apr 19. PMID: 15096528; PMCID: PMC2172045
50. Konishi H, Kuroda S, Tanaka M, Matsuzaki H, Ono Y, Kameyama K, Haga T, Kikkawa U (1995) Molecular cloning and characterization of a new member of the RAC protein kinase family: association of the pleckstrin homology domain of three types of RAC protein kinase with protein kinase C subspecies and beta gamma subunits of G proteins. Biochem Biophys Res Commun. 216(2):526–534. https://doi.org/10.1006/bbrc.1995.2654. PMID: 7488143
51. Krebs EG, Beavo JA (1979) Phosphorylation-dephosphorylation of enzymes. Annu Rev Biochem. 48:923–959. https://doi.org/10.1146/annurev.bi.48.070179.004423. PMID: 38740
52. Kriplani N, Hermida MA, Brown ER, Leslie NR (2015) Class I PI 3-kinases: Function and evolution. Adv Biol Regul. 59:53–64. https://doi.org/10.1016/j.jbior.2015.05.002. PMID: 26159297
53. Lee AS (2001) The glucose-regulated proteins: stress induction and clinical applications. Trends Biochem Sci. 26(8):504–510. https://doi.org/10.1016/s0968-0004(01)01908-9. PMID: 11504627
54. Leitges M, Sanz L, Martin P, Duran A, Braun U, García JF, Camacho F, Diaz-Meco MT, Rennert PD, Moscat J (2001) Targeted disruption of the zetaPKC gene results in the impairment of the NF-kappaB pathway. Mol Cell. 8(4):771–780. https://doi.org/10.1016/s1097-2765(01)00361-6. PMID: 11684013
55. Libich DS, Schwalbe M, Kate S, Venugopal H, Claridge JK, Edwards PJ, Dutta K, Pascal SM (2009) Intrinsic disorder and coiled-coil formation in prostate apoptosis response factor 4. FEBS J. 276(14):3710–3728. https://doi.org/10.1111/j.1742-4658.2009.07087.x. Epub 2009 May 28. PMID: 19490121
56. Mabe NW, Fox DB, Lupo R, Decker AE, Phelps SN, Thompson JW, Alvarez JV (2018) Epigenetic silencing of tumor suppressor Par-4 promotes chemoresistance in recurrent breast cancer. J Clin Invest. 128(10):4413–4428. https://doi.org/10.1172/JCI99481. Epub 2018 Aug 27. PMID: 30148456; PMCID: PMC6159989.
57. Manni S, Brancalion A, Tubi LQ, Colpo A, Pavan L, Cabrelle A, Ave E, Zaffino F, Di Maira G, Ruzzene M, Adami F, Zambello R, Pitari MR, Tassone P, Pinna LA, Gurrieri C, Semenzato G, Piazza F (2012) Protein kinase CK2 protects multiple myeloma cells from ER stress-induced apoptosis and from the cytotoxic effect of HSP90 inhibition through regulation of the unfolded protein response. Clin Cancer Res. 18(7):1888–1900. https://doi.org/10.1158/1078-0432.CCR-11-1789. Epub 2012 Feb 20. PMID: 22351691
58. Matsumoto G, Wada K, Okuno M, Kurosawa M, Nukina N (2011) Serine 403 phosphorylation of p62/SQSTM1 regulates selective autophagic clearance of ubiquitinated proteins. Mol Cell. 44(2):279–289. https://doi.org/10.1016/j.molcel.2011.07.039. PMID: 22017874
59. Mattson MP, Duan W, Chan SL, Camandola S (1999) Par-4: an emerging pivotal player in neuronal apoptosis and neurodegenerative disorders. J Mol Neurosci. 13(1-2):17–30. https://doi.org/10.1385/JMN:13:1-2:17. PMID: 10691289
60. MacDonald JA, Borman MA, Murányi A, Somlyo AV, Hartshorne DJ, Haystead TA (2001) Identification of the endogenous smooth muscle myosin phosphatase-associated kinase. Proc Natl Acad Sci U S A. 98(5):2419–2424. https://doi.org/10.1073/pnas.041331498. PMID: 11226254; PMCID: PMC30153
61. MacDonald JA, Moffat LD, Al-Ghabkari A, Sutherland C, Walsh MP (2013) Prostate-apoptosis response-4 phosphorylation in vascular smooth muscle. Arch Biochem Biophys. 535(1):84–90. https://doi.org/10.1016/j.abb.2012.11.009. Epub 2012 Dec 3. PMID: 23219599
62. McDonnell MA, Abedin MJ, Melendez M, Platikanova TN, Ecklund JR, Ahmed K, Kelekar A (2008) Phosphorylation of murine caspase-9 by the protein kinase casein kinase 2 regulates its cleavage by caspase-8. J Biol Chem. 283(29):20149–20158. https://doi.org/10.1074/jbc.M802846200. Epub 2008 May 8. PMID: 18467326; PMCID: PMC2459295
63. Moreno-Bueno G, Fernandez-Marcos PJ, Collado M, Tendero MJ, Rodriguez-Pinilla SM, Garcia-Cao I, Hardisson D, Diaz-Meco MT, Moscat J, Serrano M, Palacios J (2007) Inactivation

of the candidate tumor suppressor par-4 in endometrial cancer. Cancer Res. 67(5):1927–1934. https://doi.org/10.1158/0008-5472.CAN-06-2687. PMID: 17332319

64. Münstermann U, Fritz G, Seitz G, Lu YP, Schneider HR, Issinger OG (1990) Casein kinase II is elevated in solid human tumours and rapidly proliferating non-neoplastic tissue. Eur J Biochem. 189(2):251–257. https://doi.org/10.1111/j.1432-1033.1990.tb15484.x. PMID: 2159876
65. Nalca A, Qiu SG, El-Guendy N, Krishnan S, Rangnekar VM (1999) Oncogenic Ras sensitizes cells to apoptosis by Par-4. J Biol Chem. 274(42):29976–29983. https://doi.org/10.1074/jbc.274.42.29976. PMID: 10514481
66. Ni M, Zhang Y, Lee AS (2011) Beyond the endoplasmic reticulum: atypical GRP78 in cell viability, signalling and therapeutic targeting. Biochem J. 434(2):181–188. https://doi.org/10.1042/BJ20101569. PMID: 21309747; PMCID: PMC3353658
67. Niiro N, Ikebe M (2001) Zipper-interacting protein kinase induces Ca(2+)-free smooth muscle contraction via myosin light chain phosphorylation. J Biol Chem. 276(31):29567–29574. https://doi.org/10.1074/jbc.M102753200. Epub 2001 May 30. PMID: 11384979
68. Ortega CE, Seidner Y, Dominguez I (2014) Mining CK2 in cancer. PLoS One. 9(12):e115609. https://doi.org/10.1371/journal.pone.0115609. Erratum in: PLoS One. 2015; 10(3):e0120224. PMID: 25541719; PMCID: PMC4277308
69. Page G, Lödige I, Kögel D, Scheidtmann KH (1999a) AATF, a novel transcription factor that interacts with Dlk/ZIP kinase and interferes with apoptosis. FEBS Lett. 462(1-2):187–191. https://doi.org/10.1016/s0014-5793(99)01529-x. PMID: 10580117
70. Page G, Kögel D, Rangnekar V, Scheidtmann KH (1999b) Interaction partners of Dlk/ZIP kinase: co-expression of Dlk/ZIP kinase and Par-4 results in cytoplasmic retention and apoptosis. Oncogene. 18(51):7265–7273. https://doi.org/10.1038/sj.onc.1203170. PMID: 10602480
71. Parker BL, Shepherd NE, Trefely S, Hoffman NJ, White MY, Engholm-Keller K, Hambly BD, Larsen MR, James DE, Cordwell SJ (2014) Structural basis for phosphorylation and lysine acetylation cross-talk in a kinase motif associated with myocardial ischemia and cardioprotection. J Biol Chem. 289(37):25890–25906. https://doi.org/10.1074/jbc.M114.556035. Epub 2014 Jul 9. Erratum in: J Biol Chem. 2014 Dec 5; 289(49):33875. PMID: 25008320; PMCID: PMC4162189
72. Payrastre B, Cocco L (2015) Foreword: "The PI3-kinase/Akt pathway: From signaling to diseases". Adv Biol Regul. 59:1–3. https://doi.org/10.1016/j.jbior.2015.07.001. Epub 2015 Jul 18. PMID: 26208875
73. Pruitt K, Ulkü AS, Frantz K, Rojas RJ, Muniz-Medina VM, Rangnekar VM, Der CJ, Shields JM (2005) Ras-mediated loss of the pro-apoptotic response protein Par-4 is mediated by DNA hypermethylation through Raf-independent and Raf-dependent signaling cascades in epithelial cells. J Biol Chem. 280(24):23363–23370. https://doi.org/10.1074/jbc.M503083200. Epub 2005 Apr 14. PMID: 15831492
74. Qiao Y, Chen T, Yang H, Chen Y, Lin H, Qu W, Feng F, Liu W, Guo Q, Liu Z, Sun H (2019) Small molecule modulators targeting protein kinase CK1 and CK2. Eur J Med Chem. 181:111581. https://doi.org/10.1016/j.ejmech.2019.111581. Epub 2019 Aug 1. PMID: 31400711
75. Ranganathan P, Rangnekar VM (2005) Regulation of cancer cell survival by Par-4. Ann N Y Acad Sci. 1059:76–85. https://doi.org/10.1196/annals.1339.046. PMID: 16382046
76. Ravi R, Bedi A (2002) Sensitization of tumor cells to Apo2 ligand/TRAIL-induced apoptosis by inhibition of casein kinase II. Cancer Res. 62(15):4180–4185. PMID: 12154014
77. Romieu-Mourez R, Landesman-Bollag E, Seldin DC, Sonenshein GE (2002) Protein kinase CK2 promotes aberrant activation of nuclear factor-kappaB, transformed phenotype, and survival of breast cancer cells. Cancer Res. 62(22):6770–6778. PMID: 12438279
78. Ruzzene M, Pinna LA (2010) Addiction to protein kinase CK2: a common denominator of diverse cancer cells? Biochim Biophys Acta. 1804(3):499–504. https://doi.org/10.1016/j.bbapap.2009.07.018. Epub 2009 Aug 6. PMID: 19665589

79. Sarker D, Reid AH, Yap TA, de Bono JS (2009) Targeting the PI3K/AKT pathway for the treatment of prostate cancer. Clin Cancer Res. 15(15):4799–4805. https://doi.org/10.1158/1078-0432.CCR-08-0125. Epub 2009 Jul 28. PMID: 19638457
80. Sells SF, Wood DP Jr, Joshi-Barve SS, Muthukumar S, Jacob RJ, Crist SA, Humphreys S, Rangnekar VM (1994) Commonality of the gene programs induced by effectors of apoptosis in androgen-dependent and -independent prostate cells. Cell Growth Differ. 5(4):457–466. PMID: 8043520
81. Sells SF, Han SS, Muthukkumar S, Maddiwar N, Johnstone R, Boghaert E, Gillis D, Liu G, Nair P, Monnig S, Collini P, Mattson MP, Sukhatme VP, Zimmer SG, Wood DP Jr, McRoberts JW, Shi Y, Rangnekar VM (1997) Expression and function of the leucine zipper protein Par-4 in apoptosis. Mol Cell Biol. 17(7):3823–3832. https://doi.org/10.1128/mcb.17.7.3823. PMID: 9199316; PMCID: PMC232234
82. Shoval Y, Pietrokovski S, Kimchi A (2007) ZIPK: a unique case of murine-specific divergence of a conserved vertebrate gene. PLoS Genet. 3(10):1884–1893. https://doi.org/10.1371/journal.pgen.0030180. Epub 2007 Sep 7. PMID: 17953487; PMCID: PMC2041995
83. Skalhegg BS, Tasken K (2000) Specificity in the cAMP/PKA signaling pathway. Differential expression, regulation, and subcellular localization of subunits of PKA. Front Biosci. 5:D678–D693. https://doi.org/10.2741/skalhegg. PMID: 10922298
84. Song MS, Salmena L, Pandolfi PP (2012) The functions and regulation of the PTEN tumour suppressor. Nat Rev Mol Cell Biol. 13(5):283–296. https://doi.org/10.1038/nrm3330. PMID: 22473468
85. Staal J, Beyaert R (2018) Inflammation and NF-κB Signaling in Prostate Cancer: Mechanisms and Clinical Implications. Cells. 7(9):122. https://doi.org/10.3390/cells7090122. PMID: 30158439; PMCID: PMC6162478
86. Torres J, Pulido R (2001) The tumor suppressor PTEN is phosphorylated by the protein kinase CK2 at its C terminus. Implications for PTEN stability to proteasome-mediated degradation. J Biol Chem 276(2):993–998. https://doi.org/10.1074/jbc.M009134200. PMID: 11035045
87. Turner KM, Sun Y, Ji P, Granberg KJ, Bernard B, Hu L, Cogdell DE, Zhou X, Yli-Harja O, Nykter M, Shmulevich I, Yung WK, Fuller GN, Zhang W (2015) Genomically amplified Akt3 activates DNA repair pathway and promotes glioma progression. Proc Natl Acad Sci U S A. 112(11):3421–3426. https://doi.org/10.1073/pnas.1414573112. Epub 2015 Mar 3. PMID: 25737557; PMCID: PMC4371922
88. Vetterkind S, Morgan KG (2009) The pro-apoptotic protein Par-4 facilitates vascular contractility by cytoskeletal targeting of ZIPK. J Cell Mol Med. 13(5):887–895. https://doi.org/10.1111/j.1582-4934.2008.00374.x. Epub 2008 May 24. PMID: 18505470; PMCID: PMC2700217
89. Vivanco I, Sawyers CL (2002) The phosphatidylinositol 3-Kinase AKT pathway in human cancer. Nat Rev Cancer. 2(7):489–501. https://doi.org/10.1038/nrc839. PMID: 12094235
90. Wang G, Unger G, Ahmad KA, Slaton JW, Ahmed K (2005) Downregulation of CK2 induces apoptosis in cancer cells--a potential approach to cancer therapy. Mol Cell Biochem. 274(1-2):77–84. https://doi.org/10.1007/s11010-005-3077-1. PMID: 16342410
91. Wang G, Ahmad KA, Ahmed K (2006) Role of protein kinase CK2 in the regulation of tumor necrosis factor-related apoptosis inducing ligand-induced apoptosis in prostate cancer cells. Cancer Res 66(4):2242–2249. https://doi.org/10.1158/0008-5472.CAN-05-2772. PMID: 16489027
92. Wang L, Yang JK, Kabaleeswaran V, Rice AJ, Cruz AC, Park AY, Yin Q, Damko E, Jang SB, Raunser S, Robinson CV, Siegel RM, Walz T, Wu H (2010) The Fas-FADD death domain complex structure reveals the basis of DISC assembly and disease mutations. Nat Struct Mol Biol 17(11):1324–1329. https://doi.org/10.1038/nsmb.1920. Epub 2010 Oct 10. PMID: 20935634; PMCID: PMC2988912
93. Welch EJ, Jones BW, Scott JD (2010) Networking with AKAPs: context-dependent regulation of anchored enzymes. Mol Interv 10(2):86–97. https://doi.org/10.1124/mi.10.2.6. PMID: 20368369; PMCID: PMC2895371

94. Wong KK, Engelman JA, Cantley LC (2010) Targeting the PI3K signaling pathway in cancer. Curr Opin Genet Dev 20(1):87–90. https://doi.org/10.1016/j.gde.2009.11.002. Epub 2009 Dec 16. PMID: 20006486; PMCID: PMC2822054
95. Yang XJ, Seto E (2008) Lysine acetylation: codified crosstalk with other posttranslational modifications. Mol Cell 31(4):449–461. https://doi.org/10.1016/j.molcel.2008.07.002. PMID: 18722172; PMCID: PMC2551738
96. Yap TA, Garrett MD, Walton MI, Raynaud F, de Bono JS, Workman P (2008) Targeting the PI3K-AKT-mTOR pathway: progress, pitfalls, and promises. Curr Opin Pharmacol 8(4):393–412. https://doi.org/10.1016/j.coph.2008.08.004. Epub 2008 Aug 27. PMID: 18721898
97. Yenice S, Davis AT, Goueli SA, Akdas A, Limas C, Ahmed K (1994) Nuclear casein kinase 2 (CK-2) activity in human normal, benign hyperplastic, and cancerous prostate. Prostate 24(1):11–16. https://doi.org/10.1002/pros.2990240105. PMID: 7507238

Role of Par-4 in GRP78 Translocation

Daniel Bastida-Ruiz, Alexia Childress, and Marie Cohen

Abstract GRP78 is a classical endoplasmic reticulum (ER)-localized protein, acting as a chaperone and master regulator of the unfolding protein response in this location. Interestingly, under certain conditions, GRP78 is also located at the cell surface. More concretely, it was observed at the cell surface of several types of cancer cells, but also of trophoblastic cells and stressed cells in general. Cell surface GRP78 can bind to extracellular proteins and promote the activation of different signaling pathways. One of these proteins is prostate apoptosis response 4 (Par-4), a pro-apoptotic protein which is secreted by cancer cells. It was observed that the secretion of Par-4 was increased in parallel with GRP78 cell-surface localization, suggesting a relationship between these events. In this review, we discuss the role of Par-4 in GRP78 relocation at the cell surface.

Keywords GRP78 · Par-4 · Endoplasmic reticulum · Cell surface · Relocation · Cancer cell · Unfolded protein response · Apoptosis · Cell survival · Cell proliferation · Cell fusion · Receptor

1 Introduction

1.1 GRP78 Structure

Glucose-regulated protein (GRP78), also known as BiP [1] or HSPA5 [2, 3], is a 78-kilodalton chaperone protein of the heat shock protein 70 (HSP70) superfamily [4]. The structure of GRP78 is complex as it contains 654 amino acids [5]. The protein contains a KDEL amino acid sequence at the C-terminus with four residues that functions as an ER retention motif. At the N-terminus, 18 amino acid residues

D. Bastida-Ruiz · A. Childress · M. Cohen (✉)
Department of Pediatrics, Gynecology and Obstetrics, Faculty of Medicine, University of Geneva, Geneva, Switzerland
e-mail: marie.cohen@unige.ch

V. M. Rangnekar (ed.), *Tumor Suppressor Par-4*,
https://doi.org/10.1007/978-3-030-73572-2_7

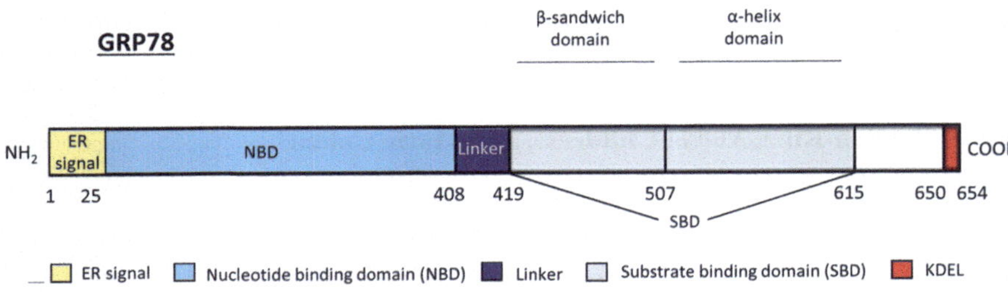

Fig. 1 Schematic representation of the GRP78 structure. At the N-terminus, in yellow, is represented the ER signaling peptide, which targets the protein to the ER. The nucleotide-binding domain (NBD) of GRP78 is represented in light blue. In grey is illustrated the substrate-binding domain (SBD), which is divided in the β-sandwich domain and the α-helix domain. Between the NBD and the SBD there is a linker, which is designed in purple. The KDEL retention amino acid sequence, serving for the protein ER retention, is represented in red at the C-terminus. The numbers under the GRP78 protein sequence describe the amino acids that mark the start and the end of the different domains

function as an ER signaling peptide to locate the protein to the ER [2, 5]. The protein is separated by the N-terminal and C-terminal domains that are connected by a short flexible linker. In the C-terminal domain—also known as the substrate-binding domain (SBD)—there are the β-sandwich and α-helical lid domains. The β-sandwich domain allows GRP78 to act as a chaperone by binding hydrophobic regions of unfolded proteins. In the N-terminal domain—also known as the nucleotide-binding domain (NBD)—there are four subdomains (IA, IB, IIA, IIB) that form an ATP-binding clef and have a dynamic purpose during adenosine triphosphate (ATP) hydrolysis. After ATP-adenosine diphosphate (ADP) exchange in the NBD, a conformational change in the GRP78 protein occurs that seals the protein and its information [6–8]. Overall, the structure of GRP78 allows it to be modifiable as it can translocate the protein from one cellular or multicellular location to another to perform new functions [5, 9–24] (Fig. 1).

2 Role of GRP78 in UPR Regulation Under ER Stress

As an ER chaperone protein, the main functions of GRP78 include its ability to bind hydrophobic regions on nascent polypeptides [6–8] and regulate the unfolded protein response (UPR) [25–29]. GRP78 acts as a key regulator of the UPR, an adaptive mechanism developed in response to ER stress, by activating inositol-requiring enzyme 1 α (IRE1α), activating transcription factor 6 (ATF6), and protein kinase RNA-like endoplasmic reticulum kinase (PERK). The active UPR affects cell fate by controlling autophagy, cell survival, and apoptosis. During homeostatic conditions, GRP78 is complexed with IRE1α, ATF6, and PERK, preventing their activation and, consequently, avoiding the UPR initiation [30–35] (Fig. 2).

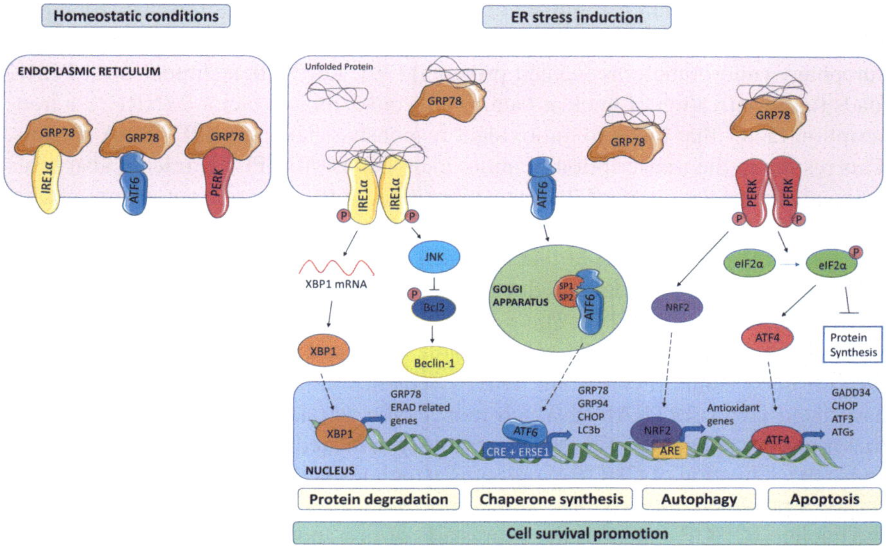

Fig. 2 Role of GRP78 in UPR regulation under ER stress. Under homeostatic conditions in the ER lumen, ER chaperone protein GRP78 is bound to proteins ATF6, IRE1α, and PERK that regulate the ER stress response. When unfolded proteins accumulate in the ER lumen, GRP78 releases ATF6, IRE1α, and PERK to regulate the folding of proteins. Next, the release of ATF6 causes it to translocate to the Golgi apparatus where enzymes SP1 and SP2 cleave it. The cleaved N-terminal cytosolic domain of ATF6 is further translocated into the nucleus where it binds to DNA regions CRE and ERSE1. This leads to the transcription of genes that code for chaperone proteins and genes that lead to autophagy activation (LC3b) and cell fate (CHOP). Then, the release of IRE1α causes the processing of multiple mRNA, such as XBP-1 mRNA. The XBP-1 protein stimulates the transcription of GRP78 and genes that cause ER-associated degradation (ERAD). IRE1α also activates JNK, which leads to the production of autophagic marker Beclin-1 by controlling Bcl2. Lastly, the release of PERK causes the activation of transcription factor NRF2, which binds to the antioxidant responsive element (ARE) regions of DNA. This binding leads to the transcription of antioxidant genes. PERK also phosphorylates eIF2α that activates the transcription factor ATF4 and enhances the transcription of genes implicated in autophagy, namely ATGs genes, and genes implicated in cell fate control, such as CHOP and GADD34. Phosphorylated eIF2α also inhibits total protein synthesisfull stop is missing

During ER stress, unfolded proteins accumulate in the ER lumen and cause GRP78 to assist in protein folding and separate from IRE1α, ATF6, and PERK [25, 36] (Fig. 2). Detachment from IRE1α leads to the processing of several mRNA such as X-box-binding protein-1 (XBP-1) mRNA, which promotes the transcription of GRP78 and genes responsible for ER-associated degradation (ERAD) [37]. IRE1α also activates the c-Jun N-terminal kinase (JNK) pathway to induce autophagy by controlling B-cell lymphoma 2 (Bcl-2) and Beclin-1 [38]. Detachment from ATF6 causes ATF6 to translocate into the Golgi apparatus where it is cleaved by serine proteases 1 and 2 (SP1 and SP2). Cleaved ATF6 N-terminal cytosolic domain then binds to cyclic adenosine monophosphate response element (CRE) and ER stress element-1 (ERSE1) DNA regions to transcribe genes that code for chaperone proteins and are involved in cell fate (C/EBP homologous protein or CHOP) and autophagy (microtubule-associated proteins) [39]. Lastly, detachment from PERK leads to the activation of nuclear factor erythroid 2-related factor 2 (NRF2), a transcription factor that binds to antioxidant responsive element (ARE) DNA regions. This results in the transcription of antioxidant genes [40]. PERK release also leads to the phosphorylation of eIF2α that activates ATF4 to transcribe autophagy-related genes (ATGs), CHOP, and other transcription factors like ATF3 to promote autophagy and regulate cell fate. Additionally, eiF2α phosphorylation leads to total protein synthesis inhibition in order to reduce the number of proteins in the lumen of the ER and diminish stress [41] (Fig. 2).

In prolonged ER stress, the cell is unable to overcome its stressful situation and the cell fate results in cell death. IRE1α activates tumor necrosis factor receptor-associated factor 2 (TRAF2), which induces the activation of a signaling cascade that promotes apoptosis [42]. Next, previously activated PERK and ATF6 pathways cause the activation of CHOP expression, which leads to growth arrest and the synthesis of pro-apoptotic proteins. At the same time, after translocation of CHOP into the nucleus, CHOP enhances the expression of pro-apoptotic proteins [43]. One of these proteins is the ER oxidoreductase 1α (Ero1α)—a cell death promoter—that causes the movement of Ca^{2+} from the ER into the mitochondria, generating the release of cytochrome *c* out of the mitochondria [44]. As a final consequence, these events lead to the caspase pathway activation, which ultimately results in apoptosis [45].

Poor vascularization of the tumor is accompanied with hypoxia, low glucose concentration, and low pH, conditions known to induce ER stress and consequently GRP78 expression [15]. As mentioned before, GRP78 regulates UPR, but also caspase-7 activity in ER [17, 46]. As a consequence, it may modulate the balance between survival and apoptosis and confer a survival advantage of tumor. Interestingly, it has been reported that the knockdown of GRP78 in mice caused a decreased tumor growth [47]. Additionally, GRP78 overexpression in breast tumors seem to be associated with therapeutic resistance [48–50]. Altogether, it was suggested that the ER-located GRP78 is implicated in tumor growth and survival, and could be used as a target for cancer treatment [51].

While GRP78 is often regarded as an ER protein, there has been new evidence that this protein is also located in the cytosol [14, 15], mitochondria [19–21], nucleus [16–18], at the surface of the cell [5, 9–13], and in the extracellular medium after being secreted [15, 22–24, 52]. GRP78 placed in various locations allows the cell to adapt to stress to promote survival [16]. In the cytosol, alternative splicing and retention of intron 1 in GRP78 result in an untargeted ER protein named GRP78va [14–16]. GRP78va regulates the UPR under ER stress by forming the active protein RNA-like ER kinase (PERK) signaling pathway [16]. The function of GRP78va is to promote cell survival and prevent ER stress-induced apoptosis [14]. In the mitochondria, GRP78 interacts with the fibrosarcoma-1 (Raf-1) protein to inhibit apoptotic signals under ER stress [53]. Mitochondrial GRP78 also regulates the balance of mitochondrial energy under ER stress [54]. In the nucleus, GRP78 can cross-link to DNA [18, 55] and is hypothesized to protect from DNA damage-induced apoptosis [56]. Additionally, GRP78 secreted by tumor cells is known to have anti-carcinogenic effects by facilitating the movement of macrophages into tumors [57].

3 GRP78 Translocates to the Surface

Forthcoming evidence has revealed that GRP78 can localize to the surface of some cancer cells [11, 25, 58–71], trophoblastic cells [72–74], and more generally stressed cells [5]. GRP78 at the cell surface was firstly identified by Arap et al., using filamentous phage clones displaying GRP78-binding peptides in their plasma membrane. These phages were able to bind to DU125 cells, a prostate cancer cell line, 30 times more than control phages. These results suggested that GRP78 was expressed at the cell surface, and that it could bind to other peptides at this location [71]. Cell surface GRP78 is now known to interact with multiple extracellular ligands and acts as a co-receptor for the entry of viruses [75–79] (Fig. 3). The interaction with ligands can trigger signal transduction, which leads to the activation of different cellular processes including cell fate, cell growth, and cell fusion [80–83].

First of all, it has been reported that the interaction of GRP78 with several ligands, namely Kringle 5 [81, 82], secreted GRP78 [83], Par-4 [74, 84, 85], or Isthmin 1 [86, 87], leads to apoptosis, and ultimately, cell death in epithelial and cancer cells (Fig. 3a). The interaction of Kringle 5 with cell surface GRP78 promotes the activation of caspase-7 [81, 82]. Similarly, the interaction of cell surface GRP78 with secreted GRP78 also activates caspase-7, and additionally, caspase-3 [83]. Moreover, Par-4 binding to cell surface GRP78 promotes the activation of the Fas-associated protein with death domain (FADD)/Caspase-8/Caspase-3 signaling pathway [84]. On the other hand, Isthmin 1 interaction with cell surface GRP78 blocks the ATP/ADP exchange between mitochondria and the cytoplasm, drastically reducing the cytoplasmic levels of ATP and, eventually, leading to cell death [86]. Finally, the binding of cell surface GRP78 with some GRP78 autoantibodies such as PAT-SM6 [62, 88, 89], Mab159 [90, 91], C38, or C107 [92, 93], also leads to apoptosis [94] (Fig. 3a). Additionally, the interaction of cell surface GRP78 with antibodies directed against its COOH-terminal domain also promotes apoptosis in prostate cancer and melanoma cell lines by upregulating p53 expression [95].

Interestingly, the opposite outcome has also been reported to be activated by the interaction of cell surface GRP78 and other ligands, such as some GRP78 autoantibodies [96–98], integrin β-1 [99], α2-macroglobulin (α2M) [63, 64, 66, 100, 101], and CRIPTO [68, 69, 102, 103], in cancer cells. The interaction of some GRP78 autoantibodies (which are currently uncharacterized) with cell surface-localized GRP78 is implicated in cell growth [96–98]. Integrin β-1/GRP78 interaction leads to the regulation of the focal adhesion kinase (FAK) signaling pathway, promoting cell proliferation [99]. The most complex interactor of cell surface-localized GRP78 to date is α2M, which is known to activate multiple signaling pathways with

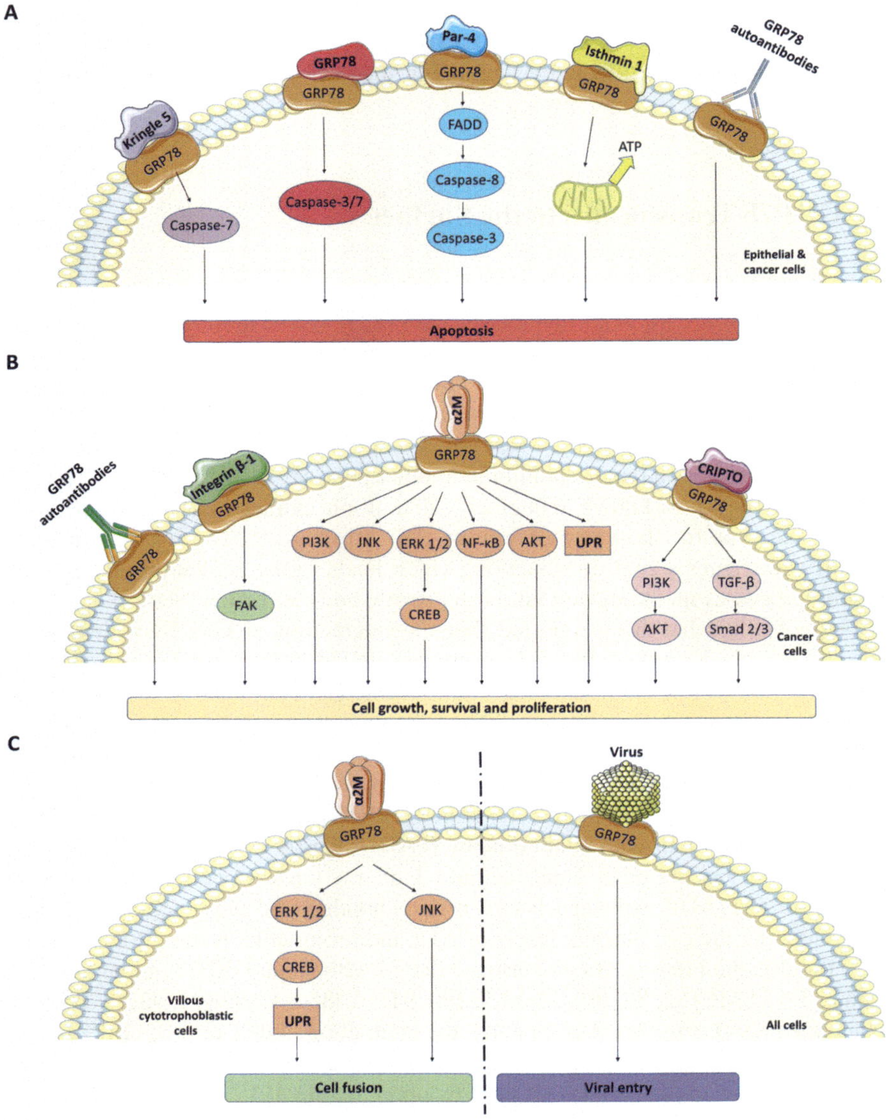

Fig. 3 Cell surface GRP78 serves as a receptor for different extracellular proteins and viruses. (**a**) Proapoptotic ligands of cell surface GRP78 and their mechanism of action in epithelial and cancer cells. Kringle 5 binds to cell surface GRP78 and activates caspase-7. Secreted extracellular GRP78 binds to cell surface GRP78 and promotes caspase-3/7 activation. Secreted Par-4 interacts with cell surface GRP78 and activates the Fas-associated protein with death domain (FADD)/Caspase-8/Caspase-3 signaling pathway. Interaction of Isthmin-1 with cell surface GRP78 blocks the ATP/ADP exchange between the mitochondria and the cytosol, triggering apoptosis due to the lack of cytosolic ATP. Some GRP78 autoantibodies binding to GRP78 lead to apoptosis. (**b**) Signaling pathways leading to cell growth, survival, and proliferation in cancer cells. The interaction between

different cellular outcomes [63, 64, 66, 100, 101]. It has been demonstrated that α2M/GRP78 interaction promotes the activation of phosphatidylinositol-4,5-biphosphate 3-kinase (PI3K) and JNK pathways, leading to cell growth of prostate cancer cells [63, 66]. Furthermore, extracellular signal-related kinase 1/2 (ERK1/2) and cAMP response element-binding protein (CREB) can also be promoted by this interaction, enhancing the proliferation of tumor cells [66]. Similarly, the activation of nuclear factor kappa-light-chain enhancer of activated B cells (NFkB) and protein kinase B (or Akt) by the binding between α2M and cell surface-localized GRP78 leads to cell survival in prostate cancer cells [66]. Interestingly, it was also reported that the α2M/GRP78 interaction could stimulate UPR activation, promoting cell survival [66]. Finally, CRIPTO promotes cell growth by binding to cell surface GRP78 and regulating the PI3K/AKT signaling pathway, and by activating the growth factor-B (TGF-β) molecular cascade in prostate and embryonic cancer cells [68, 69, 103] (Fig. 3b).

In villous cytotrophoblasts, the interaction of GRP78 with α2M activates ERK1/2 and CREB, which in turn promotes the UPR activation and JNK signaling pathways. The activation of these two molecular cascades is involved in the fusion of these cells, demonstrating the versatility of GRP78 as a receptor at the cell surface [80] (Fig. 3c).

Lastly, cell surface GRP78 is known to have the ability to interact with pathogens such as Herpes simplex virus [78], coxsackievirus [75], dengue virus serotype 2 [76], and Tembusu virus [78] and internalize them, promoting the viral entry. More recently, a modelization study suggested that GRP78 can also be involved in the viral entry of SARS-CoV-2 [79], pointing toward GRP78 as an important potential target against viral infection [16] (Fig. 3c).

For all these reasons, GRP78 has emerged as an important target for anti-cancer therapy since it is capable of regulating oncogenic signaling pathways in cancer cells. Indeed, several studies have already investigated the use of small molecules

Fig. 3 (continued) some GRP78 auto-antibodies and cell surface-localized GRP78 leads to cell proliferation. Integrin β-1 interaction with cell surface GRP78 promotes focal adhesion kinase (FAK) activation and cell proliferation. GRP78 binding to α2-macroglobulin (α2M) promotes phosphatidylinositol-4,5-biphosphate 3-kinase (PI3K) and JNK pathways activation, leading to cell growth. α2M/GRP78 interaction stimulates extracellular signal-related kinase 1/2 (ERK1/2) and cAMP response element-binding protein (CREB) activation, enhancing the proliferation of tumor cells. The activation of nuclear factor kappa-light-chain enhancer of activated B cells (NFkB) and protein kinase B (or Akt) is also promoted by the binding of α2M to cell surface GRP78, leading to cell survival. The unfolded protein response (UPR) is enhanced by the α2M/GRP78 interaction and promotes cell survival. Binding of CRIPTO to cell surface GRP78 activates PI3K/AKT signaling pathway and the growth factor-B (TGF-β) molecular cascade and leads to cell growth. (**c**) Signaling pathways leading to cell fusion and viral entry. In villous cytotrophoblastic cells, the interaction of α2M and cell surface GRP78 activates ERK1/2 and CREB signaling pathway, which activates UPR and leads to cell fusion. This binding also activates the JNK signaling pathway and leads to cell fusion. Cell surface GRP78 can also interact with viral particles and facilitate viral entry and cellular infection by these pathogens

capable of interacting with GRP78 and activate apoptosis to be used as anti-cancer therapy [104]. In first place, a study in mice transplanted with DU145 or EF43-fgf4 cell line-derived tumors, and then treated with phage peptides constructed against GRP78 was conducted. These peptides were able to bind GRP78, before being internalized and suppressed tumor growth [71]. More recently, a prototype drug called bone metastasis targeting peptidomimetic-78 (BMPT-78) was developed. This drug consists of a GRP78-binding peptide fused to a death-inducing domain. This molecule reduced cell viability in multiple human lymphome and leukemia cell lines. However, toxicology studies revealed dose-dependent toxicity of this molecule in rodents, fatal cardiac dysfunction in non-human primates was also encountered in non-human primates, forcing the discard of this molecule as an GRP78 targeting anti-cancer therapy [105]. Almost parallelly, a study of Kao et al. found out a peptide, known as BC71, capable of targeting GRP78, reducing tumor angiogenesis and increasing apoptosis, which altogether caused tumor growth suppression. Interestingly, no toxicity was detected in mice, suggesting that BC71 could emerge as an important potential GRP78-targeting anti-cancer molecule [106]. However, a better understanding of the relocation of GRP78 at the cell surface would be very helpful to block it at this specific location and/or inhibit its relocalization to the cell membrane, thus controlling the apoptotic response in cancer cells.

The relocation process of GRP78 at the cell surface is not currently well understood. Even if several studies have been performed to elucidate potential mechanisms that are involved in the transport of GRP78 to the cell surface, the contribution of each one of them is not completely enlightened. Multiple processes linked to the molecular characteristics of GRP78 have been hypothesized and demonstrated to be involved in its relocation to the cell surface.

Localizing GRP78 to the cell surface can occur in multiple ways, including the self-insertion of transmembrane four-pass GRP78 into the cell membrane [5, 107, 108], or the association of GRP78 with a transmembrane protein [107] or a glycosylphosphatidylinositol (GPI) molecule [107], enabling the protein attachment to the cell membrane and allowing the signal transfer to the cytoplasm. An examination of the hydropathicity profile of GRP78 revealed that this protein is highly hydrophilic. However, it contains four different hydrophobic domains (I, II, III, and IV) that can form transmembrane helices [108, 109]. In this manner, GRP78 can be inserted into the lipid bilayer by crossing the membrane four times. It was demonstrated that the N-terminal and C-terminal regions along with the middle segment of GRP78 are exposed to the extracellular space [5, 61, 84, 110]. Nevertheless, this conformation of GRP78 at the cell surface is not compatible with the hydropathicity profile of GRP78. It has been thus hypothesized that the hydrophobic domain I is not inserted in the cell membrane but could be actually exposed to the extracellular space. An additional hydrophobic domain V may exist in the last segment of the protein, getting inserted in the cell membrane [5]. In this way, GRP78 would have the 4 transmembrane passes while permitting the observed GRP78 configuration at the cell surface (Fig. 4). GRP78 also exists as a peripheral protein on the cell surface and as a subpopulation anchored by GPI proteins. The small portion of endogenous

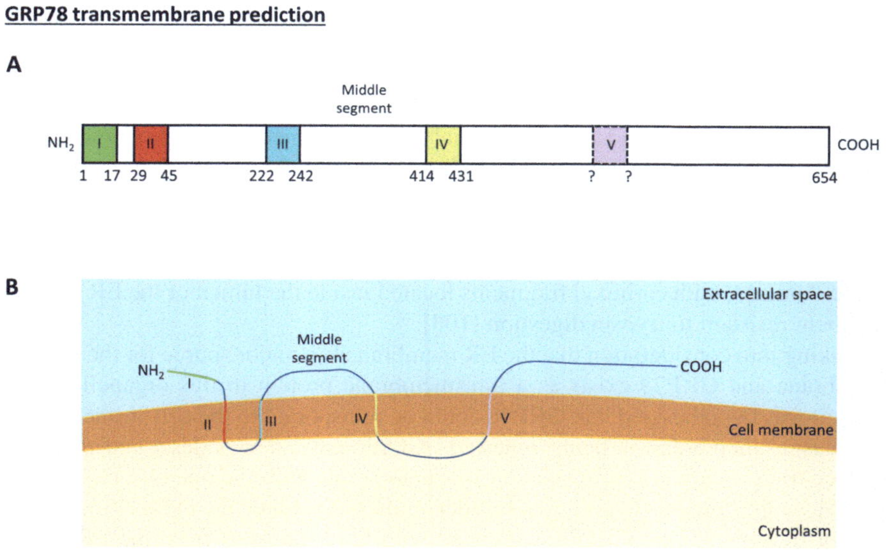

Fig. 4 Schematic representation of transmembrane GRP78. (**a**) GRP78 hydrophobic domains prediction. The hydrophobic region I is represented in green at the N-terminus. Then, the hydrophobic regions II and III are selectively represented in red and blue. The middle segment is between the hydrophobic regions II and IV, which is designed in yellow. Finally, the hypothetical hydrophobic region V is illustrated in pink. The numbers under the GRP78 protein sequence describe the amino acids that mark the start and the end of these different domains, except for hydrophobic domain V, which starting and ending amino acids are unknown. (**b**) GRP78 representation at the cell surface. At the N-terminal part, the hydrophobic domain I does not get inserted in the cell membrane and is exposed to the extracellular space, together with the middle segment and the C-terminal part of the protein. The hydrophobic regions II, III, IV, and V get inserted in the cell membrane, generating a 4-pass transmembrane protein, with 2 cytosolic segments

GRP78 attached to the cell surface by association with GPI proteins may be due to a decrease in GPI-anchored proteins from the plasma membrane under stress [107]. Several hypotheses have been proposed for the translocation of GRP78 to the cell surface and we are going to detail them in the following section.

4 Endoplasmic Reticulum Transmembrane GRP78 Origin

It was hypothesized that GRP78 in the ER could exist not only in the lumen of this organelle but also as an ER transmembrane protein since the ER membrane is a lipid bilayer where GRP78 could be inserted [111]. Some results obtained in our laboratory showed no insertion of GRP78 at the ER membrane in the ovarian cell line SKOV-3 and the trophoblastic cell line HIPEC-65 [112]. Nevertheless, Reedy et al.

demonstrated that in the GRP78 overexpressing ovarian-derived cell line C.1 cells, GRP78 could be encountered as an ER transmembrane protein. By performing a sodium carbonate protein extraction of microsomes GRP78 was detected in both the pellet and the supernatant of microsomes. These results led them to determine that GRP78 could be found as a luminal and transmembrane protein in the ER. Additionally, they demonstrated that three different GRP78 fragments of 78KDa, 50KDa, and 35KDa were generated after a trypsin digestion treatment of the microsomes. These results led them to hypothesize a GRP78 transmembrane conformation in the ER where the III and IV domains serve as putative transmembrane domains, with carboxyl fragments located inside the lumen of the ER rendering them resistant to trypsin digestion [108].

Taking into consideration that the ER membrane is a major source for the plasma membrane and GRP78 exists as a transmembrane protein in this organelle [108, 113], it was hypothesized that GRP78 could be transported to the cell surface while the plasma membrane is being renewed. Additionally, it was described that cell surface transmembrane conformation was compatible with the one observed in the ER [5], supporting this hypothesis. Further studies to validate the indirect translocation of transmembrane GRP78 associated with lipid transport from the ER to the cell surface may be needed.

5 KDEL Motif-Related Scape of GRP78 from the ER

As it was previously mentioned, GRP78 is a protein containing 654 amino acids and a structure divided into an ER signal peptide, ATPase domain, peptide-binding domain, and KDEL motif [2, 5–8] (Fig. 11.1). The preferential location of GRP78 in the ER is accomplished due to the ER signal peptide and the KDEL motif [2, 5]. The ER signal peptide consists of 18 amino acids placed at the N-terminus of the protein that have an objective to drive GRP78 to the ER [5]. On the other side, the KDEL motif is located at the C-terminus of GRP78 and is a highly conserved 4 amino acid sequence that prevents the transport of GRP78 from the ER and facilitates its return to this organelle [114].

The ER signal peptide and the KDEL motif should prevent the transport of GRP78 to other cellular compartments [114–119] (Fig. 5a). The observation of GRP78 at the cell surface suggests that this protein escapes the ER retention machinery. Several hypotheses have been proposed for this, including: (i) overloaded ER retention machinery due to a GRP78 overexpression, (ii) KDEL motif deletion from the GRP78 sequence, (iii) KDEL receptors dispersing from the Golgi apparatus (GA), and (iv) KDEL motif masking derived from an *O*-glycosylation at a close proximity to the retention sequence.

(i) Several studies have demonstrated that during ER stress conditions, the total expression of chaperone proteins—among them GRP78—is increased. At the same time, it has been reported that together with the increased GRP78 expres-

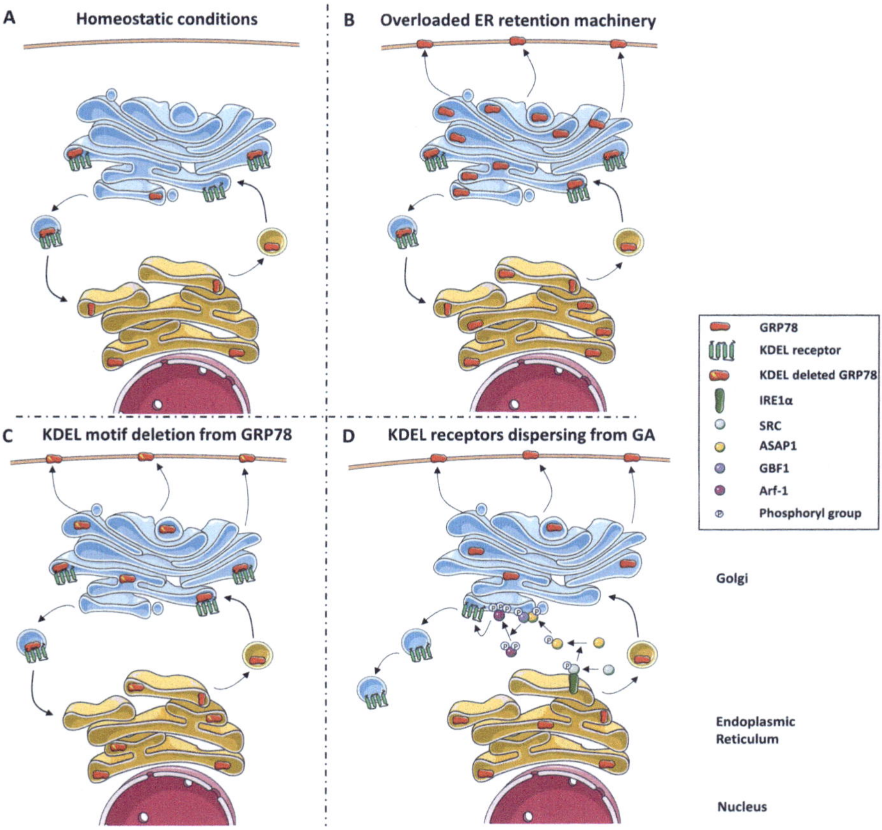

Fig. 5 Hypothesis about GRP78 transport to the cell surface depending on KDEL sequence or KDEL receptors. (**a**) Homeostatic conditions. GRP78 is residing in the lumen of the endoplasmic reticulum (ER) under homeostatic conditions. The few GRP78 molecules that escape the ER retention machinery are detected in the Golgi apparatus (GA) by the KDEL receptors, ensuring its return to the ER via vesicle retrotransport. The GRP78 molecules cannot reach the cell surface. (**b**) Overloaded ER retention machinery. Under ER stress conditions, the expression of GRP78 is enhanced in the ER. The high expression of GRP78 could lead to an overloaded ER retention machinery and, consequently, the escape of a high amount of GRP78 molecules to the GA. In the GA, the KDEL receptors may intercept GRP78; however, the uncommonly high presence of GRP78 in this organelle could lead to insufficient KDEL receptors retrotransporting GRP78 molecules to the ER. The remaining GRP78 molecules in the GA could use the vesicles transporting proteins from the GA to the cell surface, reaching this final location. (**c**) KDEL motif deletion from GRP78. The modification of the KDEL sequence of GRP78 allows this protein to be not detected by the KDEL receptors. After reaching the GA, the GRP78 copies that are unmodified would be detected by the KDEL receptors and retrotransported to the ER. On the contrary, the KDEL deleted GRP78 proteins that reach the GA would not be detected by the ER retention machinery and could be transported to the cell surface. (**d**) KDEL receptors dispersing from the GA. Upon ER stress activation, several proteins get activated, as it is the case of IRE1α. This kinase protein is capable of phosphorylating, and hence activating other proteins in the cytosol, as it is the case of SRC. The active SRC leads to the phosphorylation of ASAP1, which subsequently phosphorylates GBF1. The former one is able to activate Arf1, by exchanging its ADP molecule to an ATP molecule. Finally, active Arf1 acts in the GA recruiting the KDEL receptors and dispersing them in vesicles in the cytosol. The absence of KDEL receptors in the GA, together with the high expression of GRP78 molecules caused by the ER stress, allow the escape of GRP78 from the ER retention machinery and the relocation of GRP78 at the cell surface

sion, cell surface GRP78 is enhanced [5, 29, 107, 120]. It was speculated that the overexpression of GRP78 may overload the ER retention machinery, allowing part of the GRP78 proteins to escape from the ER, achieving relocation to different cell compartments such as the cell surface [16] (Fig. 5b).

Some studies determined that the overexpression of GRP78 generated by an increased ER stress impacted the relocation of GRP78 at the cell membrane [5, 29, 107, 120]. Additionally, Zhang et al. demonstrated that the overexpression of GRP78, independent of ER stress, was enough to accomplish the GRP78 relocation to the cell surface [5]. This suggests that the overexpression of GRP78 is a strong signal for cell surface relocation, most probably due to an overload of the ER retention machinery.

The ER retention machinery is dependent on the KDEL motif. This short sequence ensures the location of GRP78 at the ER and facilitates its return to this organelle in case it has been transported to other cellular locations. In the GA, lysosomes and secretory vesicles, reside KDEL receptors, which are seven-transmembrane-domain proteins that recognize the KDEL motif and activate the retrotransport of GRP78 to the ER [115, 121]. In a scenario where the ER retention machinery is overloaded, the KDEL receptors would be overwhelmed by the high expression of GRP78 and the retrotransport could not be achieved.

Lenin et al. recently studied the impact of insufficient expression of KDEL receptors on the relocation of GRP78 at the cell surface. The silencing of KDEL receptor 1 in primary human retinal microvascular endothelial cells was accompanied by an increase in GRP78 detection at the cell surface [13]. Overall, it has been demonstrated that GRP78 overexpression is involved in its relocation to the cell surface by escaping the KDEL retention system.

(ii) Another hypothesis regarding the relocation of GRP78 and its escape from the KDEL retrieval system is dependent on the GRP78 sequence. In the absence of the KDEL motif in the GRP78 sequence, the KDEL receptors could be transported to other cellular locations but could not recognize GRP78 as an ER residing protein (Fig. 5c).

This hypothesis was studied by Zhang et al. by transfecting 293 T cells with a plasmid expressing GRP78 that lacks the KDEL motif. At low dosages of transfected plasmid, about 5% of truncated GRP78 is expressed at the cell surface, while only 1–2% of the full GRP78 protein could be observed at this cellular location. They concluded that the deletion of the C-terminal KDEL motif affects cell surface presentation of GRP78 [5].

(iii) The third hypothesis is directly related to the KDEL receptors absence in the GA. Tsai et al. discovered that IRE1α, activates SRC, a nonreceptor protein-tyrosine kinase protein which plays a role in multiple cell mechanisms. The SRC activation in the GA leads to the phosphorylation of Arf-GAP with SH3 domain, ANK repeat, and PH domain-containing protein 1 (ASAP1), which promotes its interaction with Golgi-specific brefeldin A-resistance guanine nucleotide exchange factor 1 (GBF1) and the activation of the former one [120]. A major substrate of GBF1 is ADP ribosylation factor 1 (Arf1) protein

[120], a small GTPase that coordinates the assembly of retrograde vesicles from the GA to the ER [118]. The interaction of active GBF1 with Arf1 causes the exchange of Arf1 GDP into GTP, generating active Arf1-GTP. The activation of Arf1 generates the insertion of this protein into the GA lipid bilayer. Finally, the insertion of Arf1 leads to the dispersion of KDEL receptors from the GA and subsequently, GRP78 is capable of escaping the retrograde KDEL transport [120] (Fig. 5d).

(iv) The last hypothesis related to the KDEL motif as a cause of GRP78 location at the cell membrane is the least accepted by the scientific community with the latest results tending to be rejected. It has been reported that GRP78 is capable of undergoing post-translational modifications as *O*-linked glycosylation [5, 16, 61, 62]. An analysis of the potential sites—which could be glycosylated—revealed that a threonine located at amino acid 648 was the strongest probable site [5]. It was suggested that its glycosylation may mask or interfere with the KDEL motif that is in close proximity, leading to the escape of GRP78 from the KDEL retrieval system and its relocation to the cell surface [5].

Nevertheless, this last hypothesis is controversial since the mutation of the threonine at the 648 amino acid—and hence the elimination of the glycosylation site—did not affect GRP78 relocation to the cell surface in three different cell lines (HeLa, MCF-7, and 293 T) [5]. These results suggest that glycosylation at the vicinity of the KDEL motif did not affect masking the retention sequence and, therefore, did not affect GRP78 relocation at the cell surface [5].

6 Secreted GRP78

As we previously mentioned, GRP78 can be detected in plasma after being secreted [15, 22–24, 52]. At the same time, it is known that GRP78 at the cell surface can be found associated with GPI [107] or with other client proteins [107]. It was hypothesized that GRP78 could be linked to the cell surface after secretion by interacting with a GPI molecule or a client protein already located at the cell surface. Nevertheless, a study by Zhang et al. discarded this hypothesis by using 293 T cells and different types of conditioned medium containing GRP78 [5]. Further experiments may need to be performed in different cell types where the expression of GPI and/or GRP78 client proteins is increased after transfection—or in cells where the expression of these molecules is endogenously high—to ensure the total rejection of this hypothesis.

7 Translocation from the ER Together with Client Proteins

Multiple proteins are transported from the ER to the cell surface, mainly transmembrane proteins that serve as receptors or constituents of the cell membrane, but also proteins directed to be secreted [122–124]. These proteins are generated in the ER before its transport to the GA and its subsequent translocation to the cell surface by vesicle transport [122–125]. GRP78 is an abundant protein in the ER [126] (especially during ER stress conditions [127, 128]) and interacts with several proteins in this location [30–35, 38, 41–45, 84, 129]. It has been hypothesized that a possible translocation mechanism used by GRP78 is through its interaction with client proteins, whose final destination is the cell surface or the secretion pathway [84, 129]. In this way, GRP78 could take advantage of the intrinsic characteristics of these proteins in order to be transported to the cell membrane.

The potential GRP78 interactor proteins are numerous since GRP78 contains a peptide-binding domain with not only an affinity to bind to nascent polypeptides, novel peptides, and unfolded proteins, but also to several mature proteins such as the UPR initiator proteins (IRE1a, PERK, and ATF6), transmembrane proteins or secreted proteins [30–35, 38, 41–45, 74, 81–87, 129]. Among the transmembrane proteins that GRP78 has been shown to interact with, we can find MTJ-1, which is expressed at the cell surface [84, 130]. MTJ-1 is a protein that binds to GRP78 in the ER, modulating its activity as a chaperone [131]. Additionally, Misra et al. demonstrated that in macrophages, MTJ-1 and GRP78 coimmunoprecipitated from plasma membrane lysates. The silencing of MTJ-1 expression also abolished the localization of GRP78 to the cell surface [130]. On the other hand, Par-4—a spontaneously secreted protein in normal and cancer cells [84, 132, 133]—has also been reported to interact with GRP78 [84]. Interestingly, it was demonstrated that in the prostate adenocarcinoma cell line PC-3, Par-4 was essential for GRP78 relocation to the cell membrane. Burikhanov et al. proved that the knockdown of endogenous Par-4 by siRNA resulted in a reduction of cell membrane located GRP78 [84]. These results were later validated by Meynier et al., who demonstrated that the overexpression of Par-4 increased the membranous level of GRP78 in the ovarian cancer cell line SKOV-3, while a decrease in Par-4 expression caused a decrease in cell surface expression of GRP78 [85]. Additionally, it was demonstrated that in a non-cancerous extravillous trophoblastic cell line, concretely HIPEC-65, Par-4 overexpression or silencing had similar effects in GRP78 relocation at the cell surface to those observed by Meynier et al. and Burikhanov et al. [74]. Altogether these studies demonstrated that Par-4 may serve as an important client protein involved in GRP78 translocation to the cell membrane.

There is no consensus currently for which mechanism determines the GRP78 relocation to the cell surface. The most probable scenario includes several of the previously mentioned processes taking part in the GRP78 cell surface relocation at the same time, displaying different importance depending on the cell type and the cellular state.

Most of the mechanisms previously described are unspecific processes that could serve to relocate other proteins. More concretely, they could be used by the cell to relocate proteins residing at the ER, containing a KDEL sequence in their structure or being secreted and displaying a binding region to other cell surface located proteins. However, the translocation due to the interaction with Par-4 [74, 84, 85, 130] seems like a quite specific mechanism and an interest in a possible GRP78-relocation control has recently awakened interest in the scientific community.

8 Par-4

Par-4 is a proapoptotic tumor suppressor protein that was primarily identified in prostate cancer cells [134, 135]. Since its discovery, this protein has been demonstrated to have similar effects in numerous other human cancer cells such as colon, renal, lung, skin, breast, and pancreatic cancer cells [136–142]. Par-4 structure consists of 340 amino acids organized in several highly conserved domains, namely two nuclear localization sequences (NLS1 and NLS2) at N-terminal, a nuclear export sequence, a leucine zipper (LZ) domain at C-terminal, and the selective for apoptosis of cancer cells (SAC) domain [143–147]. The NLS serves to the protein to be imported into the nucleus through the nuclear import machinery [143, 148]. On the other side, the SAC domain, which is a short sequence of 59 amino acids long (146–203 in human Par-4), is the main functional unit for apoptosis induction in cancer cells [144, 149, 148]. The NLS and SAC domains have been reported to be essential for the apoptosis activation through Par-4 [148] (Fig. 6).

Par-4 is ubiquitously expressed among the various tissues and has been generally recognized as a protein residing in both the cytoplasm and the nucleus [150, 143]. Through the interaction with different proteins, Par-4 controls the activation of apoptosis [151, 152]. Cytoplasmic Par-4 has been reported to interact and inhibit the activation of atypical protein kinase C (ζPKC). The inhibition of ζPKC prevents the activation of nuclear factor-κB (NF-κB) and the subsequent transcription of X-chromosome-linked inhibitor of apoptosis (XIAP) [151]. Moreover, cytoplasmic Par-4 is known to interact with Akt, preventing NF-κB activation and promoting apoptosis [152].

In order to enter the nucleus and perform its proapoptotic functions in this location, Par-4 needs to be cleaved. Cleavage of this protein is performed by caspase 3. The cleaved fragment of Par-4 contains the NLS2 domain, allowing its transportation to the nucleus, the SAC domain (which confers the proapoptotic function), and

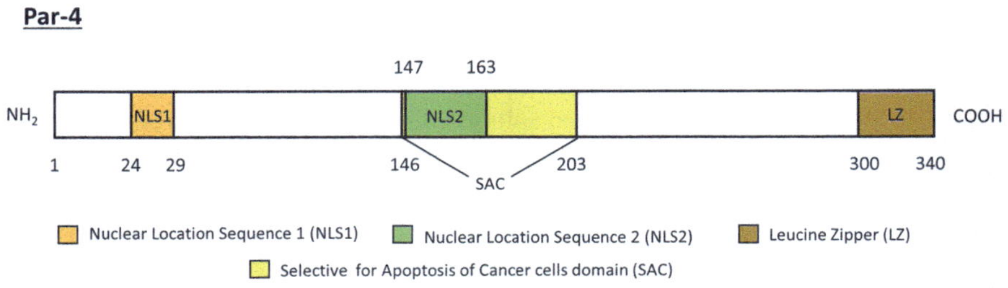

Fig. 6 Schematic representation of the Par-4 structure. At the N-terminal part of the protein (in orange) is represented the nuclear location sequence 1 (NLS1). The NLS2 is illustrated in green in the middle part of the region. In the same region (in yellow) is represented the selective for apoptosis of cancer cells (SAC) domain, which overlaps part of its sequence with NLS2. At the C-terminus is designed in light brown the leucine zipper (LZ) domain. The numbers under the Par-4 protein sequence describe the amino acids that mark the start and the end of the different domains

the leucine zipper domain at C-terminal [153]. Once the cleaved fragment of Par-4 has been transported into the nucleus, this protein is capable of interacting with different proapoptotic proteins, regulating several processes, and causing cell death [135, 154]. It was described that Par-4 interacts with the transcription factor Wilms tumor 1 (WT1) in the nucleus, repressing the transcription of the antiapoptotic gene B-cell lymphoma-2 (bcl-2) and thus promoting cell death [135]. The interaction of Par-4 with topoisomerase 1 (TOP1) leads to the sequestration of this protein and attenuates its DNA relaxation activity and its contribution to the full apoptosis response [154].

The cytoplasm and the nucleus are the classical locations of Par-4; however, several studies have recently demonstrated that this protein is also located in the ER and, importantly, has also been observed as a secreted protein [84, 132, 133]. The secretion of Par-4 has been largely investigated since it is involved in the activation of extrinsic apoptosis, reinforcing its proapoptotic role in cancer cells [84, 132, 133]. The secretion of Par-4 was surprising since it does not contain a signal peptide sequence that drives this protein to the secretion pathway [84]. Nevertheless, many other proteins that are secreted into the extracellular environment do not contain a discernible signal peptide sequence, suggesting the existence of unraveled signal peptide sequences. It has been hypothesized that Par-4 may contain atypical short signal peptide sequence(s) (with amino acid compositions that have not been elucidated yet) that could drive Par-4 into the secretory pathway [84].

Even if the primary signal peptide targeting Par-4 to the secretory pathway is not deciphered, the mechanisms leading to the secretion of this protein have been, at least partially, identified. The classical pathway involved in protein secretion starts with the translocation of the protein from the ER to the GA. The protein may suffer some posttranslational modifications in the GA before its transport to the extracellular matrix via vesicle trafficking [132, 84]. Firstly, Wang et al. discovered that Par-4 colocalized with calnexin (an ER marker) in the nephroblastoma cell line SK-NEP-1 [155]. This observation was confirmed in other cell types such as PC-3 and BPH-1 and suggested the possible trafficking of Par-4 to the extracellular matrix through the classical secretory pathway [84]. A study conducted by Burikhanov et al. demonstrated that treating PC-3 cells with brefeldin A (BFA)—a drug blocking the anterograde transport of proteins between the ER and the AG—led to a reduction in Par-4 secretion. These results confirmed that Par-4 secretion is conducted by the classical secretory pathway [84].

Par-4 secretion was observed in normal and cancer cells, but interestingly, its secretion was enhanced in hypoxic conditions [156]. The hypoxic environment is capable of triggering different processes in the cell, among them the ER stress response [157–159]. It was reported that two ER stress inducers, Thapsigargin and Tunicamycin, increased the Par-4 secretion in PC-3 cells. This study concluded that ER stress is associated with Par-4 secretion [84]. Moreover, ER stress response is known to increase the expression of chaperone proteins, among them GRP78 [30–35, 38, 42–45]. In the previously mentioned investigation, it was described that the increased Par-4 secretion caused by ER stress response induction was accompanied by an enhanced GRP78 expression [84].

We formerly discussed that GRP78 under stressful conditions is located at the cell surface, acting as a receptor of several extracellular proteins [63–69, 74, 81–87, 96–103]. The relocation of GRP78 [5, 16, 107, 120] and Par-4 [84, 129] was increased under ER stress conditions in cancer cells, increasing the interest of the potential relationship between these two proteins.

9 GRP78/Par-4 Interaction

The observations demonstrating an increased Par-4 secretion under ER stress conditions [84, 129]—paralleled to an increased GRP78 relocation to the cell surface [5, 16, 107, 120]—led to the hypothesis that these two proteins may interact in their new locations and activate some intracellular responses. GRP78 at the cell surface is capable of binding to different secreted proteins, such as α2M and CRIPTO [63–69, 74, 81–87, 96–103]. Burikhanov et al. decided to study the potential interaction of secreted Par-4 and cell surface located GRP78 in PC-3 cells. In this study, they used TNF-related apoptosis-inducing ligand (TRAIL) as an ER stress inducer. Under these conditions, it was shown that Par-4 binds to GRP78 at the cell membrane [84]. A more detailed study of the potential protein regions implicated in the

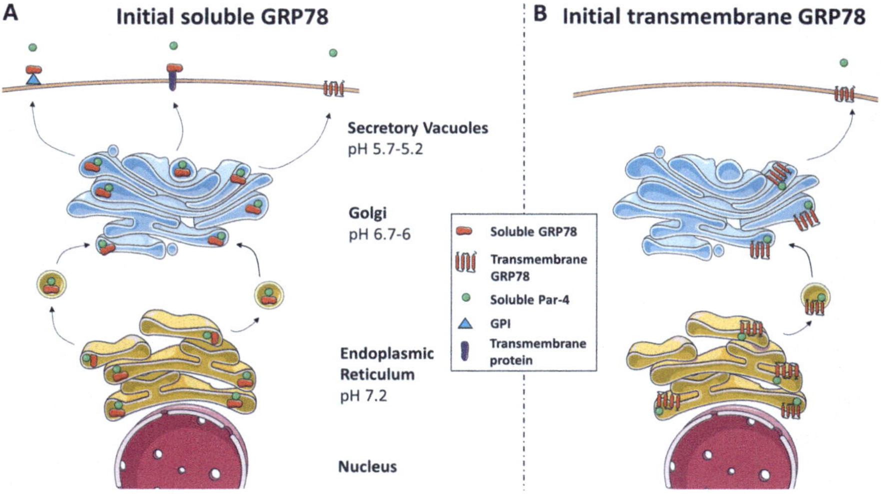

Fig. 7 Hypothesis about Par-4-dependent relocation of GRP78 to the cell surface depending on GRP78 localization in ER. (**a**) GRP78 in the lumen of ER. GRP78 and Par-4 are located in the lumen of the ER. GRP78/Par-4 complex is transported to the GA. Following the classical secretion pathway, Par-4 and GRP78 get into secretory vacuoles that transport them to the cell surface. GRP78 could be expressed at the cell membrane bound to a glycosylphosphatidylinositol (GPI) molecule or another transmembrane protein. GRP78 might also be inserted at the cell membrane and the insertion is probably achieved during the transportation due to the pH level decrease. (**b**) Transmembrane GRP78. GRP78 is inserted in the membrane of the ER, and Par-4 is bound to it in the luminal region. Following the classical secretory pathway, both proteins are co-transported to the AG, and from there to the cell membrane by vesicles. GRP78 is expressed at the cell surface as a transmembrane protein and Par-4 is secreted

interaction revealed that the SAC domain of Par-4 and the N-terminal region of GRP78 are directly involved in this binding. The intracellular effects generated by this interaction led to caspase-8 and -3 activation. The caspases activation was FADD-dependent, and the final cellular outcome was apoptosis as previously described in Fig. 3a [84]. In this way, this investigation highlighted a novel manner by which Par-4 is implicated in activating apoptosis in cancer cells where the translocation of GRP78 to the cell surface is high.

Additionally, the same study demonstrated that the interaction between cell surface located GRP78 and Par-4 produces a robust ER stress response involving GRP78 upregulation. Moreover, ER stress activation led to PERK activation, one of the main components of the ER stress response, and this protein was implicated in the caspase-8 activation, reinforcing the apoptotic signaling. Finally, the Par-4/GRP78 interaction at the cell surface increased the trafficking of GRP78 to the cell surface, enhancing the loop between ER stress, GRP78 at the cell surface, and apoptosis [84].

The ability of GRP78 and Par-4 to interact at the cell surface [84], together with the already mentioned effects of Par-4 silencing on GRP78 relocation to the cell membrane [74, 84, 85], reinforced the hypothesis suggesting that Par-4 may serve as an important client protein involved in GRP78 translocation to the cell membrane.

The missing link at this point was a demonstration of Par-4 interacting with GRP78 in the ER and GA, which would allow the transport of the two proteins at the same time to the cell surface.

Burikhanov et al. proved that Par-4 and GRP78 co-localize intracellularly in PC-3 cells and BPH-1 cells [84]. Moreover, Meynier et al. validated these results in SKOV-3 cells, indicating that the colocalization of these two proteins was taking place in the cytoplasm, mainly in the perinuclear region [85]. Finally, another study on the intracellular location of GRP78 and Par-4 proteins led to the observation of the colocalization of these two proteins with calnexin: an ER marker [155]. This final observation conducted the validation of the interaction of GRP78 and Par-4 in the ER, strengthening the hypothesis that Par-4 serves as a client protein, supporting the GRP78 translocation to the cell surface.

The domains and exact amino acids that serve for Par-4 and GRP78 interaction in the ER have not been deciphered yet. However, considering all the information available at this moment, a probable hypothesis on GRP78 translocation to the cell surface via Par-4 client protein has been proposed. Soluble GRP78 may be residing in the lumen of the ER, folded in a manner that several of its binding sequences are accessible, allowing the binding of GRP78 with other proteins. On the other hand, Par-4 might be located in its correctly folded conformation in the lumen of the ER. As it was demonstrated at the cell surface, the SAC domain of Par-4 and the N-terminal region of GRP78 could interact [160], leading to Par-4 acting as a client protein used by the cell to relocate GRP78 to the cell surface. Once Par-4 and GRP78 are bound, the signal peptide of Par-4 suggested to be present in the protein sequence [84] might be recognized by the secretory pathway, leading to Par-4 secretion. Consequently, as GRP78 may be bound to Par-4 [84, 85, 155], the transportation of Par-4 might take place together. It is known that Par-4 secretion is performed through the classical pathway, therefore, Par-4 and GRP78 should be firstly co-transported to the GA [84, 132]. Some unpublished results obtained in our lab confirmed that blockage of ER-GA transport with Brefeldin A or monensin decreased the transport of GRP78 to the cell membrane in Par-4 overexpressing SKOV-3 and HIPEC cells.

The relocation of GRP78 at the cell surface can take place in three different ways as was previously mentioned: it could be associated with a GPI molecule [107], associated with a transmembrane protein [107], or directly inserted in the cell membrane as a 4-pass transmembrane protein [5, 107, 108]. In the case of GRP78 complexed with GPI or transmembrane proteins, we would expect that it exits the GA already complexed with them [161]. In the case of GRP78 present at the cell surface as a transmembrane protein, it means that it should be already inserted in the bilipid layer before its arrival to the cell surface [162]. The transport of the Par-4/GRP78 complex would take place via the secretory pathway, and at some point during the transport of these proteins to the cell surface, GRP78 might be inserted into the Golgi membrane or the final secretory vesicle membrane. One of the variables that could promote the insertion of GRP78 to the membrane—as it happens with other proteins [163]—is the pH. It has been reported that the lumen content and pH of the ER, Golgi, and secretory vesicles are not the same. In fact, it is known that while the secretory pathway advances, the pH acidifies, with a maximum value of 7.2 at the ER lumen and a minimum of 5.7–5.2 at the secretory vesicles. Meanwhile, the Golgi pH shows intermediate values varying from 6.7 at the ER proximal part to 6.0 at the ER most distal part [164]. The exposure to different pH concentrations might promote the GRP78 insertion into the Golgi membrane or vesicle membrane during its transport to the cell surface together with Par-4. This hypothesis seems less plausible for the following reasons. Firstly, the insertion of GRP78 should take place while it is still bound to Par-4, complicating the process. Another option might be that GRP78 separates from Par-4 to get inserted into the cell membrane, and in this way, Par-4 is already free to be secreted and GRP78 is in its transmembrane form (Fig. 7a).

Another hypothesis explaining the transmembrane GRP78 relocation from the ER to the cell surface has been proposed. It has been reported that GRP78 could be located as a transmembrane protein in the ER [108]. The structure of the cell membrane inserted GRP78 is a 4-pass transmembrane protein, with the N-terminal and C-terminal sides of the protein at the extracellular side [5]. Reddy et al. previously described the transmembrane conformation of GRP78 at the ER as a multiple-pass transmembrane protein, with the C-terminal at the ER lumen [108]. With this information, we suppose that the conformation of GRP78 at the cell membrane and the membrane of the ER might be the same: a 4-pass transmembrane protein with the terminal sides in the lumen of the ER [5108]. In this way, Par-4 could bind to the N-terminal side of GRP78 through its SAC domain [160], allowing the transport of GRP78 to the GA through the secretory pathway [84, 107]. Consecutively, Par-4 together with GRP78 may be transported to the cell surface through vesicles, allowing the secretion of Par-4 and the relocation of GRP78 to the cell surface inserted in the membrane (Fig. 7b).

Finally, Lee suggested that cytoplasmic Par-4 may interact with GRP78 through transmembrane ER protein complexes [129]. This hypothesis could partially explain the transport of GRP78 from the ER to the GA. However, Par-4 has to be in the lumen of the GA to be secreted and not in the cytoplasmic side. The observation that GRP78 relocation to the cell surface was accompanied by an increment in Par-4 secretion [74, 84, 85] would not fit this hypothesis, making it an unplausible option.

The exact mechanism leading to GRP78 relocation to cell surface has not been completely unraveled; however, the information that has been collected during the past years start to fit, allowing the formulation of a possible hypothesis. It seems that Par-4 is a client protein that facilitates GRP78 translocation to the cell surface by forming a protein complex that is transported altogether. The observation that an increased Par-4 secretion was accompanied by an increment of cell surface-localized GRP78 reinforced this hypothesis [74, 84, 85]. The co-localization of both proteins in the ER [84, 155] and the discovery of the implication of the classical secretion pathway in the relocalization of these proteins [84, 107] partially clarified the mechanism used by the cell to transport these proteins. More studies are needed to clarify the exact mechanism leading to GRP78 relocation.

The secretion of Par-4 and relocation of GRP78 are enhanced under ER stress conditions [5, 16, 71, 84, 107, 120, 129]. It is interesting to observe how the cell promotes different mechanisms to self-destroy the cell under stressful circumstances. In cancer cells, the ER stress is known to be overactivated [165–171], demonstrating the importance of GRP78 translocation and Par-4 secretion and their potential interaction at the cell surface to promote apoptosis and cell death in these abnormal cells [160, 84, 85].

References

1. Haas IG, Wabl M (1983) Immunoglobulin heavy chain binding protein. Nature 306(5941):387–389. https://doi.org/10.1038/306387a0
2. Munro S, Pelham HR (1986) An Hsp70-like protein in the ER: identity with the 78 kd glucose-regulated protein and immunoglobulin heavy chain binding protein. Cell 46(2):291–300. https://doi.org/10.1016/0092-8674(86)90746-4
3. Wang J, Lee J, Liem D, Ping P (2017) HSPA5 gene encoding Hsp70 chaperone BiP in the endoplasmic reticulum. Gene 618:14–23. https://doi.org/10.1016/j.gene.2017.03.005
4. Radons J (2016) The human HSP70 family of chaperones: where do we stand? Cell Stress Chaperones 21(3):379–404. https://doi.org/10.1007/s12192-016-0676-6
5. Zhang Y, Liu R, Ni M, Gill P, Lee AS (2010) Cell surface relocalization of the endoplasmic reticulum chaperone and unfolded protein response regulator GRP78/BiP. J Biol Chem 285(20):15065–15075. https://doi.org/10.1074/jbc.M109.087445
6. Mayer MP (2013) Hsp70 chaperone dynamics and molecular mechanism. Trends Biochem Sci 38(10):507–514. https://doi.org/10.1016/j.tibs.2013.08.001
7. Yang J, Nune M, Zong Y, Zhou L, Liu Q (2015) Close and allosteric opening of the polypeptide-binding site in a human Hsp70 chaperone BiP. Structure 23(12):2191–2203. https://doi.org/10.1016/j.str.2015.10.012

8. Gurusinghe K, Mishra A, Mishra S (2018) Glucose-regulated protein 78 substrate-binding domain alters its conformation upon EGCG inhibitor binding to nucleotide-binding domain: molecular dynamics studies. Sci Rep 8(1):5487. https://doi.org/10.1038/s41598-018-22905-6
9. Gonzalez-Gronow M, Selim MA, Papalas J, Pizzo SV (2009) GRP78: a multifunctional receptor on the cell surface. Antioxid Redox Signal 11(9):2299–2306. https://doi.org/10.1089/ARS.2009.2568
10. Quinones QJ, de Ridder GG, Pizzo SV (2008) GRP78: a chaperone with diverse roles beyond the endoplasmic reticulum. Histol Histopathol 23(11):1409–1416. https://doi.org/10.14670/HH-23.1409
11. Berger CL, Dong Z, Hanlon D, Bisaccia E, Edelson RL (1997) A lymphocyte cell surface heat shock protein homologous to the endoplasmic reticulum chaperone, immunoglobulin heavy chain binding protein BIP. Int J Cancer 71(6):1077–1085. https://doi.org/10.1002/(sici)1097-0215(19970611)71:6<1077::aid-ijc26>3.0.co;2-9
12. Kang BR, Yang SH, Chung BR, Kim W, Kim Y (2016) Cell surface GRP78 as a biomarker and target for suppressing glioma cells. Sci Rep 6:34922. https://doi.org/10.1038/srep34922
13. Lenin R, Nagy PG, Jha KA, Gangaraju R (2019) GRP78 translocation to the cell surface and O-GlcNAcylation of VE-Cadherin contribute to ER stress-mediated endothelial permeability. Sci Rep 9(1):10783. https://doi.org/10.1038/s41598-019-47246-w
14. Ni M, Zhou H, Wey S, Baumeister P, Lee AS (2009) Regulation of PERK signaling and leukemic cell survival by a novel cytosolic isoform of the UPR regulator GRP78/BiP. PLoS One 4(8):e6868. https://doi.org/10.1371/journal.pone.0006868
15. Casas C (2017) GRP78 at the Centre of the Stage in Cancer and Neuroprotection. Front Neurosci 11:177. https://doi.org/10.3389/fnins.2017.00177
16. Ni M, Zhang Y, Lee AS (2011) Beyond the endoplasmic reticulum: atypical GRP78 in cell viability, signalling and therapeutic targeting. Biochem J 434(2):181–188. https://doi.org/10.1042/BJ20101569
17. Rao RV, Peel A, Logvinova A, del Rio G, Hermel E, Yokota T, Goldsmith PC, Ellerby LM, Ellerby HM, Bredesen DE (2002) Coupling endoplasmic reticulum stress to the cell death program: role of the ER chaperone GRP78. FEBS Lett 514(2–3):122–128. https://doi.org/10.1016/s0014-5793(02)02289-5
18. Matsumoto A, Hanawalt PC (2000) Histone H3 and heat shock protein GRP78 are selectively cross-linked to DNA by photoactivated gilvocarcin V in human fibroblasts. Cancer Res 60(14):3921–3926
19. Prasad M, Pawlak KJ, Burak WE, Perry EE, Marshall B, Whittal RM, Bose HS (2017) Mitochondrial metabolic regulation by GRP78. Sci Adv 3(2):e1602038. https://doi.org/10.1126/sciadv.1602038
20. Sun FC, Wei S, Li CW, Chang YS, Chao CC, Lai YK (2006) Localization of GRP78 to mitochondria under the unfolded protein response. Biochem J 396(1):31–39. https://doi.org/10.1042/BJ20051916
21. Teng Y, Ren X, Li H, Shull A, Kim J, Cowell JK (2016) Mitochondrial ATAD3A combines with GRP78 to regulate the WASF3 metastasis-promoting protein. Oncogene 35(3):333–343. https://doi.org/10.1038/onc.2015.86
22. Takemoto H, Yoshimori T, Yamamoto A, Miyata Y, Yahara I, Inoue K, Tashiro Y (1992) Heavy chain binding protein (BiP/GRP78) and endoplasmin are exported from the endoplasmic reticulum in rat exocrine pancreatic cells, similar to protein disulfide-isomerase. Arch Biochem Biophys 296(1):129–136. https://doi.org/10.1016/0003-9861(92)90554-a
23. Kern J, Untergasser G, Zenzmaier C, Sarg B, Gastl G, Gunsilius E, Steurer M (2009) GRP-78 secreted by tumor cells blocks the antiangiogenic activity of bortezomib. Blood 114(18):3960–3967. https://doi.org/10.1182/blood-2009-03-209668
24. Marin-Briggiler CI, Gonzalez-Echeverria MF, Munuce MJ, Ghersevich S, Caille AM, Hellman U, Corrigall VM, Vazquez-Levin MH (2010) Glucose-regulated protein 78 (Grp78/BiP) is secreted by human oviduct epithelial cells and the recombinant protein modulates

sperm-zona pellucida binding. Fertil Steril 93(5):1574–1584. https://doi.org/10.1016/j.fertnstert.2008.12.132
25. Wang M, Wey S, Zhang Y, Ye R, Lee AS (2009) Role of the unfolded protein response regulator GRP78/BiP in development, cancer, and neurological disorders. Antioxid Redox Signal 11(9):2307–2316. https://doi.org/10.1089/ARS.2009.2485
26. Ni M, Lee AS (2007) ER chaperones in mammalian development and human diseases. FEBS Lett 581(19):3641–3651. https://doi.org/10.1016/j.febslet.2007.04.045
27. Hendershot LM (2004) The ER function BiP is a master regulator of ER function. Mt Sinai J Med 71(5):289–297
28. Lee AS (2005) The ER chaperone and signaling regulator GRP78/BiP as a monitor of endoplasmic reticulum stress. Methods 35(4):373–381. https://doi.org/10.1016/j.ymeth.2004.10.010
29. Li J, Lee AS (2006) Stress induction of GRP78/BiP and its role in cancer. Curr Mol Med 6(1):45–54. https://doi.org/10.2174/156652406775574523
30. Ma Y, Hendershot LM (2004) ER chaperone functions during normal and stress conditions. J Chem Neuroanat 28(1–2):51–65. https://doi.org/10.1016/j.jchemneu.2003.08.007
31. Chakrabarti A, Chen AW, Varner JD (2011) A review of the mammalian unfolded protein response. Biotechnol Bioeng 108(12):2777–2793. https://doi.org/10.1002/bit.23282
32. Hetz C (2012) The unfolded protein response: controlling cell fate decisions under ER stress and beyond. Nat Rev Mol Cell Biol 13(2):89–102. https://doi.org/10.1038/nrm3270
33. Hwang J, Qi L (2018) Quality control in the endoplasmic reticulum: crosstalk between ERAD and UPR pathways. Trends Biochem Sci 43(8):593–605. https://doi.org/10.1016/j.tibs.2018.06.005
34. Sano R, Reed JC (2013) ER stress-induced cell death mechanisms. Biochim Biophys Acta 1833(12):3460–3470. https://doi.org/10.1016/j.bbamcr.2013.06.028
35. Novoa I, Zeng H, Harding HP, Ron D (2001) Feedback inhibition of the unfolded protein response by GADD34-mediated dephosphorylation of eIF2alpha. J Cell Biol 153(5):1011–1022. https://doi.org/10.1083/jcb.153.5.1011
36. Bastida-Ruiz D, Aguilar E, Ditisheim A, Yart L, Cohen M (2017) Endoplasmic reticulum stress responses in placentation - A true balancing act. Placenta 57:163–169. https://doi.org/10.1016/j.placenta.2017.07.004
37. Yoshida H, Matsui T, Yamamoto A, Okada T, Mori K (2001) XBP1 mRNA is induced by ATF6 and spliced by IRE1 in response to ER stress to produce a highly active transcription factor. Cell 107(7):881–891. https://doi.org/10.1016/s0092-8674(01)00611-0
38. Ogata M, Hino S, Saito A, Morikawa K, Kondo S, Kanemoto S, Murakami T, Taniguchi M, Tanii I, Yoshinaga K, Shiosaka S, Hammarback JA, Urano F, Imaizumi K (2006) Autophagy is activated for cell survival after endoplasmic reticulum stress. Mol Cell Biol 26(24):9220–9231. https://doi.org/10.1128/MCB.01453-06
39. Bailey D, O'Hare P (2007) Transmembrane bZIP transcription factors in ER stress signaling and the unfolded protein response. Antioxid Redox Signal 9(12):2305–2321. https://doi.org/10.1089/ars.2007.1796
40. Cullinan SB, Diehl JA (2004) PERK-dependent activation of Nrf2 contributes to redox homeostasis and cell survival following endoplasmic reticulum stress. J Biol Chem 279(19):20108–20117. https://doi.org/10.1074/jbc.M314219200
41. Senft D, Ronai ZA (2015) UPR, autophagy, and mitochondria crosstalk underlies the ER stress response. Trends Biochem Sci 40(3):141–148. https://doi.org/10.1016/j.tibs.2015.01.002
42. Nishitoh H, Matsuzawa A, Tobiume K, Saegusa K, Takeda K, Inoue K, Hori S, Kakizuka A, Ichijo H (2002) ASK1 is essential for endoplasmic reticulum stress-induced neuronal cell death triggered by expanded polyglutamine repeats. Genes Dev 16(11):1345–1355. https://doi.org/10.1101/gad.992302
43. Puthalakath H, O'Reilly LA, Gunn P, Lee L, Kelly PN, Huntington ND, Hughes PD, Michalak EM, McKimm-Breschkin J, Motoyama N, Gotoh T, Akira S, Bouillet P, Strasser A (2007) ER

stress triggers apoptosis by activating BH3-only protein Bim. Cell 129(7):1337–1349. https://doi.org/10.1016/j.cell.2007.04.027

44. Li G, Mongillo M, Chin KT, Harding H, Ron D, Marks AR, Tabas I (2009) Role of ERO1-alpha-mediated stimulation of inositol 1,4,5-triphosphate receptor activity in endoplasmic reticulum stress-induced apoptosis. J Cell Biol 186(6):783–792. https://doi.org/10.1083/jcb.200904060
45. Ott M, Robertson JD, Gogvadze V, Zhivotovsky B, Orrenius S (2002) Cytochrome c release from mitochondria proceeds by a two-step process. Proc Natl Acad Sci U S A 99(3):1259–1263. https://doi.org/10.1073/pnas.241655498
46. Rao RV, Hermel E, Castro-Obregon S, del Rio G, Ellerby LM, Ellerby HM, Bredesen DE (2001) Coupling endoplasmic reticulum stress to the cell death program. Mechanism of caspase activation. J Biol Chem 276(36):33869–33874. https://doi.org/10.1074/jbc.M102225200
47. Wang M, Ye R, Barron E, Baumeister P, Mao C, Luo S, Fu Y, Luo B, Dubeau L, Hinton DR, Lee AS (2010) Essential role of the unfolded protein response regulator GRP78/BiP in protection from neuronal apoptosis. Cell Death Differ 17(3):488–498. https://doi.org/10.1038/cdd.2009.144
48. Fernandez PM, Tabbara SO, Jacobs LK, Manning FC, Tsangaris TN, Schwartz AM, Kennedy KA, Patierno SR (2000) Overexpression of the glucose-regulated stress gene GRP78 in malignant but not benign human breast lesions. Breast Cancer Res Treat 59(1):15–26. https://doi.org/10.1023/a:1006332011207
49. Scriven P, Coulson S, Haines R, Balasubramanian S, Cross S, Wyld L (2009) Activation and clinical significance of the unfolded protein response in breast cancer. Br J Cancer 101(10):1692–1698. https://doi.org/10.1038/sj.bjc.6605365
50. Cook KL, Clarke R (2015) Role of GRP78 in promoting therapeutic-resistant breast cancer. Future Med Chem 7(12):1529–1534. https://doi.org/10.4155/FMC.15.80
51. Madden E, Logue SE, Healy SJ, Manie S, Samali A (2019) The role of the unfolded protein response in cancer progression: from oncogenesis to chemoresistance. Biol Cell 111(1):1–17. https://doi.org/10.1111/boc.201800050
52. Laverriere A, Landau R, Charvet I, Irion O, Bischof P, Morales M, Cohen M (2009) GRP78 as a marker of pre-eclampsia: an exploratory study. Mol Hum Reprod 15(9):569–574. https://doi.org/10.1093/molehr/gap037
53. Shu CW, Sun FC, Cho JH, Lin CC, Liu PF, Chen PY, Chang MD, Fu HW, Lai YK (2008) GRP78 and Raf-1 cooperatively confer resistance to endoplasmic reticulum stress-induced apoptosis. J Cell Physiol 215(3):627–635. https://doi.org/10.1002/jcp.21340
54. Ye R, Jung DY, Jun JY, Li J, Luo S, Ko HJ, Kim JK, Lee AS (2010) Grp78 heterozygosity promotes adaptive unfolded protein response and attenuates diet-induced obesity and insulin resistance. Diabetes 59(1):6–16. https://doi.org/10.2337/db09-0755
55. Barker S, Weinfeld M, Zheng J, Li L, Murray D (2005) Identification of mammalian proteins cross-linked to DNA by ionizing radiation. J Biol Chem 280(40):33826–33838. https://doi.org/10.1074/jbc.M502477200
56. Huang SP, Chen JC, Wu CC, Chen CT, Tang NY, Ho YT, Lo C, Lin JP, Chung JG, Lin JG (2009) Capsaicin-induced apoptosis in human hepatoma HepG2 cells. Anticancer Res 29(1):165–174
57. La X, Zhang L, Yang Y, Li H, Song G, Li Z (2019) Tumor-secreted GRP78 facilitates the migration of macrophages into tumors by promoting cytoskeleton remodeling. Cell Signal 60:1–16
58. Xiao G, Chung TF, Pyun HY, Fine RE, Johnson RJ (1999) KDEL proteins are found on the surface of NG108-15 cells. Brain Res Mol Brain Res 72(2):121–128. https://doi.org/10.1016/s0169-328x(99)00188-6
59. Shin BK, Wang H, Yim AM, Le Naour F, Brichory F, Jang JH, Zhao R, Puravs E, Tra J, Michael CW, Misek DE, Hanash SM (2003) Global profiling of the cell surface proteome of cancer cells uncovers an abundance of proteins with chaperone function. J Biol Chem 278(9):7607–7616. https://doi.org/10.1074/jbc.M210455200

60. Liu Y, Steiniger SC, Kim Y, Kaufmann GF, Felding-Habermann B, Janda KD (2007) Mechanistic studies of a peptidic GRP78 ligand for cancer cell-specific drug delivery. Mol Pharm 4(3):435–447. https://doi.org/10.1021/mp060122j
61. Jakobsen CG, Rasmussen N, Laenkholm AV, Ditzel HJ (2007) Phage display derived human monoclonal antibodies isolated by binding to the surface of live primary breast cancer cells recognize GRP78. Cancer Res 67(19):9507–9517. https://doi.org/10.1158/0008-5472.CAN-06-4686
62. Rauschert N, Brandlein S, Holzinger E, Hensel F, Muller-Hermelink HK, Vollmers HP (2008) A new tumor-specific variant of GRP78 as target for antibody-based therapy. Lab Investig 88(4):375–386. https://doi.org/10.1038/labinvest.2008.2
63. Misra UK, Gonzalez-Gronow M, Gawdi G, Wang F, Pizzo SV (2004) A novel receptor function for the heat shock protein Grp78: silencing of Grp78 gene expression attenuates alpha2M*-induced signalling. Cell Signal 16(8):929–938. https://doi.org/10.1016/j.cellsig.2004.01.003
64. Misra UK, Deedwania R, Pizzo SV (2005) Binding of activated alpha2-macroglobulin to its cell surface receptor GRP78 in 1-LN prostate cancer cells regulates PAK-2-dependent activation of LIMK. J Biol Chem 280(28):26278–26286. https://doi.org/10.1074/jbc.M414467200
65. Gonzalez-Gronow M, Cuchacovich M, Llanos C, Urzua C, Gawdi G, Pizzo SV (2006) Prostate cancer cell proliferation in vitro is modulated by antibodies against glucose-regulated protein 78 isolated from patient serum. Cancer Res 66(23):11424–11431. https://doi.org/10.1158/0008-5472.CAN-06-1721
66. Misra UK, Deedwania R, Pizzo SV (2006) Activation and cross-talk between Akt, NF-kappaB, and unfolded protein response signaling in 1-LN prostate cancer cells consequent to ligation of cell surface-associated GRP78. J Biol Chem 281(19):13694–13707. https://doi.org/10.1074/jbc.M511694200
67. Al-Hashimi AA, Caldwell J, Gonzalez-Gronow M, Pizzo SV, Aboumrad D, Pozza L, Al-Bayati H, Weitz JI, Stafford A, Chan H, Kapoor A, Jacobsen DW, Dickhout JG, Austin RC (2010) Binding of anti-GRP78 autoantibodies to cell surface GRP78 increases tissue factor procoagulant activity via the release of calcium from endoplasmic reticulum stores. J Biol Chem 285(37):28912–28923. https://doi.org/10.1074/jbc.M110.119107
68. Shani G, Fischer WH, Justice NJ, Kelber JA, Vale W, Gray PC (2008) GRP78 and Cripto form a complex at the cell surface and collaborate to inhibit transforming growth factor beta signaling and enhance cell growth. Mol Cell Biol 28(2):666–677. https://doi.org/10.1128/MCB.01716-07
69. Kelber JA, Panopoulos AD, Shani G, Booker EC, Belmonte JC, Vale WW, Gray PC (2009) Blockade of Cripto binding to cell surface GRP78 inhibits oncogenic Cripto signaling via MAPK/PI3K and Smad2/3 pathways. Oncogene 28(24):2324–2336. https://doi.org/10.1038/onc.2009.97
70. Liu C, Bhattacharjee G, Boisvert W, Dilley R, Edgington T (2003) In vivo interrogation of the molecular display of atherosclerotic lesion surfaces. Am J Pathol 163(5):1859–1871. https://doi.org/10.1016/S0002-9440(10)63545-6
71. Arap MA, Lahdenranta J, Mintz PJ, Hajitou A, Sarkis AS, Arap W, Pasqualini R (2004) Cell surface expression of the stress response chaperone GRP78 enables tumor targeting by circulating ligands. Cancer Cell 6(3):275–284. https://doi.org/10.1016/j.ccr.2004.08.018
72. Arnaudeau S, Arboit P, Bischof P, Shin-ya K, Tomida A, Tsuruo T, Irion O, Cohen M (2009) Glucose-regulated protein 78: a new partner of p53 in trophoblast. Proteomics 9(23):5316–5327. https://doi.org/10.1002/pmic.200800865
73. Fradet S, Pierredon S, Ribaux P, Epiney M, Shin Ya K, Irion O, Cohen M (2012) Involvement of membrane GRP78 in trophoblastic cell fusion. PLoS One 7(8):e40596. https://doi.org/10.1371/journal.pone.0040596
74. Cohen M, Ribaux P, Epiney M, Irion O (2013) Role of prostate apoptosis response 4 in translocation of GRP78 from the endoplasmic reticulum to the cell surface of trophoblastic cells. PLoS One 8(11):e80231. https://doi.org/10.1371/journal.pone.0080231

75. Triantafilou K, Fradelizi D, Wilson K, Triantafilou M (2002) GRP78, a coreceptor for coxsackievirus A9, interacts with major histocompatibility complex class I molecules which mediate virus internalization. J Virol 76(2):633–643. https://doi.org/10.1128/jvi.76.2.633-643.2002
76. Jindadamrongwech S, Thepparit C, Smith DR (2004) Identification of GRP 78 (BiP) as a liver cell expressed receptor element for dengue virus serotype 2. Arch Virol 149(5):915–927. https://doi.org/10.1007/s00705-003-0263-x
77. Honda T, Horie M, Daito T, Ikuta K, Tomonaga K (2009) Molecular chaperone BiP interacts with Borna disease virus glycoprotein at the cell surface. J Virol 83(23):12622–12625. https://doi.org/10.1128/JVI.01201-09
78. Zhao D, Liu Q, Han K, Wang H, Yang J, Bi K, Liu Y, Liu N, Tian Y, Li Y (2018) Identification of glucose-regulated protein 78 (GRP78) as a receptor in BHK-21 cells for duck Tembusu virus infection. Front Microbiol 9:694. https://doi.org/10.3389/fmicb.2018.00694
79. Ibrahim IM, Abdelmalek DH, Elshahat ME, Elfiky AA (2020) COVID-19 spike-host cell receptor GRP78 binding site prediction. J Infect 80(5):554–562. https://doi.org/10.1016/j.jinf.2020.02.026
80. Bastida-Ruiz D, Wuillemin C, Pederencino A, Yaron M, Martinez de Tejada B, Pizzo SV, Cohen M (2020) Activated alpha2-macroglobulin binding to cell surface GRP78 induces trophoblastic cell fusion. Sci Rep 10(1):9666. https://doi.org/10.1038/s41598-020-66554-0
81. Davidson DJ, Haskell C, Majest S, Kherzai A, Egan DA, Walter KA, Schneider A, Gubbins EF, Solomon L, Chen Z, Lesniewski R, Henkin J (2005) Kringle 5 of human plasminogen induces apoptosis of endothelial and tumor cells through surface-expressed glucose-regulated protein 78. Cancer Res 65(11):4663–4672. https://doi.org/10.1158/0008-5472.CAN-04-3426
82. McFarland BC, Stewart J Jr, Hamza A, Nordal R, Davidson DJ, Henkin J, Gladson CL (2009) Plasminogen kringle 5 induces apoptosis of brain microvessel endothelial cells: sensitization by radiation and requirement for GRP78 and LRP1. Cancer Res 69(13):5537–5545. https://doi.org/10.1158/0008-5472.CAN-08-4841
83. Vig S, Buitinga M, Rondas D, Crevecoeur I, van Zandvoort M, Waelkens E, Eizirik DL, Gysemans C, Baatsen P, Mathieu C, Overbergh L (2019) Cytokine-induced translocation of GRP78 to the plasma membrane triggers a pro-apoptotic feedback loop in pancreatic beta cells. Cell Death Dis 10(4):309. https://doi.org/10.1038/s41419-019-1518-0
84. Burikhanov R, Zhao Y, Goswami A, Qiu S, Schwarze SR, Rangnekar VM (2009) The tumor suppressor Par-4 activates an extrinsic pathway for apoptosis. Cell 138(2):377–388. https://doi.org/10.1016/j.cell.2009.05.022
85. Meynier S, Kramer M, Ribaux P, Tille JC, Delie F, Petignat P, Cohen M (2015) Role of PAR-4 in ovarian cancer. Oncotarget 6(26):22641–22652. https://doi.org/10.18632/oncotarget.4010
86. Chen M, Qiu T, Wu J, Yang Y, Wright GD, Wu M, Ge R (2018) Extracellular anti-angiogenic proteins augment an endosomal protein trafficking pathway to reach mitochondria and execute apoptosis in HUVECs. Cell Death Differ 25(11):1905–1920. https://doi.org/10.1038/s41418-018-0092-9
87. Chen M, Zhang Y, Yu VC, Chong YS, Yoshioka T, Ge R (2014) Isthmin targets cell-surface GRP78 and triggers apoptosis via induction of mitochondrial dysfunction. Cell Death Differ 21(5):797–810. https://doi.org/10.1038/cdd.2014.3
88. Hensel F, Eckstein M, Rosenwald A, Brandlein S (2013) Early development of PAT-SM6 for the treatment of melanoma. Melanoma Res 23(4):264–275. https://doi.org/10.1097/CMR.0b013e328362cbc8
89. Rasche L, Duell J, Morgner C, Chatterjee M, Hensel F, Rosenwald A, Einsele H, Topp MS, Brandlein S (2013) The natural human IgM antibody PAT-SM6 induces apoptosis in primary human multiple myeloma cells by targeting heat shock protein GRP78. PLoS One 8(5):e63414. https://doi.org/10.1371/journal.pone.0063414
90. Liu R, Li X, Gao W, Zhou Y, Wey S, Mitra SK, Krasnoperov V, Dong D, Liu S, Li D, Zhu G, Louie S, Conti PS, Li Z, Lee AS, Gill PS (2013) Monoclonal antibody against cell surface

GRP78 as a novel agent in suppressing PI3K/AKT signaling, tumor growth, and metastasis. Clin Cancer Res 19(24):6802–6811. https://doi.org/10.1158/1078-0432.CCR-13-1106
91. Lee AS (2014) Glucose-regulated proteins in cancer: molecular mechanisms and therapeutic potential. Nat Rev Cancer 14(4):263–276. https://doi.org/10.1038/nrc3701
92. de Ridder GG, Ray R, Pizzo SV (2012) A murine monoclonal antibody directed against the carboxyl-terminal domain of GRP78 suppresses melanoma growth in mice. Melanoma Res 22(3):225–235. https://doi.org/10.1097/CMR.0b013e32835312fd
93. Mo L, Bachelder RE, Kennedy M, Chen PH, Chi JT, Berchuck A, Cianciolo G, Pizzo SV (2015) Syngeneic murine ovarian Cancer model reveals that ascites enriches for ovarian Cancer stem-like cells expressing membrane GRP78. Mol Cancer Ther 14(3):747–756. https://doi.org/10.1158/1535-7163.MCT-14-0579
94. Ge R, Kao C (2019) Cell surface GRP78 as a death receptor and an anticancer drug target. Cancers (Basel) 11(11). https://doi.org/10.3390/cancers11111787
95. Misra UK, Mowery Y, Kaczowka S, Pizzo SV (2009) Ligation of cancer cell surface GRP78 with antibodies directed against its COOH-terminal domain up-regulates p53 activity and promotes apoptosis. Mol Cancer Ther 8(5):1350–1362. https://doi.org/10.1158/1535-7163.MCT-08-0990
96. Al-Hashimi AA, Lebeau P, Majeed F, Polena E, Lhotak S, Collins CAF, Pinthus JH, Gonzalez-Gronow M, Hoogenes J, Pizzo SV, Crowther M, Kapoor A, Rak J, Gyulay G, D'Angelo S, Marchio S, Pasqualini R, Arap W, Shayegan B, Austin RC (2017) Autoantibodies against the cell surface-associated chaperone GRP78 stimulate tumor growth via tissue factor. J Biol Chem 292(51):21180–21192. https://doi.org/10.1074/jbc.M117.799908
97. Crane ED, Al-Hashimi AA, Chen J, Lynn EG, Won KD, Lhotak S, Naeim M, Platko K, Lebeau P, Byun JH, Shayegan B, Krepinsky JC, Rayner KJ, Marchio S, Pasqualini R, Arap W, Austin RC (2018) Anti-GRP78 autoantibodies induce endothelial cell activation and accelerate the development of atherosclerotic lesions. JCI Insight 3(24). https://doi.org/10.1172/jci.insight.99363
98. Dadey DYA, Kapoor V, Hoye K, Khudanyan A, Collins A, Thotala D, Hallahan DE (2017) Antibody targeting GRP78 enhances the efficacy of radiation therapy in human glioblastoma and non-small cell lung Cancer cell lines and tumor models. Clin Cancer Res 23(10):2556–2564. https://doi.org/10.1158/1078-0432.CCR-16-1935
99. Li Z, Zhang L, Zhao Y, Li H, Xiao H, Fu R, Zhao C, Wu H, Li Z (2013) Cell-surface GRP78 facilitates colorectal cancer cell migration and invasion. Int J Biochem Cell Biol 45(5):987–994. https://doi.org/10.1016/j.biocel.2013.02.002
100. Misra UK, Pizzo SV (2014) Activated alpha2-macroglobulin binding to cell surface GRP78 induces T-loop phosphorylation of Akt1 by PDK1 in association with raptor. PLoS One 9(2):e88373. https://doi.org/10.1371/journal.pone.0088373
101. Misra UK, Pizzo SV (2015) Activated alpha2-macroglobulin binding to human prostate cancer cells triggers insulin-like responses. J Biol Chem 290(15):9571–9587. https://doi.org/10.1074/jbc.M114.617837
102. Zoni E, Chen L, Karkampouna S, Granchi Z, Verhoef EI, La Manna F, Kelber J, Pelger RCM, Henry MD, Snaar-Jagalska E, van Leenders G, Beimers L, Kloen P, Gray PC, van der Pluijm G, Kruithof-de Julio M (2017) CRIPTO and its signaling partner GRP78 drive the metastatic phenotype in human osteotropic prostate cancer. Oncogene 36(33):4739–4749. https://doi.org/10.1038/onc.2017.87
103. Gray PC, Vale W (2012) Cripto/GRP78 modulation of the TGF-beta pathway in development and oncogenesis. FEBS Lett 586(14):1836–1845. https://doi.org/10.1016/j.febslet.2012.01.051
104. Araujo N, Hebbar N, Rangnekar VM (2018) GRP78 is a targetable receptor on Cancer and stromal cells. EBioMedicine 33:2–3. https://doi.org/10.1016/j.ebiom.2018.06.030
105. Staquicini DI, D'Angelo S, Ferrara F, Karjalainen K, Sharma G, Smith TL, Tarleton CA, Jaalouk DE, Kuniyasu A, Baze WB, Chaffee BK, Hanley PW, Barnhart KF, Koivunen E, Marchio S, Sidman RL, Cortes JE, Kantarjian HM, Arap W, Pasqualini R (2018) Therapeutic targeting of membrane-associated GRP78 in leukemia and lymphoma: preclinical efficacy

in vitro and formal toxicity study of BMTP-78 in rodents and primates. Pharmacogenomics J 18(3):436–443. https://doi.org/10.1038/tpj.2017.46
106. Kao C, Chandna R, Ghode A, Dsouza C, Chen M, Larsson A, Lim SH, Wang M, Cao Z, Zhu Y, Anand GS, Ge R (2018) Proapoptotic cyclic peptide BC71 targets cell-surface GRP78 and functions as an anticancer therapeutic in mice. EBioMedicine 33:22–32. https://doi.org/10.1016/j.ebiom.2018.06.004
107. Tsai YL, Zhang Y, Tseng CC, Stanciauskas R, Pinaud F, Lee AS (2015) Characterization and mechanism of stress-induced translocation of 78-kilodalton glucose-regulated protein (GRP78) to the cell surface. J Biol Chem 290(13):8049–8064. https://doi.org/10.1074/jbc.M114.618736
108. Reddy RK, Mao C, Baumeister P, Austin RC, Kaufman RJ, Lee AS (2003) Endoplasmic reticulum chaperone protein GRP78 protects cells from apoptosis induced by topoisomerase inhibitors: role of ATP binding site in suppression of caspase-7 activation. J Biol Chem 278(23):20915–20924. https://doi.org/10.1074/jbc.M212328200
109. Wooden SK, Lee AS (1992) Comparison of the genomic organizations of the rat grp78 and hsc73 gene and their evolutionary implications. DNA Seq 3(1):41–48. https://doi.org/10.3109/10425179209039694
110. Hardy B, Battler A, Weiss C, Kudasi O, Raiter A (2008) Therapeutic angiogenesis of mouse hind limb ischemia by novel peptide activating GRP78 receptor on endothelial cells. Biochem Pharmacol 75(4):891–899. https://doi.org/10.1016/j.bcp.2007.10.008
111. Alberts B, Johnson A, Lewis J, Raff M, Roberts K, Walter P (2002) Intracellular compartments and protein sorting. In: Molecular Biology of the Cell. Garland Science, New York
112. Meynier S (2016) Mécanismes de relocalisation de la Glucose-regulated protein 78 à la surface des cellules du cancer de l'ovaire., University of Geneva
113. Chen WT, Lee AS (2011) Measurement and modification of the expression level of the chaperone protein and signaling regulator GRP78/BiP in mammalian cells. Methods Enzymol 490:217–233. https://doi.org/10.1016/B978-0-12-385114-7.00013-1
114. Munro S, Pelham HR (1987) A C-terminal signal prevents secretion of luminal ER proteins. Cell 48(5):899–907. https://doi.org/10.1016/0092-8674(87)90086-9
115. Capitani M, Sallese M (2009) The KDEL receptor: new functions for an old protein. FEBS Lett 583(23):3863–3871. https://doi.org/10.1016/j.febslet.2009.10.053
116. Denecke J, De Rycke R, Botterman J (1992) Plant and mammalian sorting signals for protein retention in the endoplasmic reticulum contain a conserved epitope. EMBO J 11(6):2345–2355
117. Wiersma VR, Michalak M, Abdullah TM, Bremer E, Eggleton P (2015) Mechanisms of translocation of ER chaperones to the cell surface and immunomodulatory roles in cancer and autoimmunity. Front Oncol 5:7. https://doi.org/10.3389/fonc.2015.00007
118. D'Souza-Schorey C, Chavrier P (2006) ARF proteins: roles in membrane traffic and beyond. Nat Rev Mol Cell Biol 7(5):347–358. https://doi.org/10.1038/nrm1910
119. Raykhel I, Alanen H, Salo K, Jurvansuu J, Nguyen VD, Latva-Ranta M, Ruddock L (2007) A molecular specificity code for the three mammalian KDEL receptors. J Cell Biol 179(6):1193–1204. https://doi.org/10.1083/jcb.200705180
120. Tsai YL, Ha DP, Zhao H, Carlos AJ, Wei S, Pun TK, Wu K, Zandi E, Kelly K, Lee AS (2018) Endoplasmic reticulum stress activates SRC, relocating chaperones to the cell surface where GRP78/CD109 blocks TGF-beta signaling. Proc Natl Acad Sci U S A 115(18):E4245–E4254. https://doi.org/10.1073/pnas.1714866115
121. Stornaiuolo M, Lotti LV, Borgese N, Torrisi MR, Mottola G, Martire G, Bonatti S (2003) KDEL and KKXX retrieval signals appended to the same reporter protein determine different trafficking between endoplasmic reticulum, intermediate compartment, and Golgi complex. Mol Biol Cell 14(3):889–902. https://doi.org/10.1091/mbc.e02-08-0468
122. Lodish H, Berk A, Zipursky SL (2000) Overview of the secretory pathway. In: WH F (ed) Molecular cell biology. 4th edn., New York

123. Kaiser C, Gimeno RE, Shaywitz DA (1997) Protein secretion, membrane biogenesis and endocytosis. In: The Molecular and Cellular Biology of the Yeast, Saccharomyces cerevisiae, 1st edn. Cold Spring Harbor Laboratory Press, Cold Spring Harbor, pp 91–227
124. Tokarev AA, Alfonso A, Segev N (2000-2013) Overview of intracellular compartments and trafficking pathways. Madame Curie Bioscience Database, Landes Bioscience
125. Cooper GM (2000) The cell: A molecular approach. In: The Mechanism of Vesicular Transport. 2nd edn. Sinauer Associate, Sunderland (MA)
126. Pfaffenbach KT, Lee AS (2011) The critical role of GRP78 in physiologic and pathologic stress. Curr Opin Cell Biol 23(2):150–156. https://doi.org/10.1016/j.ceb.2010.09.007
127. Ibrahim IM, Abdelmalek DH, Elfiky AA (2019) GRP78: a cell's response to stress. Life Sci 226:156–163. https://doi.org/10.1016/j.lfs.2019.04.022
128. Lee AS (1987) Coordinated regulation of a set of genes by glucose and calcium ionophores in mammalian cells. Trends Biochem Sci 12:20–23
129. Lee AS (2009) The Par-4-GRP78 TRAIL, more twists and turns. Cancer Biol Ther 8(22):2103–2105. https://doi.org/10.4161/cbt.8.22.10140
130. Misra UK, Gonzalez-Gronow M, Gawdi G, Pizzo SV (2005) The role of MTJ-1 in cell surface translocation of GRP78, a receptor for alpha 2-macroglobulin-dependent signaling. J Immunol 174(4):2092–2097. https://doi.org/10.4049/jimmunol.174.4.2092
131. Chevalier M, Rhee H, Elguindi EC, Blond SY (2000) Interaction of murine BiP/GRP78 with the DnaJ homologue MTJ1. J Biol Chem 275(26):19620–19627. https://doi.org/10.1074/jbc.M001333200
132. Burikhanov R, Hebbar N, Noothi SK, Shukla N, Sledziona J, Araujo N, Kudrimoti M, Wang QJ, Watt DS, Welch DR, Maranchie J, Harada A, Rangnekar VM (2017) Chloroquine-inducible Par-4 secretion is essential for tumor cell apoptosis and inhibition of metastasis. Cell Rep 18(2):508–519. https://doi.org/10.1016/j.celrep.2016.12.051
133. Hebbar N, Wang C, Rangnekar VM (2012) Mechanisms of apoptosis by the tumor suppressor Par-4. J Cell Physiol 227(12):3715–3721. https://doi.org/10.1002/jcp.24098
134. Sells SF, Wood DP Jr, Joshi-Barve SS, Muthukumar S, Jacob RJ, Crist SA, Humphreys S, Rangnekar VM (1994) Commonality of the gene programs induced by effectors of apoptosis in androgen-dependent and -independent prostate cells. Cell Growth Differ 5(4):457–466
135. Johnstone RW, See RH, Sells SF, Wang J, Muthukkumar S, Englert C, Haber DA, Licht JD, Sugrue SP, Roberts T, Rangnekar VM, Shi Y (1996) A novel repressor, par-4, modulates transcription and growth suppression functions of the Wilms' tumor suppressor WT1. Mol Cell Biol 16(12):6945–6956. https://doi.org/10.1128/mcb.16.12.6945
136. Lucas T, Pratscher B, Krishnan S, Fink D, Gunsberg P, Wolschek M, Wacheck V, Muster T, Romirer I, Wolff K, Pehamberger H, Eichler HG, Rangnekar VM, Jansen B (2001) Differential expression levels of Par-4 in melanoma. Melanoma Res 11(4):379–383. https://doi.org/10.1097/00008390-200108000-00008
137. Gurumurthy S, Goswami A, Vasudevan KM, Rangnekar VM (2005) Phosphorylation of Par-4 by protein kinase A is critical for apoptosis. Mol Cell Biol 25(3):1146–1161. https://doi.org/10.1128/MCB.25.3.1146-1161.2005
138. Cook J, Krishnan S, Ananth S, Sells SF, Shi Y, Walther MM, Linehan WM, Sukhatme VP, Weinstein MH, Rangnekar VM (1999) Decreased expression of the pro-apoptotic protein Par-4 in renal cell carcinoma. Oncogene 18(5):1205–1208. https://doi.org/10.1038/sj.onc.1202416
139. Zhao Y, Burikhanov R, Qiu S, Lele SM, Jennings CD, Bondada S, Spear B, Rangnekar VM (2007) Cancer resistance in transgenic mice expressing the SAC module of Par-4. Cancer Res 67(19):9276–9285. https://doi.org/10.1158/0008-5472.CAN-07-2124
140. Kline CL, Shanmugavelandy SS, Kester M, Irby RB (2009) Delivery of PAR-4 plasmid in vivo via nanoliposomes sensitizes colon tumor cells subcutaneously implanted into nude mice to 5-FU. Cancer Biol Ther 8(19):1831–1837. https://doi.org/10.4161/cbt.8.19.9592
141. Lee JW, Lee KF, Hsu HY, Hsu LP, Shih WL, Chu YC, Hsiao WT, Liu PF (2007) Protein expression and intracellular localization of prostate apoptosis response-4 (Par-4) are associated with

apoptosis induction in nasopharyngeal carcinoma cell lines. Cancer Lett 257(2):252–262. https://doi.org/10.1016/j.canlet.2007.08.004
142. Lee TJ, Lee JT, Kim SH, Choi YH, Song KS, Park JW, Kwon TK (2008) Overexpression of Par-4 enhances thapsigargin-induced apoptosis via down-regulation of XIAP and inactivation of Akt in human renal cancer cells. J Cell Biochem 103(2):358–368. https://doi.org/10.1002/jcb.21642
143. El-Guendy N, Zhao Y, Gurumurthy S, Burikhanov R, Rangnekar VM (2003) Identification of a unique core domain of par-4 sufficient for selective apoptosis induction in cancer cells. Mol Cell Biol 23(16):5516–5525. https://doi.org/10.1128/mcb.23.16.5516-5525.2003
144. Shrestha-Bhattarai T, Rangnekar VM (2010) Cancer-selective apoptotic effects of extracellular and intracellular Par-4. Oncogene 29(27):3873–3880. https://doi.org/10.1038/onc.2010.141
145. Sato M, Yao VJ, Arap W, Pasqualini R (2010) GRP78 signaling hub a receptor for targeted tumor therapy. Adv Genet 69:97–114. https://doi.org/10.1016/S0065-2660(10)69006-2
146. Irby RB, Kline CL (2013) Par-4 as a potential target for cancer therapy. Expert Opin Ther Targets 17(1):77–87. https://doi.org/10.1517/14728222.2013.731047
147. Thayyullathil F, Pallichankandy S, Rahman A, Kizhakkayil J, Chathoth S, Patel M, Galadari S (2013) Caspase-3 mediated release of SAC domain containing fragment from Par-4 is necessary for the sphingosine-induced apoptosis in Jurkat cells. J Mol Signal 8(1):2. https://doi.org/10.1186/1750-2187-8-2
148. Sledziona J, Rangnekar VM (2017) Regulation of caspase-mediated apoptosis by the tumor suppressor Par-4. In: Micheau O (ed) TRAIL, Fas Ligand, TNF and TLR3 in Cancer. Springer, Cham, pp 91–106. https://doi.org/10.1007/978-3-319-56805-8_5
149. Sarkar S, Jain S, Rai V, Sahoo DK, Raha S, Suklabaidya S, Senapati S, Rangnekar VM, Maiti IB, Dey N (2015) Plant-derived SAC domain of PAR-4 (prostate apoptosis response 4) exhibits growth inhibitory effects in prostate cancer cells. Front Plant Sci 6:822. https://doi.org/10.3389/fpls.2015.00822
150. Sells SF, Han SS, Muthukkumar S, Maddiwar N, Johnstone R, Boghaert E, Gillis D, Liu G, Nair P, Monnig S, Collini P, Mattson MP, Sukhatme VP, Zimmer SG, Wood DP Jr, McRoberts JW, Shi Y, Rangnekar VM (1997) Expression and function of the leucine zipper protein Par-4 in apoptosis. Mol Cell Biol 17(7):3823–3832. https://doi.org/10.1128/mcb.17.7.3823
151. Garcia-Cao I, Duran A, Collado M, Carrascosa MJ, Martin-Caballero J, Flores JM, Diaz-Meco MT, Moscat J, Serrano M (2005) Tumour-suppression activity of the proapoptotic regulator Par4. EMBO Rep 6(6):577–583. https://doi.org/10.1038/sj.embor.7400421
152. Diaz-Meco MT, Abu-Baker S (2009) The Par-4/PTEN connection in tumor suppression. Cell Cycle 8(16):2518–2522. https://doi.org/10.4161/cc.8.16.9384
153. Chaudhry P, Singh M, Parent S, Asselin E (2012) Prostate apoptosis response 4 (Par-4), a novel substrate of caspase-3 during apoptosis activation. Mol Cell Biol 32(4):826–839. https://doi.org/10.1128/MCB.06321-11
154. Goswami A, Qiu S, Dexheimer TS, Ranganathan P, Burikhanov R, Pommier Y, Rangnekar VM (2008) Par-4 binds to topoisomerase 1 and attenuates its DNA relaxation activity. Cancer Res 68(15):6190–6198. https://doi.org/10.1158/0008-5472.CAN-08-0831
155. Wang J, Li Y, Ma F, Zhou H, Ding R, Lu B, Zou L, Li J, Lu R (2017) Inhibitory effect of Par-4 combined with cisplatin on human Wilms' tumor cells. Tumour Biol 39(7):1010428317716689. https://doi.org/10.1177/1010428317716689
156. Garcia SA, Robovnitchaia IG, Mangone FR, Nagai MA (2018) Prostate apoptosis response-4 (PAR4) secretion in breast tumor and normal cell lines cultured in hypoxic conditions. Clin Cancer Res
157. Chipurupalli S, Kannan E, Tergaonkar V, D'Andrea R, Robinson N (2019) Hypoxia induced ER stress response as an adaptive mechanism in Cancer. Int J Mol Sci 20(3). https://doi.org/10.3390/ijms20030749
158. Delbrel E, Soumare A, Naguez A, Label R, Bernard O, Bruhat A, Fafournoux P, Tremblais G, Marchant D, Gille T, Bernaudin JF, Callard P, Kambouchner M, Martinod E, Valeyre D,

Uzunhan Y, Planes C, Boncoeur E (2018) HIF-1alpha triggers ER stress and CHOP-mediated apoptosis in alveolar epithelial cells, a key event in pulmonary fibrosis. Sci Rep 8(1):17939. https://doi.org/10.1038/s41598-018-36063-2
159. Bartoszewska S, Collawn JF (2020) Unfolded protein response (UPR) integrated signaling networks determine cell fate during hypoxia. Cell Mol Biol Lett 25:18. https://doi.org/10.1186/s11658-020-00212-1
160. Nayak D, Katoch A, Sharma D, Faheem MM, Chakraborty S, Sahu PK, Chikan NA, Amin H, Gupta AP, Gandhi SG, Mukherjee D, Goswami A (2019) Indolylkojyl methane analogue IKM5 potentially inhibits invasion of breast cancer cells via attenuation of GRP78. Breast Cancer Res Treat 177(2):307–323. https://doi.org/10.1007/s10549-019-05301-0
161. Zurzolo C, Simons K (2016) Glycosylphosphatidylinositol-anchored proteins: membrane organization and transport. Biochim Biophys Acta 1858(4):632–639. https://doi.org/10.1016/j.bbamem.2015.12.018
162. Viotti C (2016) ER to Golgi-dependent protein secretion: the conventional pathway. In: Pompa A, De Marchis F (eds) Unconventional protein secretion. Methods in molecular biology. Humana Press, New York
163. Munguira ILB, Takahashi H, Casuso I, Scheuring S (2017) Lysenin toxin membrane insertion is pH-dependent but independent of Neighboring Lysenins. Biophys J 113(9):2029–2036. https://doi.org/10.1016/j.bpj.2017.08.056
164. Paroutis P, Touret N, Grinstein S (2004) The pH of the secretory pathway: measurement, determinants, and regulation. Physiology (Bethesda) 19:207–215. https://doi.org/10.1152/physiol.00005.2004
165. Yadav RK, Chae SW, Kim HR, Chae HJ (2014) Endoplasmic reticulum stress and cancer. J Cancer Prev 19(2):75–88. https://doi.org/10.15430/JCP.2014.19.2.75
166. Lee E, Nichols P, Spicer D, Groshen S, Yu MC, Lee AS (2006) GRP78 as a novel predictor of responsiveness to chemotherapy in breast cancer. Cancer Res 66(16):7849–7853. https://doi.org/10.1158/0008-5472.CAN-06-1660
167. Nagelkerke A, Bussink J, Mujcic H, Wouters BG, Lehmann S, Sweep FC, Span PN (2013) Hypoxia stimulates migration of breast cancer cells via the PERK/ATF4/LAMP3-arm of the unfolded protein response. Breast Cancer Res 15(1):R2. https://doi.org/10.1186/bcr3373
168. Gupta S, McGrath B, Cavener DR (2009) PERK regulates the proliferation and development of insulin-secreting beta-cell tumors in the endocrine pancreas of mice. PLoS One 4(11):e8008. https://doi.org/10.1371/journal.pone.0008008
169. Su R, Li Z, Li H, Song H, Bao C, Wei J, Cheng L (2010) Grp78 promotes the invasion of hepatocellular carcinoma. BMC Cancer 10:20. https://doi.org/10.1186/1471-2407-10-20
170. Delie F, Petignat P, Cohen M (2012) GRP78 protein expression in ovarian Cancer patients and perspectives for a drug-targeting approach. J Oncol 2012:468615. https://doi.org/10.1155/2012/468615
171. Lee HK, Xiang C, Cazacu S, Finniss S, Kazimirsky G, Lemke N, Lehman NL, Rempel SA, Mikkelsen T, Brodie C (2008) GRP78 is overexpressed in glioblastomas and regulates glioma cell growth and apoptosis. Neuro-Oncology 10(3):236–243. https://doi.org/10.1215/15228517-2008-006

PAR-4 in the Regulation of Stem Cell Death and Embryo Development

Ahmed Elsherbini and Erhard Bieberich

Abstract Prostate apoptosis response-4 (PAR-4) is widely known to induce apoptosis in cancer cells, suggesting that it may also regulate cell death in other rapidly dividing cell types, particularly stem cells. In fact, the stem cell niche was one of the first tissues that were found to express high levels of PAR-4, suggesting that PAR-4 regulates cell death during stem cell differentiation and embryo morphogenesis. In the last two decades, embryonic stem (ES) cells or induced pluripotent stem (iPS) cells emerged as a source for tissue regeneration or repair, using stem cell therapy. However, stem cell therapy is hampered by unwanted transformation of ES or iPS cells into tumorigenic tissue. Currently, there are only limited intervention strategies available to prevent tumor growth from pluripotent stem cells, which is a major obstacle in stem cell therapy using ES or iPS cell-derived progenitor cell transplants. Research from our laboratory demonstrated that PAR-4 mediated apoptosis regulates differentiation of neural progenitor cells (NPCs) during brain development and activation of PAR-4 is a viable strategy to eliminate tumorigenic stem cells from NPC transplants. We discovered that activation of PAR-4 in these cells is achieved by upregulation of a sphingolipid, ceramide, that partners with PAR-4 to induce apoptosis. Simultaneous increase of ceramide and PAR-4 levels eliminates tumorigenic stem cells and regulates stem cell survival during normal development. This chapter discusses mechanisms underlying apoptosis induction by ceramide and PAR-4 during stem cell differentiation and tissue development, as well as novel strategies to utilize these mechanisms as a means to make stem cell therapy safer.

Keywords Stem cells · Neural progenitor cells · Differentiation · Development · Apoptosis · Sphingolipids · Ceramide · Asymmetric cell division · Ectoderm · Pluripotency · Induced pluripotent stem cells · Embryonic stem cells · Fetal alcohol syndrome · Teratoma · Stem cell therapy

A. Elsherbini · E. Bieberich (✉)
Department of Physiology, University of Kentucky School of Medicine, Lexington, KY, USA

Veterans Affairs Medical Center, Lexington, KY, USA
e-mail: Erhard.bieberich@uky.edu

V. M. Rangnekar (ed.), *Tumor Suppressor Par-4*,
https://doi.org/10.1007/978-3-030-73572-2_8

1 Stem Cell Differentiation and the Role of PAR-4 in the Elimination of Unwanted Cells in Embryogenesis

The zygote is the ultimate totipotent stem cell that develops into embryonic and extra-embryonic (e.g., placenta) tissues. Ongoing cell division produces more stem cells than needed for morphogenesis of the embryo. Cells that do not differentiate or form intermediary tissues are removed by apoptosis, which is initiated and regulated by extracellular and intracellular cues, or a combination of both. One predominant mechanism that orchestrates extracellular and intracellular signals during embryonic morphogenesis is cell polarity. Polarization into spatially distinct portions of the embryo begins with the zygote (*C. elegans*) or the blastomere, generally guided by cell surfaces contacting other cells or pointing toward cell-free and luminal spaces [1]. These contact-free cell surfaces establish the apical membrane of differentiating stem cells, which is maintained by accumulation of apical protein kinase Cζ (aPKC) and associated cell polarity proteins (Par6, Par3) as well as small GTPases (Cdc42) [2–5]. This apical polarity protein complex hooks the cell membrane up to the cytoskeleton and establishes the apicobasal "body axis" of the cell, a master regulator of epithelial cell polarity in the developing embryo. Stem cells that do not establish or lose cell polarity (e.g., due to dissociation or detachment) are prone to elimination mediated by apoptosis. Hence, the task at hand is to understand the molecular mechanism that links cell polarity (or lack thereof) to apoptosis in differentiating stem cells.

After the discovery of Par-4 in 1994 by Vivek Rangnekar's group in Lexington, Kentucky, Moscat's and Pfizenmeier's groups reported that PAR-4 selectively binds to aPKC and that aPKC is dose-dependently activated (low concentration) or inhibited (high concentration) by ceramide, respectively [6–8]. Moscat hypothesized that with a high concentration of ceramide, aPKC changes its confirmation and becomes susceptible to inhibition by PAR-4. In 1997, Rangnekar's group found that PAR-4 is highly expressed in stem cell compartments of adult rat tissues [9]. About 3 years later, we discovered that ceramide-induced interaction of aPKC with PAR-4 regulates apoptosis in differentiating embryonic stem (ES) cells [10, 11]. Based on this initial observation, our research suggests that ceramide is a "morphogenetic lipid" that is confined to the apical membrane of primitive ectoderm cells, a pluripotent stem cell layer in the early embryo, and that interaction of ceramide with aPKC either initiates complex formation with other cell polarity proteins such as Par6-Cdc42 or the apoptosis inducer PAR-4 [12–17] (Fig. 1). This discovery was possible due to our ability to generate a ceramide specific antibody and a novel technology, lipid-mediated affinity chromatography (LIMAC), to use this antibody to pull down cell polarity and pro-apoptotic protein complexes from differentiating stem cells [18, 19]. We also have shown the conversion of ceramide to glycosphingolipids and asymmetric distribution of PAR-4 are instrumental in the cell fate decision, apoptosis or further differentiation, of stem cells. It is known that other lipids critical for the regulation of cell polarity such as phosphatidylinositol phosphates (PIPs) bind to Annexin2/Cdc42 when associated with Par6 and aPKC (Fig. 1); however, it is not

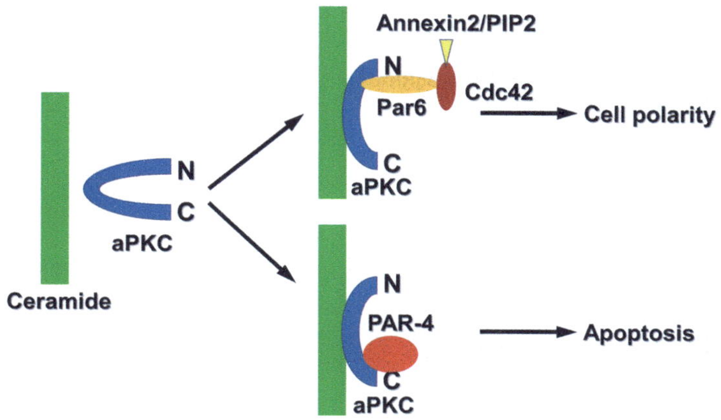

Fig. 1 Ceramide-mediated binding of aPKC and sphingolipid-induced protein complexes (SLIPs) regulating cell polarity (top) or apoptosis (bottom)
aPKC, atypical protein kinase Cζ/λ; Cdc42, cell division control protein 42; Par6, portioning-defective 6; PIP2, phosphatidylinositol 4,5-bisphosphate

clear how this binding affects cell polarity vs. apoptosis during stem cell differentiation. We suggested that ceramide and PIPs simultaneously bind to the aPKC complex for induction of cell polarity, while ceramide binding alone induces interaction of aPKC with PAR-4 and apoptosis [15, 16, 20–22].

2 PAR-4 and the Regulation of the Cytoskeleton in Dissociation-Induced Apoptosis of Stem Cells

The induction of apoptosis by PAR-4 is a conserved mechanism throughout embryo development, and it is tightly associated with cell attachment and polarity, two processes critically relying on the regulation of the cytoskeleton. While studies in our group were targeted toward the interaction of ceramide with aPKC, PAR-4, and downstream protein complexes that regulate the cytoskeleton, studies in other groups took a non-biased approach when CRISPR/Cas9 screening of stem cells became available. The general principle of these loss-of-function screens is that a biological process is interrupted when after transfection with an inducible Cas9-sgRNA library, a gene is deleted that is critical for regulation of this process. Most recently, Ihry et al. performed such a genome-wide screen on human pluripotent stem cells (hPSCs) dissociated from clusters into single cell clones (Fig. 2) [23]. Single cell cloning of hPSCs led to poor survival unless the Rho-kinase (ROCK) pathway was pharmacologically inhibited or PAR-4 (*PAWR*) was deleted in the absence of ROCK inhibitors. The group found that in addition to PAR-4, deletion of its binding partner aPKC (*PRKCZ*) also sustained survival of dissociated hPSCs, suggesting that dissociation-induced cell death was mediated by interaction of

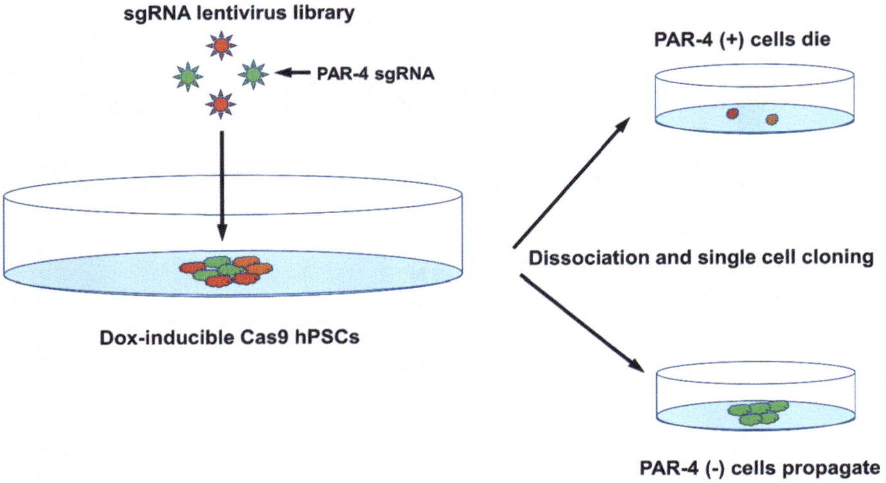

Fig. 2 CRISPR/Cas9 screening identifies PAR-4 in dissociation-induced apoptosis of human PSCs Loss of function due to deletion of PAR-4 with doxycycline-induced Cas9 and lentiviral transduction of PAR-4-specific sgRNA prevents apoptosis of human PSCs after dissociation and cloning of single cells

PAR-4-aPKC with the ROCK cell signaling pathway inducing actomyosin contraction and blebbing of the membrane. This conclusion was supported by immunolabeling studies showing co-localization of PAR-4 and aPKC with actin-labeled blebs of the plasma membrane of dissociated hPSC cells [23]. The critical role of PAR-4 in dissociation-induced cell death of hPSCs was confirmed by another group using a similar screen [24].

It is noteworthy that dissociation-induced cell death is specific for hPSCs and it is not observed with mouse embryonic stem (ES) cells, which can be expanded from single cells [25]. It has been suggested that this difference results from acquisition of cell polarity in clusters of hPSC cells, which resemble epiblast cells [25]. These cells undergo apoptosis if not incorporated into the polarized epithelium of the epiblast. Hence, the study from Ihry et al. supports the hypothesis that the interaction of PAR-4 with aPKC is important for the regulation of cell polarity vs. apoptosis in embryo development. The function of aPKC binding to ceramide may constitute an upstream regulator positioning a cell polarity (with Par6-Cdc42) vs. pro-apoptotic (with PAR-4) complex at the plasma membrane (Fig. 3). This proposed mechanism is consistent with a study showing that the aPKC-Par6 complex (although independently of Cdc42) negatively regulates Rho/ROCK [26], while others show activation of Rho/ROCK by aPKC and Cdc42 [27]. To date, it is not entirely clear how these aPKC complexes affect actomyosin in stem cell polarity and if ceramide is involved in this regulation.

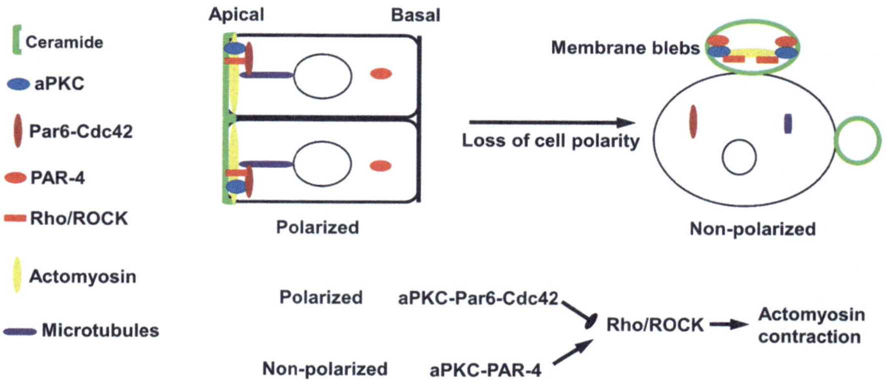

Fig. 3 Proposed mechanism for regulation of actomyosin by PAR-4
In polarized human PSCs, a cell polarity complex of aPKC with Par6 and presumably Cdc42 prevents actomyosin contraction, while in non-polarized (e.g., dissociated) cells, a complex of aPKC with PAR-4 activates Rho/ROCK and induces actomyosin contraction, which leads to membrane blebbing concurrent with apoptosis. The regulatory function of ceramide in these processes is not known

3 Ceramide Metabolism and Asymmetric Distribution of PAR-4 During Stem Cell Differentiation

Over the last two decades, our research uncovered a remarkable synchronization between ceramide metabolism and PAR-4 expression during neural stem cell differentiation and embryonic brain development. While expression of PAR-4 is typically associated with apoptosis, upregulation of ceramide is conditionally pro-apoptotic or pro-differentiation depending on its intracellular distribution and metabolic conversion to phosphosphingolipids and glycosphingolipids [14–16, 20, 22]. Using the ceramide antibody developed in our laboratory, we found that in embryonic bodies (EBs) derived from pluripotent ES cells, ceramide is concentrated in the apical plasma membrane of primitive ectoderm cells [13]. During gastrulation, the primitive ectoderm layer gives rise to all three germ layers in the embryo, ectoderm, mesoderm, and endoderm. To form the primitive ectoderm layer in EBs, cells in the center are eliminated by apoptosis giving rise to a lumen or cavity that is enclosed by the apical membrane of primitive ectoderm cells. While ceramide is upregulated in all cells of the EBs, the primitive ectoderm layer is protected from ceramide-induced apoptosis due to confinement of ceramide to the apical membrane and its interaction with the cell polarity complex consisting of aPKC and Cdc42-Par6 (Figs. 1 and 3).

To further differentiate primitive ectoderm cells, EBs are attached to the surface of tissue culture dishes which leads to flattening of the EBs and specification of the pluripotent stem cells to multipotent neuroprogenitor cells (NPCs), which migrate out of the periphery of the attached EBs. These in vitro differentiated NPCs correspond to neural precursor cells in the ventricular and subventricular zone that give rise to glia and neurons of the developing brain and spinal cord. In contrast to residual pluripotent stem cells in the center of the attached EB, these NPCs no longer express the transcription factor and pluripotency marker Oct-4, but the intermediate filament protein nestin [11, 13]. Interestingly, during neural differentiation to NPCs, ceramide levels are continuously upregulated, but apoptosis induction is limited to PAR-4 expressing cells. We discovered that this remarkable selectivity is due to the asymmetric distribution of PAR-4 during stem cell division.

We showed that immediately prior to or during cell division, NPCs upregulate PAR-4 and ceramide and give rise to two daughter cells, one that inherits PAR-4 and undergoes apoptosis, and another one that lacks PAR-4 and continues to proliferate and undergo neural differentiation (Fig. 1). When ceramide biosynthesis is blocked by incubation with fumonisin B1 (FB1), an inhibitor of ceramide synthases, PAR-4 (+) cells survive and continue to proliferate as well. Apoptosis of PAR-4 (+) cells is restored by incubation of FB1-treated cells with the novel ceramide analog N-oleoyl serinol (S18). Likewise, when PAR-4 expression is knocked down with anti-sense morpholinos, apoptosis is prevented despite the presence of ceramide [11]. Consistently, transfection of NPCs with a plasmid encoding RFP-tagged PAR-4 and simultaneous incubation with S18 leads to apoptosis in the PAR-4 expressing cells. These results clearly demonstrated that during asymmetric division of NPCs, compartmentalization of PAR-4 to one of the daughter cells leads to ceramide-induced apoptosis of the PAR-4 (+) cell or otherwise put as "quarantining cell death" (Fig. 4), while the other NPC differentiates into neurons and glia. Whether this compartmentalization of PAR-4 is due to asymmetric distribution of transcription factors, mRNA, or the protein itself, is not yet known. We have shown that in EB-derived

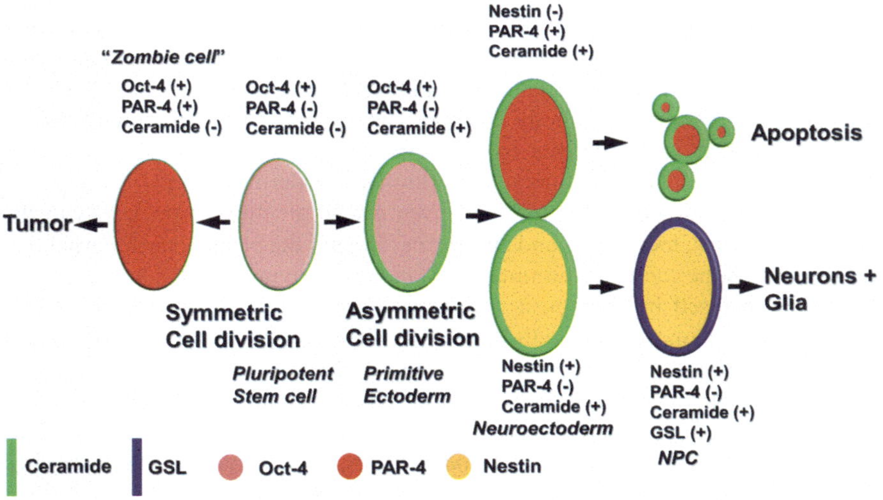

Fig. 4 Asymmetric cell division of neuroectodermal stem cells leading to apoptosis (top) or further differentiation into neural progenitor cells (NPCs) and then neurons and glia (bottom) GSL, glycosphingolipid; Oct-4, octamer-binding transcription factor 4

stem cells, a short, 33 kDa isoform of PAR-4 (p33) is co-expressed with pro-apoptotic 38 kDa PAR-4 (p38). The p33 isoform binds to PAR-4 (p38) and F-actin, which suppresses induction of apoptosis [28]. Since F-actin is primarily formed at the apical cell membrane, one may speculate that this releases aPKC from being inhibited in a pro-apoptotic complex with PAR-4 and instead allows formation of the cell polarity complex guiding the microtubule spindle axis required for asymmetric cell division. More clarity is needed to determine if upregulation of PAR-4 isoforms and ceramide in NPCs and asymmetric distribution of PAR-4 are only initiated in the beginning of NPC proliferation or continue for several generations of progeny cells.

When NPCs enter a phase of symmetric cell division to eventually give rise to precursor cells that terminally differentiate into neurons and glia, ceramide levels reach their peak in the developing mouse brain (embryonic age E16.5-E18.5) [10, 11]. In this developmental stage, a monocellular layer of NPCs lining the ventricles – radial glial cells – undergoes asymmetric cell division into one daughter cell that remains a self-renewing radial glia cell and another daughter cell that differentiates into a neural precursor cell and undergoes a limited number of symmetric cell divisions while radially and tangentially migrating to populate the layers of the brain during a process termed "lamination" [29–36]. During lamination, ceramide is intensively converted to glycosphingolipids, particularly gangliosides, which prevents unwanted apoptosis and promotes neuronal differentiation in neural precursor cells [10]. We found that gangliosides of a particular type, so-called b-series gangliosides, suppress expression of PAR-4 and increase levels of its counterplayer, the anti-apoptotic protein B-cell lymphoma (Bcl-2), in differentiating neural precursor cells of the intermediate zone in embryonic mouse brain [10]. PAR-4 is a transcriptional co-repressor of Bcl-2, suggesting that downregulation of PAR-4 leads to upregulation of Bcl-2 and inhibition of apoptosis [37–39]. Our experimental evidence indicated that b-series gangliosides activate the insulin growth factor (IGF)-1-to-PI3K/Akt-to-ERK/MAPK cell signaling pathway, which downregulates PAR-4 and upregulates Bcl-2 [10, 40] (Fig. 5).

Several years later, this cell signaling pathway for downregulation of PAR-4 expression was confirmed by other groups [41, 42]. B-series gangliosides were also shown to activate other growth factor receptors for neuronal differentiation, such as EGF-1, which is still a research area intensively studied in the laboratory of the author's former mentor, Dr. Robert K. Yu [43]. While up- and downregulation of PAR-4 in NPCs and neural precursor cells, respectively, is one way of regulating apoptosis during brain development, it should be noted that knocking out PAR-4 did not lead to overt brain developmental phenotypes [44]. This may be disappointing from a research perspective, but it is not unusual, as several compensatory mechanisms often rebalance apoptosis vs. cell survival during embryonic development. Therefore, our laboratory began to focus on the role of aberrant PAR-4 expression in dysregulation of apoptosis during pathological brain development.

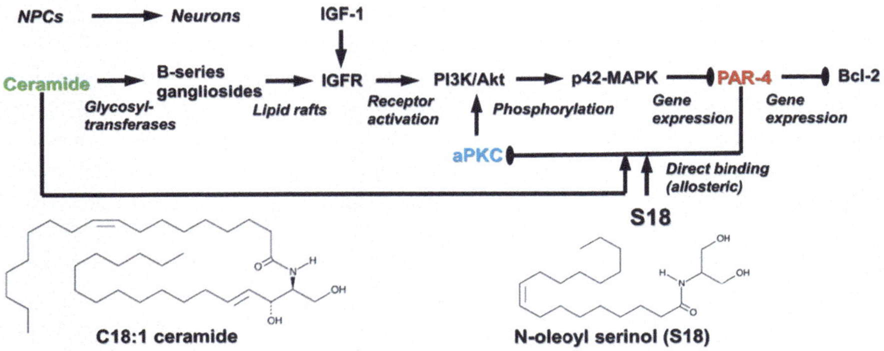

Fig. 5 Ceramide- and glycosphingolipid-induced cell signaling pathway regulating apoptosis and neural differentiation of stem cells
Bcl-2, b-cell lymphoma 2; IGF-1, insulin growth factor 1; IGFR, IGF-1 receptor; PI3K/Akt, phosphatidylinositol 3 kinase/protein kinase B; p42-MAPK, p42 mitogen-activated protein kinase

4 PAR-4 in Fetal Alcohol Syndrome and Dysregulated Stem Cell Apoptosis

Neural crest cells (NCCs) are among the earliest derivatives from neuroectodermal stem cells in the neural tube [45]. They are multipotent stem cells, but unlike NPCs, they differentiate into a variety neural and non-neural cells, including peripheral neurons and glia, melanocytes in the skin, smooth muscle cells, and facial bone and cartilage cells. Due to this versatility, they are highly migratory and populate different tissues throughout the mouse embryo shortly after neural tube closure (embryonic age E9.5-E10.5). We have shown that NCCs from mouse branchial arch explant cultures co-expressed PAR-4 and ceramide, although at a low level, which did not induce apoptosis [46]. However, PAR-4 and ceramide levels were increased by several-fold when NCCs were exposed to ethanol at a concentration corresponding to alcohol levels causing fetal alcohol syndrome (FAS). FAS is a condition occurring in about 1/1000 women, who consume alcohol during pregnancies they are often not aware of [47, 48]. FAS leads to malformation of neural crest-derived tissue, particularly facial bones, and mental retardation.

Our research showed that ethanol activates neutral sphingomyelinase 2 (nSMase2), a key cell signaling enzyme that hydrolyzes sphingomyelin in the plasma membrane to ceramide. The reason for ethanol-induced PAR-4 expression in NCCs is not known. We focused on understanding the effects of ethanol-induced NCC apoptosis due to simultaneous elevation of ceramide and PAR-4 on brain development. We discovered that one specific tissue, the meninges, was underdeveloped in mouse embryos when pregnant dams were given alcohol. The meninges is derived from the mesoderm and NCCs, and it secretes growth factors toward the developing bone and brain. The immunolabeling of laminin, an extracellular matrix protein coating the NCC-derived cell layer in the meninges of embryos from ethanol-treated dams, was severely reduced, indicating aberrant NCC differentiation, probably due to ceramide/PAR-4-induced apoptosis [46]. Dysregulated development of this cell layer leads to growth factor deficiency, ultimately resulting in facial malformation and retardation of brain development characteristic of FAS.

In the absence of PAR-4, ceramide can act as "topological lipid" as well as "cell signaling lipid" with a plethora of different functions from metabolic and cell cycle regulation to generating specialized lipid rafts, so called ceramide-rich platforms (CRPs), and sphingolipid-induced protein complexes (SLIPs) that regulate cell polarity, endocytosis and exosome biogenesis, and formation of membrane protrusions such as cilia [20, 49–53]. Our laboratory showed that one of the mechanisms by which PAR-4 modulates the function of ceramide is the aPKC switch from apoptosis induction by ceramide-mediated binding of PAR-4 to orchestration of a protein complex with Par6-CDC42 when PAR-4 is absent [10–13, 15, 54–56] (Fig. 1). Our laboratory has shown that in the absence of PAR-4, ceramide promotes migration of NPCs and their differentiation to neurons [15, 16, 21]. On the other hand, despite PAR-4 being expressed, apoptosis may not be induced when ceramide is missing, which may lead to stem cell-derived tumors.

5 Ceramide and PAR-4-Mediated Apoptosis in Stem Cell-Derived Tumor Cells

Tumors can arise from stem cells at any developmental stage: (1) pluripotent (Oct-4(+)) stem cells give rise to teratoma, a disorganized embryonic mass composed of all three germ layers, (2) multipotent (nestin (+)) NPCs can develop neurocytomas and bear the risk of forming other types of tumors after transplantation into the brain or spinal cord, and (3) in adult brain, tumor stem cells may give rise to brain tumors such as glioblastoma. Our research focused on the role of PAR-4 in inducing apoptosis in Oct-4 (+) stem cells derived from ES or iPS cells. These Oct-4 (+) cells are residual pluripotent stem cells that contaminate stem cell transplants despite extensive differentiation toward a population of homogenous NPCs. As little as 0.001% of residual Oct-4 (+) cells in a transplant of ES- or iPS cell-derived NPCs will give rise to tumors in the majority of animals receiving the transplant, a major risk in stem cell therapy [57]. Since extensive differentiation to NPCs appears not to lower the proportion of these residual, tumor-forming stem cells, the only strategies to increase safety of stem cell therapy is to eliminate Oct-4 (+) cells by cell sorting or induction of apoptosis.

During neural differentiation, stem cells downregulated Oct-4 and upregulated nestin. They undergo apoptosis when they express PAR-4 (Fig. 6A). We discovered that one of the main reasons for survival of Oct-4 (+) stem cells during differentiation to NPCs is the lack of ceramide. While these cells co-express PAR-4 (+), they are not eliminated by apoptosis and continue to proliferate, eventually forming stem cell-derived tumors. We borrowed the term "Zombie cells" as a term to characterize

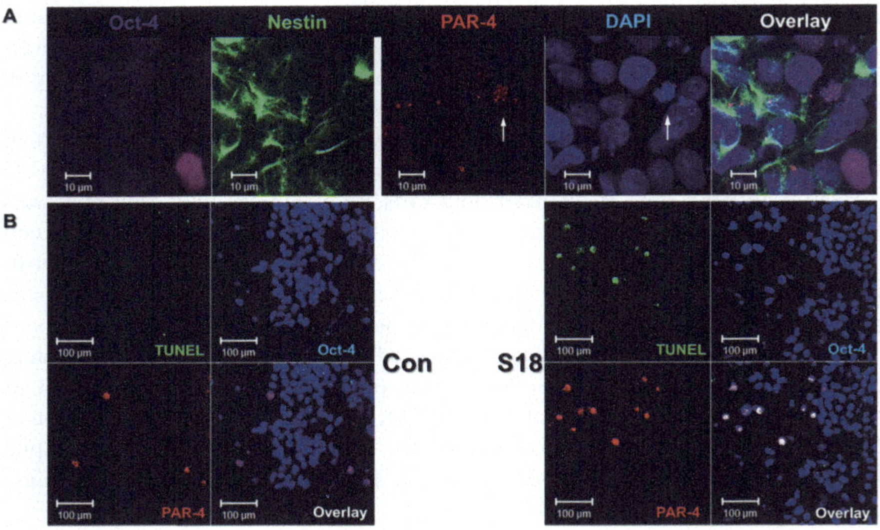

Fig. 6 PAR-4- and S18-induced apoptosis during stem cell differentiation to eliminate Oct-4 (+) tumor-forming stem cells

these "undead" stem cells reminiscent of failed apoptosis observed with other cell types. To induce apoptosis in Oct-4(+)/PAR-4(+) "Zombie" stem cells, we added the water-soluble novel ceramide analog S18 (Fig. 6B) to the culture medium of attached EBs. As expected, cells in the center of the EB underwent apoptosis, while NPCs in the periphery continued to proliferate and, eventually, differentiate into neurons and glia. When these S18-treated NPCs were injected into mouse brain, they differentiated into normal brain tissue, while non-treated control cells developed teratoma [12, 21]. These data demonstrated that incubation of stem cell transplants with ceramide analogs is a viable strategy to eliminate tumor-forming stem cells and significantly enhance safety of stem cell therapy. Several types of tumor stem cells are known to express PAR-4 [58, 59], suggesting that inducing ceramide-/PAR-4-mediated apoptosis by either increasing ceramide production through chemotherapy or administering ceramide/ceramide analogs is a novel strategy in cancer therapy.

6 Conclusions and Future Directions

PAR-4 has been firmly established as a molecular and pharmacological target in cancer therapy. Studies from our laboratory discovered an additional role for PAR-4 in rapidly dividing stem cells. Asymmetric distribution of PAR-4 and simultaneous upregulation of ceramide regulate the stem cell number by inducing apoptosis in the PAR-4(+)/ceramide(+) stem cell. The other cell converts ceramide into glycosphingolipids and continues to differentiate. However, if ceramide is not upregulated or metabolized, the PAR-4(+) cell may continue to divide and form a tumor, which is a risk in stem cell therapy. The tumor-forming stem cell can be eliminated by incubation with novel ceramide analogs such as N-oleoyl serinol (S18). In future studies, we will explore this approach to increase safety and efficacy of stem cell therapy.

Acknowledgments This work was supported by the NIH grants R01AG034389 and R01AG064234, and the VA merit award I01BX003643. We are grateful to institutional support by the Department of Physiology (Chair Dr. Alan Daugherty) at the University of Kentucky College of Medicine, Lexington, KY 40536.

References

1. Ajduk A, Zernicka-Goetz M (2016) Polarity and cell division orientation in the cleavage embryo: from worm to human. Mol Hum Reprod 22(10):691–703. Epub 2015/12/15. PubMed PMID: 26660321; PMCID: PMC5062000. https://doi.org/10.1093/molehr/gav068
2. Etienne-Manneville S, Hall A (2003) Cell polarity: Par6, aPKC and cytoskeletal crosstalk. Curr Opin Cell Biol 15(1):67–72. PubMed PMID: 12517706
3. Suzuki A, Akimoto K, Ohno S (2003) Protein kinase C lambda/iota (PKClambda/iota): a PKC isotype essential for the development of multicellular organisms. J Biochem (Tokyo) 133(1):9–16. PubMed PMID: 12761193
4. Hutterer A, Betschinger J, Petronczki M, Knoblich JA (2004) Sequential roles of Cdc42, Par-6, aPKC, and Lgl in the establishment of epithelial polarity during Drosophila embryogenesis. Dev Cell 6(6):845–854. PubMed PMID: 15177032
5. Kim M, Datta A, Brakeman P, Yu W, Mostov KE (2007) Polarity proteins PAR6 and aPKC regulate cell death through GSK-3beta in 3D epithelial morphogenesis. J Cell Sci 120(Pt 14):2309–2317. PubMed PMID: 17606986
6. Lozano J, Berra E, Municio MM, Diaz-Meco MT, Dominguez I, Sanz L, Moscat J (1994) Protein kinase C zeta isoform is critical for kappa B-dependent promoter activation by sphingomyelinase. J Biol Chem 269(30):19200–19202. PubMed PMID: 8034680
7. Diaz-Meco MT, Municio MM, Frutos S, Sanchez P, Lozano J, Sanz L, Moscat J (1996) The product of par-4, a gene induced during apoptosis, interacts selectively with the atypical isoforms of protein kinase C. Cell 86(5):777–786. PubMed PMID: 8797824
8. Muller G, Ayoub M, Storz P, Rennecke J, Fabbro D, Pfizenmaier K (1995) PKC zeta is a molecular switch in signal transduction of TNF-alpha, bifunctionally regulated by ceramide and arachidonic acid. EMBO J 14(9):1961–1969. PubMed PMID: 7744003
9. Boghaert ER, Sells SF, Walid AJ, Malone P, Williams NM, Weinstein MH, Strange R, Rangnekar VM (1997) Immunohistochemical analysis of the proapoptotic protein Par-4 in normal rat tissues. Cell Growth Differ 8(8):881–890. Epub 1997/08/01. PubMed PMID: 9269897
10. Bieberich E, MacKinnon S, Silva J, Yu RK (2001) Regulation of apoptosis during neuronal differentiation by ceramide and b-series complex gangliosides. J Biol Chem 276(48):44396–44404. PubMed PMID: 11574545
11. Bieberich E, MacKinnon S, Silva J, Noggle S, Condie BG (2003) Regulation of cell death in mitotic neural progenitor cells by asymmetric distribution of prostate apoptosis response 4 (PAR-4) and simultaneous elevation of endogenous ceramide. J Cell Biol 162(3):469–479. PubMed PMID: 12885759
12. Bieberich E, Silva J, Wang G, Krishnamurthy K, Condie BG (2004) Selective apoptosis of pluripotent mouse and human stem cells by novel ceramide analogues prevents teratoma for-

mation and enriches for neural precursors in ES cell-derived neural transplants. J Cell Biol 167(4):723–734. PubMed PMID: 15545317

13. Krishnamurthy K, Wang G, Silva J, Condie BG, Bieberich E (2007) Ceramide regulates atypical PKC{zeta}/{lambda}-mediated cell polarity in primitive ectoderm cells: a NOVEL FUNCTION OF SPHINGOLIPIDS IN MORPHOGENESIS. J Biol Chem 282(5):3379–3390. PubMed PMID: 17105725
14. Bieberich E (2004) Integration of glycosphingolipid metabolism and cell-fate decisions in cancer and stem cells: review and hypothesis. Glycoconj J 21(6):315–327. PubMed PMID: 15514480
15. Bieberich E (2010) There is more to a lipid than just being a fat: Sphingolipid-guided differentiation of oligodendroglial lineage from embryonic stem cells. Neurochem Res. Epub 2010/12/08. PubMed PMID: 21136155. https://doi.org/10.1007/s11064-010-0338-5
16. Bieberich E (2011) Ceramide in stem cell differentiation and embryo development: novel functions of a topological cell-signaling lipid and the concept of ceramide compartments. J Lipids:610306. Epub 2011/04/15. PubMed PMID: 21490805; PMCID: 3066852. https://doi.org/10.1155/2011/610306
17. Wang G, Spassieva SD, Bieberich E (1697) Ceramide and S1P signaling in embryonic stem cell differentiation. Methods Mol Biol 2018:153–171. https://doi.org/10.1007/7651_2017_43. PubMed PMID: 28540559
18. Krishnamurthy K, Dasgupta S, Bieberich E (2007) Development and characterization of a novel anti-ceramide antibody. J Lipid Res 48(4):968–975. PubMed PMID: 17210985
19. Bieberich E (2011) Lipid vesicle-mediated affinity chromatography using magnetic activated cell sorting (LIMACS): a novel method to analyze protein-lipid interaction. J Vis Exp 50. Epub 2011/05/05. doi: 2657 [pii]. https://doi.org/10.3791/2657. PubMed PMID: 21540829
20. Bieberich E (2008) Ceramide signaling in cancer and stem cells. Future Lipidol 3(3):273–300. Epub 2008/12/04. PubMed PMID: 19050750; PMCID: 2493052. https://doi.org/10.2217/17460875.3.3.273
21. Bieberich E (2008) Smart drugs for smarter stem cells: making SENSe (sphingolipid-enhanced neural stem cells) of ceramide. Neurosignals 16(2–3):124–139. Epub 2008/02/07. doi: 000111558 [pii]. 10.1159/000111558. PubMed PMID: 18253053
22. Bieberich E (2012) It's a lipid's world: bioactive lipid metabolism and signaling in neural stem cell differentiation. Neurochem Res. 37(6):1208–1229. Epub 2012/01/17. https://doi.org/10.1007/s11064-011-0698-5. PubMed PMID: 22246226; PMCID: 3343224
23. Ihry RJ, Salick MR, Ho DJ, Sondey M, Kommineni S, Paula S, Raymond J, Henry B, Frias E, Wang Q, Worringer KA, Ye C, Russ C, Reece-Hoyes JS, Altshuler RC, Randhawa R, Yang Z, McAllister G, Hoffman GR, Dolmetsch R, Kaykas A (2019) Genome-scale CRISPR screens identify human pluripotency-specific genes. Cell Rep 27(2):616–630. Epub 2019/04/11. PubMed PMID: 30970262. https://doi.org/10.1016/j.celrep.2019.03.043
24. Mair B, Aldridge PM, Atwal RS, Philpott D, Zhang M, Masud SN, Labib M, Tong AHY, Sargent EH, Angers S, Moffat J, Kelley SO (2019) High-throughput genome-wide phenotypic screening via immunomagnetic cell sorting. Nat Biomed Eng 3(10):796–805. Epub 2019/09/25. PubMed PMID: 31548591. https://doi.org/10.1038/s41551-019-0454-8
25. Takashima Y, Guo G, Loos R, Nichols J, Ficz G, Krueger F, Oxley D, Santos F, Clarke J, Mansfield W, Reik W, Bertone P, Smith A (2014) Resetting transcription factor control circuitry toward ground-state pluripotency in human. Cell 158(6):1254–1269. Epub 2014/09/13. PubMed PMID: 25215486; PMCID: PMC4162745. https://doi.org/10.1016/j.cell.2014.08.029
26. Zhang H, Macara IG (2008) The PAR-6 polarity protein regulates dendritic spine morphogenesis through p190 RhoGAP and the Rho GTPase. Dev Cell 14(2):216–226. Epub 2008/02/13. PubMed PMID: 18267090; PMCID: PMC2768076. https://doi.org/10.1016/j.devcel.2007.11.020
27. Coghlan MP, Chou MM, Carpenter CL (2000) Atypical protein kinases Clambda and -zeta associate with the GTP-binding protein Cdc42 and mediate stress fiber loss. Mol Cell Biol 20(8):2880–2889. Epub 2000/03/25. PubMed PMID: 10733591; PMCID: PMC85517. https://doi.org/10.1128/mcb.20.8.2880-2889.2000

28. Wang G, Silva J, Krishnamurthy K, Bieberich E (2006) A novel isoform of prostate apoptosis response 4 (PAR-4) that co-distributes with F-actin and prevents apoptosis in neural stem cells. Apoptosis: Int J Programm Cell Death 11(3):315–325. PubMed PMID: 16520897
29. Taverna E, Gotz M, Huttner WB (2014) The cell biology of neurogenesis: toward an understanding of the development and evolution of the neocortex. Annu Rev Cell Dev Biol 30:465–502. Epub 2014/07/09. PubMed PMID: 25000993. https://doi.org/10.1146/annurev-cellbio-101011-155801
30. Alvarez-Buylla A, Garcia-Verdugo JM, Tramontin AD (2001) A unified hypothesis on the lineage of neural stem cells. Nat Rev Neurosci 2(4):287–293. PubMed PMID: 11283751
31. Miyata T, Kawaguchi A, Okano H, Ogawa M (2001) Asymmetric inheritance of radial glial fibers by cortical neurons. Neuron 31(5):727–741. PubMed PMID: 11567613
32. Fishell G, Kriegstein AR (2003) Neurons from radial glia: the consequences of asymmetric inheritance. Curr Opin Neurobiol 13(1):34–41. PubMed PMID: 12593980
33. Gotz M, Huttner WB (2005) The cell biology of neurogenesis. Nat Rev Mol Cell Biol 6(10):777–788. Epub 2005/11/30. PubMed PMID: 16314867. https://doi.org/10.1038/nrm1739
34. Noctor SC, Martinez-Cerdeno V, Kriegstein AR (2007) Neural stem and progenitor cells in cortical development. Novartis Found Symp 288:59–73. discussion – 8, 96–8. Epub 2008/05/23. PubMed PMID: 18494252
35. Zhang Y, Niu B, Yu D, Cheng X, Liu B, Deng J (2010) Radial glial cells and the lamination of the cerebellar cortex. Brain Struct Funct 215(2):115–122. Epub 2010/09/30. PubMed PMID: 20878181. https://doi.org/10.1007/s00429-010-0278-5
36. De Juan RC, Borrell V (2015) Coevolution of radial glial cells and the cerebral cortex. Glia 63(8):1303–1319. Epub 2015/03/27. PubMed PMID: 25808466; PMCID: PMC5008138. https://doi.org/10.1002/glia.22827
37. Mattson MP, Duan W, Chan SL, Camandola S (1999) Par-4: an emerging pivotal player in neuronal apoptosis and neurodegenerative disorders. J Molecul Neurosci: MN 13(1–2):17–30. Epub 2000/02/26. doi: JMN:13:1–2:17 [pii]. 10.1385/JMN:13:1-2:17. PubMed PMID: 10691289
38. Camandola S, Mattson MP (2000) Pro-apoptotic action of PAR-4 involves inhibition of NF-kappaB activity and suppression of BCL-2 expression. J Neurosci Res 61(2):134–139. PubMed PMID: 10878585
39. Cheema SK, Mishra SK, Rangnekar VM, Tari AM, Kumar R, Lopez-Berestein G (2003) Par-4 transcriptionally regulates Bcl-2 through a WT1-binding site on the bcl-2 promoter. J Biol Chem 278(22):19995–20005. PubMed PMID: 12644474
40. Barradas M, Monjas A, Diaz-Meco MT, Serrano M, Moscat J (1999) The downregulation of the pro-apoptotic protein Par-4 is critical for Ras-induced survival and tumor progression. EMBO J 18(22):6362–6369. Epub 1999/11/24. PubMed PMID: 10562548; PMCID: PMC1171699. https://doi.org/10.1093/emboj/18.22.6362
41. Chung H, Seo S, Moon M, Park S (2007) IGF-I inhibition of apoptosis is associated with decreased expression of prostate apoptosis response-4. J Endocrinol 194(1):77–85. Epub 2007/06/27. PubMed PMID: 17592023. https://doi.org/10.1677/JOE-07-0073
42. Casolari DA, Pereira MC, de Bessa Garcia SA, Nagai MA (2011) Insulin-like growth factor-1 and 17beta-estradiol down-regulate prostate apoptosis response-4 expression in MCF-7 breast cancer cells. Int J Mol Med 28(3):337–342. Epub 2011/05/14. PubMed PMID: 21567071. https://doi.org/10.3892/ijmm.2011.691
43. Yanagisawa M, Liour SS, Yu RK (2004) Involvement of gangliosides in proliferation of immortalized neural progenitor cells. J Neurochem 91(4):804–812. Epub 2004/11/05. doi: JNC2750 [pii]. PubMed PMID: 15525334. https://doi.org/10.1111/j.1471-4159.2004.02750.x
44. Affar El B, Luke MP, Gay F, Calvo D, Sui G, Weiss RS, Li E, Shi Y (2006) Targeted ablation of Par-4 reveals a cell type-specific susceptibility to apoptosis-inducing agents. Cancer Res 66(7):3456–3462. Epub 2006/04/06. PubMed PMID: 16585168. https://doi.org/10.1158/0008-5472.CAN-05-0964
45. Teng L, Labosky PA (2006) Neural crest stem cells. Adv Exp Med Biol 589:206–212. PubMed PMID: 17076284

46. Wang G, Bieberich E (2010) Prenatal alcohol exposure triggers ceramide-induced apoptosis in neural crest-derived tissues concurrent with defective cranial development. Cell Death Dis 1(5):e46. Epub 2011/03/03. doi: cddis201022 [pii]. https://doi.org/10.1038/cddis.2010.22. PubMed PMID: 21364652; PMCID: 3032308
47. Nayak RB, Murthy P (2008) Fetal alcohol spectrum disorder. Indian Pediatr 45(12):977–983. PubMed PMID: 19129565
48. Sokol RJ, Delaney-Black V, Nordstrom B (2003) Fetal alcohol spectrum disorder. JAMA 290(22):2996–2999. PubMed PMID: 14665662
49. Burgert A, Schlegel J, Becam J, Doose S, Bieberich E, Schubert-Unkmeir A, Sauer M (2017) Characterization of plasma membrane ceramides by super-resolution microscopy. Angew Chem . PubMed PMID: 28379629. https://doi.org/10.1002/anie.201700570
50. Bieberich E (2018) Sphingolipids and lipid rafts: Novel concepts and methods of analysis. Chem Phys Lipids 216:114–131. PubMed PMID: 30194926; PMCID: 6196108. https://doi.org/10.1016/j.chemphyslip.2018.08.003
51. Jiang X, Zhu Z, Qin H, Tripathi P, Zhong L, Elsherbini A, Karki S, Crivelli SM, Zhi W, Wang G, Spassieva SD, Bieberich E (2019) Visualization of ceramide-associated proteins in ceramide-rich platforms using a cross-linkable ceramide analog and proximity ligation assays with anti-ceramide antibody. Front Cell & Develop Biol (7):166. https://doi.org/10.3389/fcell.2019.00166. PubMed PMID: 31475148; PMCID: 6706757
52. Elsherbini A, Qin H, Zhu Z, Tripathi P, Wang G, Crivelli SM, Spassieva SD, Bieberich E (2021) Extracellular vesicles containing ceramide-rich platforms: "Mobile raft" isolation and analysis. Methods Mol Biol 2187:87–98. Epub 2020/08/10. PubMed PMID: 32770502. https://doi.org/10.1007/978-1-0716-0814-2_5
53. Tripathi P, Zhu Z, Qin H, Elsherbini A, Roush EA, Crivelli SM, Spassieva SD, Bieberich E (2021) Cross-link/proximity ligation assay for visualization of lipid and protein complexes in lipid rafts. Methods Mol Biol 2187:337–348. Epub 2020/08/10. PubMed PMID: 32770517. https://doi.org/10.1007/978-1-0716-0814-2_20
54. Bieberich E, Kawaguchi T, Yu RK (2000) N-acylated serinol is a novel ceramide mimic inducing apoptosis in neuroblastoma cells. J Biol Chem 275(1):177–181. Epub 2000/01/05. PubMed PMID: 10617602
55. Wang G, Silva J, Krishnamurthy K, Tran E, Condie BG, Bieberich E (2005) Direct binding to ceramide activates protein kinase Czeta before the formation of a pro-apoptotic complex with PAR-4 in differentiating stem cells. J Biol Chem 280(28):26415–26424. PubMed PMID: 15901738
56. Wang G, Krishnamurthy K, Umapathy NS, Verin AD, Bieberich E (2009) The carboxyl-terminal domain of atypical protein kinase Czeta binds to ceramide and regulates junction formation in epithelial cells. J Biol Chem 284(21):14469–14475. Epub 2009/03/24. doi: M808909200 [pii]. https://doi.org/10.1074/jbc.M808909200. PubMed PMID: 19304661; PMCID: 2682895
57. Guan Y, Zou H, Chen X, Zhao C, Wang J, Cai Y, Chan P, Chen L, Zhang YA (2014) Ischemia, immunosuppression, and SSEA-1-negative cells all contribute to tumors resulting from mouse embryonic stem cell-derived neural progenitor transplantation. J Neurosci Res 92(1):74–85. Epub 2013/10/15. PubMed PMID: 24123213. https://doi.org/10.1002/jnr.23292
58. Jagtap JC, Dawood P, Shah RD, Chandrika G, Natesh K, Shiras A, Hegde AS, Ranade D, Shastry P (2014) Expression and regulation of prostate apoptosis response-4 (Par-4) in human glioma stem cells in drug-induced apoptosis. PLoS One 9(2):e88505. Epub 2014/02/14. PubMed PMID: 24523904; PMCID: PMC3921173. https://doi.org/10.1371/journal.pone.0088505
59. Liu Y, Gilbert MR, Kyprianou N, Rangnekar VM, Horbinski C (2014) The tumor suppressor prostate apoptosis response-4 (Par-4) is regulated by mutant IDH1 and kills glioma stem cells. Acta Neuropathol 128(5):723–732. Epub 2014/08/20. PubMed PMID: 25135281; PMCID: PMC4201966. https://doi.org/10.1007/s00401-014-1334-7

RASSF2 and the PAR-4 Connection

Howard Donninger and Geoffrey J. Clark

Abstract RASSF2 is a member of the RASSF family of tumor suppressors (Donninger H, Schmidt ML, Mezzanotte J, Barnoud T, Clark GJ, Semin Cell Dev Biol 58:86–95, 2016). RASSF proteins all lack enzymatic activity and act by serving as scaffolding and possibly transport proteins for their binding partners. The RASSF proteins can serve as apoptotic effectors of the RAS oncoproteins. They are often inactivated in tumors by epigenetic mechanisms. Indeed, RASSF2 is so frequently inactivated in prostate cancer that methylated RASSF2 promoter DNA detected in the urine is a better indication of prostate cancer than PSA levels (Donninger H, Hesson L, Vos M, Beebe K, Gordon L, Sidransky D, Liu JW, Schlegel T, Payne S, Hartmann A, Mol Cell Biol 30:2608–2620, 2010). The mechanisms by which RASSF proteins, including RASSF2, mediate their biological effects remain only partially understood. However, an important and specific effector of RASSF2 function is the PAR-4 protein. RASSF2 may be able to control both the nuclear localization and the secretion of PAR-4.

Keywords RASSF2 · K-RAS · PAR-4 · Apoptosis · Transformation · Signaling cascades · Promoter methylation · Nuclear shuttling · Protein post-translational modification

H. Donninger
Department of Medicine, University of Louisville, Louisville, KY, USA

University of Louisville School of Medicine, and Experimental Therapeutics Group, James Graham Brown Cancer Center, University of Louisville, Louisville, KY, USA

G. J. Clark (✉)
University of Louisville School of Medicine, and Experimental Therapeutics Group, James Graham Brown Cancer Center, University of Louisville, Louisville, KY, USA

Department of Pharmacology & Toxicology, University of Louisville, Louisville, KY, USA
e-mail: h0donn01@louisville.edu; gjclar01@louisville.edu

V. M. Rangnekar (ed.), *Tumor Suppressor Par-4*,
https://doi.org/10.1007/978-3-030-73572-2_9

1 RASSF2

1.1 Gene Structure

RASSF2 is a member of the RASSF family of tumor suppressors and is frequently inactivated by promoter methylation, particularly in prostate cancer [1, 2]. The *RASSF2* gene is found on the short arm of chromosome 20 (20p13) and occupies approximately 43 kb of genomic DNA [3]. It codes for three splice variants, *RASSF2A*, *RASSF2B*, and *RASSF2C* [3, 4], with *RASSF2A* being the predominant one. *RASSF2A* encodes a protein of 326 amino acids, and while *RASSF2C* contains an identical open reading frame to *RASSF2A*, it is unclear whether this isoform is actually expressed [3, 4]. *RASSF2B* mRNA is predicted to encode a protein of 157 amino acids but is expressed at extremely low levels, and to date, a protein corresponding to the predicted size of *RASSF2B* has yet to be detected [3, 4]. Thus, the protein is merely referred to as RASSF2. RASSF2 mRNA is readily detectable in a wide variety of normal tissues including the brain, lung, spleen, thymus, peripheral blood, small intestine, placenta, heart, colon, stomach, and skeletal muscle, with the highest expression found in the brain and peripheral blood [5–7]. The *RASSF2* promoter has a large CpG island encompassing the first two non-coding exons [3, 4], raising the possibility that, similar to the more well-characterized RASSF1A family member, this gene may also be inactivated by promoter hypermethylation. Indeed, the RASSF2 promoter has been shown to be methylated in numerous cancer cell lines and primary cancers, including colorectal, gastric, nasopharyngeal, lung, breast, oral squamous, thyroid, pancreatic, cervical, Ewing sarcoma, and ovarian carcinomas [4, 7–21].

1.2 RASSF2 Protein Structure and Subcellular Localization

The *RASSF2* gene encodes a 326 amino acid protein with domains that are conserved between other classical RASSF family members. These include a Ras-association domain (amino acids 175–265) and an acidic coiled-coil Sav/RASSF/Hpo (SARAH) domain (amino acids 272–319) both located toward the C-terminus (Fig. 1). In addition, RASSF2 contains a nuclear localization signal (NLS) encompassing amino acids 151–167 as well as a nuclear export signal (amino acids 240–260; Fig. 1) which is required for export of RASSF2 out of the nucleus [22]. RASSF2 shares 14% amino acid identity with RASSF1A and the RA domain of RASSF2 shares 28% and 31% identity to the RA domains of RASSF1A and NORE1A, respectively [4, 23]. RASSF2 binds directly to K-RAS in a GTP-dependent manner via the RA domain [5, 24] and interacts with other SARAH motif, containing proteins such as the MST kinases via it's SARAH domain [25]. RASSF2 is predominantly located in the nucleus, however, it can shuttle between

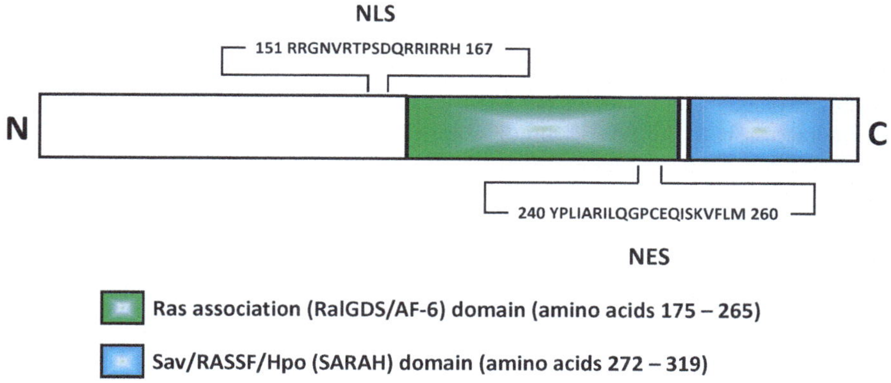

Fig. 1 RASSF2 protein structure
RASSF2 is a 326 amino acid protein containing a Ras association domain, a SARAH motif, a nuclear localization signal (NLS), and a nuclear export signal (NES)

the nucleus and cytoplasm [14, 22, 26]. Localization of RASSF2 may be cell context dependent and may be regulated by phosphorylation [22, 25].

1.3 RASSF2 Function

RASSF2 has potent tumor suppressor properties. Overexpression of RASSF2 inhibits cell proliferation and induces apoptosis in vitro and inhibits tumor growth in vivo [5, 14]. Conversely, loss of RASSF2 expression enhances transformation and promotes invasion [8, 24]. RASSF2 is a bona fide RAS effector and forms an endogenous complex preferentially with K-RAS, and the interaction occurs more favorably when K-RAS is in its active, GTP bound state [5, 24, 27]. Loss of RASSF2 expression significantly enhances the transforming potential of K-RAS [8, 24], and there appears to be a correlation between RASSF2 inactivation and K-RAS/BRAF mutations in primary tumors [8, 11, 12]. Like the better characterized RASSF1A and NORE1A family members, RASSF2 lacks the ability to bind to DNA and has no enzymatic activity and likely acts as a scaffold interacting with and targeting multiple binding partners to distinct subcellular compartments, thereby regulating multiple growth inhibitory pathways. Being a tumor suppressor, RASSF2 functions in promoting cell death and apoptosis, and the way it affects these pro-apoptotic functions occurs by interacting with various effectors. RASSF2 interacts with both the Mst1/2 kinases [25, 28], which are central regulators of the Hippo signaling pathway [29], but as yet, it is uncertain whether this interaction modulates the Hippo signaling pathway. The interaction of RASSF2 with the Mst1 and Mst2 differs in that RASSF2 appears to stabilize Mst2 [25], whereas the interaction with Mst1 stabilizes RASSF2 [28]. RASSF2 also enhances Mst1 autophosphorylation and increases its kinase activity to select substrates [28]. Additionally, RASSF2 appears to be a substrate for Mst2 and is phosphorylated by Mst2 [25], however, it is unclear how this affects RASSF2 function but does suggest that RASSF2 may be activated by Mst2 and may regulate apoptosis independently of the Hippo signaling pathway. A further mechanism by which RASSF2 exerts its pro-apoptotic activity is through its interaction with the PAR-4 tumor suppressor [2]. RASSF2 may have functions extending beyond its role as a tumor suppressor as it forms complexes with a variety of proteins that have key functions in diverse biological processes [30]. RASSF2 may have a role in metabolism since it interacts with C1QBP [30], a multifunctional, multiligand binding chaperone that is found predominantly in the mitochondria [31, 32], where it serves a critical role in regulating metabolism [32, 33]. C1QBP is also a substrate for MAP kinase and can translocate to the nucleus where it plays a role in homologous recombination [34, 35] suggesting that RASSF2 may also function in DNA damage repair. A further RASSF2 interacting partner is vimentin [30], a type III intermediate filament protein that is widely expressed across many cell types and is a marker of epithelial-mesenchymal transition [36]. It forms an integral part of the cytoskeleton and functions in crucial biological processes including cell adhesion, motility, and migration [37–39] as well as apoptosis [40, 41], thereby implicating RASSF2 in the regulation of these biological processes. Intriguingly, RASSF2 induces deacetylation of vimentin in a K-Ras-dependent fashion which is mediated by SIRT1 [30], and RASSF2 interacts with SIRT1 as well as histone deacetylase 6 (HDAC6; unpublished observations). SIRT1 is an NAD^+-dependent deacetylase, and HDAC6 is a member of a family of zinc-dependent isoenzymes, both of which modulate the acetylation of multiple targets and play critical roles in diverse biological processes including metabolism, immune response, autophagy, cell migration, and adhesion [42–53]. The interaction of RASSF2 with SIRT1 and HDAC6 implicates RASSF2 in the regulation of protein acetylation and may also connect K-Ras to regulation of the acetylome, since RASSF2 action on protein acetylation may be K-Ras-dependent [30]. Thus, RASSF2 may have functions extending beyond controlling cell proliferation and apoptosis and may couple K-Ras to biological functions not previously ascribed to this oncogene (Fig. 2). In fact, RASSF2 may be the most important member of the RASSF family as it is the only member whose loss in vivo results in severe growth retardation and subsequent post-natal mortality shortly after birth [54].

2 PAR-4

As has been described at length elsewhere in this book, PAR-4 (prostate apoptosis response protein 4) is a multifaceted tumor suppressor with an intriguing ability to selectively promote cell death in tumor cells [55]. PAR-4 is a 40 kDa protein that is widely expressed and contains three main functional domains: (1) two nuclear localization sequences, (2) a leucine-zipper domain, and (3) a SAC domain (Selective domain for Apoptosis induction in Cancer cells). It can activate the intrinsic apoptotic pathways, a process that requires its nuclear localization, and the

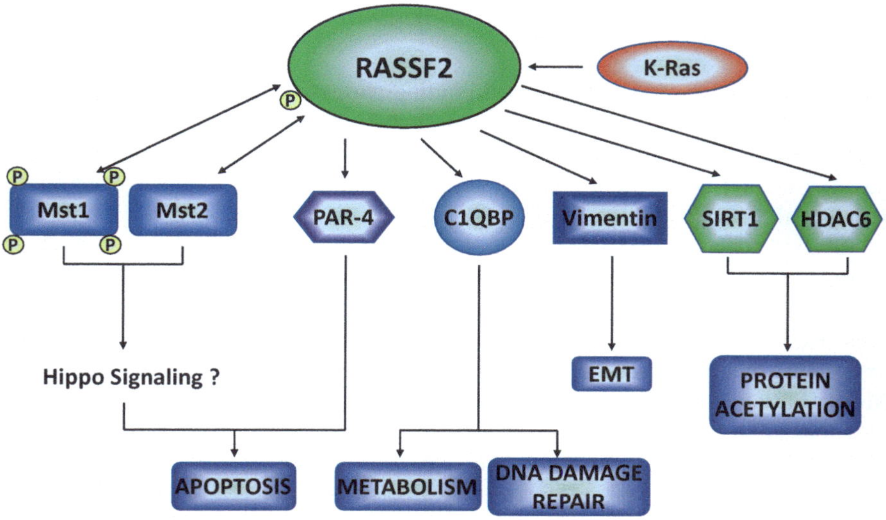

Fig. 2 Summary of the biological processes, both known and predicted, regulated by RASSF2 RASSF2 interacts with multiple binding partners to potentially regulate diverse biological processes, including apoptosis, DNA damage repair, metabolism, epithelial to mesenchymal transition (EMT), and protein acetylation

extrinsic apoptotic pathways via secretion and binding to GRP78 on the cell surface [56]. In addition to apoptosis, the PAR-4 protein has also been implicated in the induction of autophagy and senescence, as well as in the suppression of EMT [57–59]. So it has a broad range of biological functions. The factors that regulate these powerful and complex biological properties remain only partially understood.

Of particular interest to cancer is the role of PAR-4 in repressing transformation induced by the notorious K-RAS oncoprotein. K-RAS is the most frequently activated oncoprotein yet detected in human cancer. It may be responsible for driving the majority of human cancers due to mutations in the gene or by defects in RAS-regulating proteins [60]. However, in addition to promoting growth and transformation, deregulation of K-RAS can also stimulate apoptosis [61]. PAR-4 contributes to this RAS-induced apoptosis and can block RAS-induced transformation due to induced cell death. Indeed, RAS-mediated transformation appears to require the suppression of PAR-4 expression [62, 63]. This suggests there may be a link between K-RAS and PAR-4. The RAS effector protein RASSF2 serves as such a link.

2.1 The RASSF2 and PAR-4 Interaction

PAR-4 was identified as a two-hybrid binding partner in a screen with full-length RASSF2 [2]. The prey consisted of residues 68–340 of PAR-4, and so the first NLS of PAR-4 appears unnecessary for the interaction. Subsequent studies using RASSF2 deletion mutant GST-fusion proteins suggested that RASSF2 interacts at multiple points with PAR-4.

RASSF2 is bound by activated K-RAS and serves as a pro-apoptotic K-RAS effector [5]. Thus, if PAR-4 serves as part of a K-RAS/RASSF2 apoptotic signaling pathway, we might expect the interaction of RASSF2 and PAR-4 to increase in the presence of activated K-RAS. Indeed, co-transfection of activated K-RAS to an exogenous overexpression system enhanced the RASSF2/PAR-4 interaction [2]. Moreover, knocking down activated K-RAS in lung cancer cells reduced the complex formation between PAR-4 AND RASSF2.

PAR-4 has potent apoptotic potential in Ras-driven cancer cells, and overexpression of PAR-4 in such cell systems can completely abrogate cell survival in colony formation assays. However, knocking down RASSF2 suppressed this PAR-4-mediated apoptosis [2]. The mechanism may involve effects on the subcellular localization of PAR-4. Nuclear localization is critical for PAR-4 to induce the intrinsic apoptotic pathways, in part at least by its inhibitory effects on NF-κB. Activated RAS drives PAR-4 into the nucleus, and this is dependent upon the presence of RASSF2. RASSF2 binds importin alpha and exhibits a substantial nuclear localization in its resting state. RASSF proteins can shuttle in and out of the nucleus [26], thus, it makes sense that RASSF2 promotes the nuclear localization of PAR-4 [2]. Moreover, RASSF2 has been reported to suppress the NF-κB pathway [54], and the connection between RASSF2 and PAR-4 nuclear localization can probably explain this effect.

In addition to its nuclear-based functions, PAR-4 also acts as a secreted protein that binds cell surface GRP78 to stimulate apoptosis. GRP78 is usually an ER (endoplasmic reticulum) localized protein involved in the ER stress response. However, at the plasma membrane, the binding of PAR-4 causes the assembly of a FADD/caspase 8 complex that triggers the extrinsic apoptosis pathway via caspase 3 [56]. This aspect may be responsible for the cancer cell preferential killing effects of PAR-4, as GRP78 tends to be preferentially expressed on cancer cells. The secretion of PAR-4 is regulated, in part, by sequestration with vimentin, an intermediate cytoskeletal filament [64]. Activated RAS promotes the release of PAR-4 from the sequestration complex. RASSF2 may provide a mechanistic explanation for this effect as we observed complex formation between vimentin and RASSF2 [30]. This complex is suppressed by activated RAS, so this may allow the release of PAR-4 from the vimentin complex. Indeed, suppression of RASSF2 by microRNA leads to enhanced secretion of PAR-4 [65]. Thus, in addition to modulating the nuclear localization of PAR-4, RASSF2 may also modulate the secretory/GRP78 pathway of PAR-4. Perhaps RASSF2 levels may even determine the relative activity of each of these pathways.

Our proteomic studies of RASSF2 have revealed several other potential RASSF2/PAR-4 connections. PAR-4 is regulated by phosphorylation by PKA. RASSF2 can complex with the regulatory sub-unit of PKA and so may influence the phosphorylation state of PAR-4. In addition, RASSF2 and other RASSF proteins can complex with a variety of phosphatases which may also allow a level of PAR-4 regulation [30].

Although RASSF2 is one of the least studied RASSF family members, it may be the most important. Not only is it very frequently inactivated in human tumors, it is also the only family member that is essential for viability in transgenic mice [54]. In our experiments, we have not detected any obvious interaction between PAR-4 and other RASSF family members such as RASSF1A and RASSF5. Perhaps the RASSF2-specific nature of the interaction with PAR-4 contributes to the essential role of RASSF2 in life.

3 Conclusions and Future Perspectives

Activation of RAS signaling is arguably the most frequently occurring driver of human cancer. RAS exerts it proliferative effects by activating its effectors such as MAP kinase, PI3 kinase, and RALGDS. Similarly, RAS can inhibit growth and induce apoptosis via interaction with a distinct set of effectors, the best studied of which are the RASSF family of tumor suppressors. While RASSF2 is not the best characterized of this family, it may be the most important member and has potent apoptotic activity. Loss of RASSF2 expression dramatically potentiates the transforming effects of activated K-RAS. Similarly, inactivation of PAR-4 is required for the full transforming activity of activated K-RAS to manifest. The RASSF2/PAR-4 interaction is critical for PAR-4 function and for restraining the transforming potential of activated K-RAS. RASSF2 regulation of PAR-4 activity may be more complex than merely re-localizing it to the nucleus and may involve RASSF2-mediated post-translational modification of PAR-4. This may further impact the activity of PAR-4 and may be crucial for correct PAR-4 function in biological processes extending beyond its impact on K-RAS signaling, such as autophagy, senescence, and EMT, processes that RASSF2 has also been implicated in regulating. A thorough investigation of the impact of the RASSF2/PAR-4 interaction on these biological processes will provide further insight into the function of these proteins and may also provide evidence of the extent of involvement of RAS in the regulation of these pivotal processes.

References

1. Donninger H, Schmidt ML, Mezzanotte J, Barnoud T, Clark GJ (2016) Ras signaling through RASSF proteins. Semin Cell Dev Biol 58:86–95
2. Donninger H, Hesson L, Vos M, Beebe K, Gordon L, Sidransky D, Liu JW, Schlegel T, Payne S, Hartmann A et al (2010) The Ras effector RASSF2 controls the PAR-4 tumor suppressor. Mol Cell Biol 30:2608–2620

3. Hesson LB, Latif F (2008–2009) RASSF2 (Ras association (RalGDS/AF-6) domain family member 2). http://atlasgeneticsoncology.org/Genes/GC_RASSF2.html
4. Hesson LB, Wilson R, Morton D, Adams C, Walker M, Maher ER, Latif F (2005) CpG island promoter hypermethylation of a novel Ras-effector gene RASSF2A is an early event in colon carcinogenesis and correlates inversely with K-ras mutations. Oncogene 24:3987–3994
5. Vos MD, Ellis CA, Elam C, Ulku AS, Taylor BJ, Clark GJ (2003) RASSF2 is a novel K-Ras-specific effector and potential tumor suppressor. J Biol Chem 278:28045–28051
6. RASSF2. https://www.proteinatlas.org/ENSG00000101265-RASSF2/tissue
7. Maruyama R, Akino K, Toyota M, Suzuki H, Imai T, Ohe-Toyota M, Yamamoto E, Nojima M, Fujikane T, Sasaki Y et al (2008) Cytoplasmic RASSF2A is a proapoptotic mediator whose expression is epigenetically silenced in gastric cancer. Carcinogenesis 29:1312–1318
8. Akino K, Toyota M, Suzuki H, Mita H, Sasaki Y, Ohe-Toyota M, Issa JP, Hinoda Y, Imai K, Tokino T (2005) The Ras effector RASSF2 is a novel tumor-suppressor gene in human colorectal cancer. Gastroenterology 129:156–169
9. Endoh M, Tamura G, Honda T, Homma N, Terashima M, Nishizuka S, Motoyama T (2005) RASSF2, a potential tumour suppressor, is silenced by CpG island hypermethylation in gastric cancer. Br J Cancer 93:1395–1399
10. Zhang Z, Sun D, Van Do N, Tang A, Hu L, Huang G (2007) Inactivation of RASSF2A by promoter methylation correlates with lymph node metastasis in nasopharyngeal carcinoma. Int J Cancer 120:32–38
11. Park HW, Kang HC, Kim IJ, Jang SG, Kim K, Yoon HJ, Jeong SY, Park JG (2007) Correlation between hypermethylation of the RASSF2A promoter and K-ras/BRAF mutations in microsatellite-stable colorectal cancers. Int J Cancer 120:7–12
12. Nosho K, Yamamoto H, Takahashi T, Mikami M, Taniguchi H, Miyamoto N, Adachi Y, Arimura Y, Itoh F, Imai K et al (2007) Genetic and epigenetic profiling in early colorectal tumors and prediction of invasive potential in pT1 (early invasive) colorectal cancers. Carcinogenesis 28:1364–1370
13. Kaira K, Sunaga N, Tomizawa Y, Yanagitani N, Ishizuka T, Saito R, Nakajima T, Mori M (2007) Epigenetic inactivation of the RAS-effector gene RASSF2 in lung cancers. Int J Oncol 31:169–173
14. Cooper WN, Dickinson RE, Dallol A, Grigorieva EV, Pavlova TV, Hesson LB, Bieche I, Broggini M, Maher ER, Zabarovsky ER et al (2008) Epigenetic regulation of the ras effector/tumour suppressor RASSF2 in breast and lung cancer. Oncogene 27:1805–1811
15. Imai T, Toyota M, Suzuki H, Akino K, Ogi K, Sogabe Y, Kashima L, Maruyama R, Nojima M, Mita H et al (2008) Epigenetic inactivation of RASSF2 in oral squamous cell carcinoma. Cancer Sci 99:958–966
16. Huang KH, Huang SF, Chen IH, Liao CT, Wang HM, Hsieh LL (2009) Methylation of RASSF1A, RASSF2A, and HIN-1 is associated with poor outcome after radiotherapy, but not surgery, in oral squamous cell carcinoma. Clin Cancer Res 15:4174–4180
17. Schagdarsurengin U, Richter AM, Hornung J, Lange C, Steinmann K, Dammann RH (2010) Frequent epigenetic inactivation of RASSF2 in thyroid cancer and functional consequences. Mol Cancer 9:264
18. Zhao L, Cui Q, Lu Z, Chen J (2012) Aberrant methylation of RASSF2A in human pancreatic ductal adenocarcinoma and its relation to clinicopathologic features. Pancreas 41:206–211
19. Guerrero-Setas D, Perez-Janices N, Blanco-Fernandez L, Ojer A, Cambra K, Berdasco M, Esteller M, Maria-Ruiz S, Torrea N, Guarch R (2013) RASSF2 hypermethylation is present and related to shorter survival in squamous cervical cancer. Mod Pathol 26:1111–1122
20. Gharanei S, Brini AT, Vaiyapuri S, Alholle A, Dallol A, Arrigoni E, Kishida T, Hiruma T, Avigad S, Grimer R et al (2013) RASSF2 methylation is a strong prognostic marker in younger age patients with Ewing sarcoma. Epigenetics 8:893–898
21. Ren F, Wang DB, Li T, Chen YH, Li Y (2014) Identification of differentially methylated genes in the malignant transformation of ovarian endometriosis. J Ovarian Res 7:73
22. Kumari G, Mahalingam S (2009) Extracellular signal-regulated kinase 2 (ERK-2) mediated phosphorylation regulates nucleo-cytoplasmic shuttling and cell growth control of Ras-associated tumor suppressor protein, RASSF2. Exp Cell Res 315:2775–2790

23. van der Weyden L, Adams DJ (2007) The Ras-association domain family (RASSF) members and their role in human tumourigenesis. Biochim Biophys Acta 1776:58–85
24. Clark J, Freeman J, Donninger H (2012) Loss of RASSF2 enhances tumorigencity of lung cancer cells and confers resistance to chemotherapy. Mol Biol Int 2012:705948
25. Cooper WN, Hesson LB, Matallanas D, Dallol A, von Kriegsheim A, Ward R, Kolch W, Latif F (2009) RASSF2 associates with and stabilizes the proapoptotic kinase MST2. Oncogene 28:2988–2998
26. Kumari G, Singhal PK, Rao MR, Mahalingam S (2007) Nuclear transport of Ras-associated tumor suppressor proteins: different transport receptor binding specificities for arginine-rich nuclear targeting signals. J Mol Biol 367:1294–1311
27. Calvisi DF, Ladu S, Gorden A, Farina M, Conner EA, Lee JS, Factor VM, Thorgeirsson SS (2006) Ubiquitous activation of Ras and Jak/Stat pathways in human HCC. Gastroenterology 130:1117–1128
28. Song H, Oh S, Oh HJ, Lim DS (2010) Role of the tumor suppressor RASSF2 in regulation of MST1 kinase activity. Biochem Biophys Res Commun 391:969–973
29. Pan D (2010) The hippo signaling pathway in development and cancer. Dev Cell 19:491–505
30. Barnoud T, Wilkey DW, Merchant ML, Clark JA, Donninger H (2016) Proteomics analysis reveals novel RASSF2 interaction partners. Cancers (Basel) 8:37
31. Dedio J, Jahnen-Dechent W, Bachmann M, Muller-Esterl W (1998) The multiligand-binding protein gC1qR, putative C1q receptor, is a mitochondrial protein. J Immunol 160:3534–3542
32. Gotoh K, Morisaki T, Setoyama D, Sasaki K, Yagi M, Igami K, Mizuguchi S, Uchiumi T, Fukui Y, Kang D (2018) Mitochondrial p32/C1qbp is a critical regulator of dendritic cell metabolism and maturation. Cell Rep 25(1800–1815):e1804
33. Fogal V, Richardson AD, Karmali PP, Scheffler IE, Smith JW, Ruoslahti E (2010) Mitochondrial p32 protein is a critical regulator of tumor metabolism via maintenance of oxidative phosphorylation. Mol Cell Biol 30:1303–1318
34. Bai Y, Wang W, Li S, Zhan J, Li H, Zhao M, Zhou XA, Li S, Li X, Huo Y et al (2019) C1QBP promotes homologous recombination by stabilizing MRE11 and controlling the assembly and activation of MRE11/RAD50/NBS1 complex. Mol Cell 75(1299–1314):e1296
35. Majumdar M, Meenakshi J, Goswami SK, Datta K (2002) Hyaluronan binding protein 1 (HABP1)/C1QBP/p32 is an endogenous substrate for MAP kinase and is translocated to the nucleus upon mitogenic stimulation. Biochem Biophys Res Commun 291:829–837
36. Yang J, Weinberg RA (2008) Epithelial-mesenchymal transition: at the crossroads of development and tumor metastasis. Dev Cell 14:818–829
37. Ivaska J, Pallari HM, Nevo J, Eriksson JE (2007) Novel functions of vimentin in cell adhesion, migration, and signaling. Exp Cell Res 313:2050–2062
38. Mendez MG, Kojima S, Goldman RD (2010) Vimentin induces changes in cell shape, motility, and adhesion during the epithelial to mesenchymal transition. FASEB J 24:1838–1851
39. Zhu QS, Rosenblatt K, Huang KL, Lahat G, Brobey R, Bolshakov S, Nguyen T, Ding Z, Belousov R, Bill K et al (2011) Vimentin is a novel AKT1 target mediating motility and invasion. Oncogene 30:457–470
40. Su L, Pan P, Yan P, Long Y, Zhou X, Wang X, Zhou R, Wen B, Xie L, Liu D (2019) Role of vimentin in modulating immune cell apoptosis and inflammatory responses in sepsis. Sci Rep 9:5747
41. Byun Y, Chen F, Chang R, Trivedi M, Green KJ, Cryns VL (2001) Caspase cleavage of vimentin disrupts intermediate filaments and promotes apoptosis. Cell Death Differ 8:443–450
42. Haigis MC, Sinclair DA (2010) Mammalian sirtuins: biological insights and disease relevance. Annu Rev Pathol 5:253–295
43. Chang HC, Guarente L (2014) SIRT1 and other sirtuins in metabolism. Trends Endocrinol Metab 25:138–145
44. Wang Y, Bi Y, Chen X, Li C, Li Y, Zhang Z, Wang J, Lu Y, Yu Q, Su H et al (2016) Histone deacetylase SIRT1 negatively regulates the differentiation of interleukin-9-producing CD4(+) T cells. Immunity 44:1337–1349

45. Elesela S, Morris SB, Narayanan S, Kumar S, Lombard DB, Lukacs NW (2020) Sirtuin 1 regulates mitochondrial function and immune homeostasis in respiratory syncytial virus infected dendritic cells. PLoS Pathog 16:e1008319
46. Lee JY, Koga H, Kawaguchi Y, Tang W, Wong E, Gao YS, Pandey UB, Kaushik S, Tresse E, Lu J et al (2010) HDAC6 controls autophagosome maturation essential for ubiquitin-selective quality-control autophagy. EMBO J 29:969–980
47. Winkler R, Benz V, Clemenz M, Bloch M, Foryst-Ludwig A, Wardat S, Witte N, Trappiel M, Namsolleck P, Mai K et al (2012) Histone deacetylase 6 (HDAC6) is an essential modifier of glucocorticoid-induced hepatic gluconeogenesis. Diabetes 61:513–523
48. Li Y, Shin D, Kwon SH (2013) Histone deacetylase 6 plays a role as a distinct regulator of diverse cellular processes. FEBS J 280:775–793
49. Valenzuela-Fernandez A, Cabrero JR, Serrador JM, Sanchez-Madrid F (2008) HDAC6: a key regulator of cytoskeleton, cell migration and cell-cell interactions. Trends Cell Biol 18:291–297
50. Moreno-Gonzalo O, Mayor F Jr, Sanchez-Madrid F (2018) HDAC6 at crossroads of infection and innate immunity. Trends Immunol 39:591–595
51. Lee IH, Cao L, Mostoslavsky R, Lombard DB, Liu J, Bruns NE, Tsokos M, Alt FW, Finkel T (2008) A role for the NAD-dependent deacetylase Sirt1 in the regulation of autophagy. Proc Natl Acad Sci U S A 105:3374–3379
52. Zhang Y, Zhang M, Dong H, Yong S, Li X, Olashaw N, Kruk PA, Cheng JQ, Bai W, Chen J et al (2009) Deacetylation of cortactin by SIRT1 promotes cell migration. Oncogene 28:445–460
53. Tang BL (2010) Sirt1 and cell migration. Cell Adhes Migr 4:163–165
54. Song H, Kim H, Lee K, Lee DH, Kim TS, Song JY, Lee D, Choi D, Ko CY, Kim HS et al (2012) Ablation of Rassf2 induces bone defects and subsequent haematopoietic anomalies in mice. EMBO J 31:1147–1159
55. Hebbar N, Wang C, Rangnekar VM (2012) Mechanisms of apoptosis by the tumor suppressor Par-4. J Cell Physiol 227:3715–3721
56. Burikhanov R, Zhao Y, Goswami A, Qiu S, Schwarze SR, Rangnekar VM (2009) The tumor suppressor Par-4 activates an extrinsic pathway for apoptosis. Cell 138:377–388
57. Katoch A, Jamwal VL, Faheem MM, Kumar S, Senapati S, Yadav G, Gandhi SG, Goswami A (2021) Overlapping targets exist between the Par-4 and miR-200c axis which regulate EMT and proliferation of pancreatic cancer cells. Transl Oncol 14:100879
58. Thayyullathil F, Cheratta AR, Pallichankandy S, Subburayan K, Tariq S, Rangnekar VM, Galadari S (1867) Par-4 regulates autophagic cell death in human cancer cells via upregulating p53 and BNIP3. Biochim Biophys Acta, Mol Cell Res 2020:118692
59. Subburayan K, Thayyullathil F, Pallichankandy S, Rahman A, Galadari S (2018) Par-4-dependent p53 up-regulation plays a critical role in thymoquinone-induced cellular senescence in human malignant glioma cells. Cancer Lett 426:80–97
60. Prior IA, Lewis PD, Mattos C (2012) A comprehensive survey of Ras mutations in cancer. Cancer Res 72:2457–2467
61. Downward J (1998) Ras signalling and apoptosis. Curr Opin Genet Dev 8:49–54
62. Qiu SG, Krishnan S, El-Guendy N, Rangnekar VM (1999) Negative regulation of Par-4 by oncogenic Ras is essential for cellular transformation. Oncogene 18:7115–7123
63. Pruitt K, Ulku AS, Frantz K, Rojas RJ, Muniz-Medina VM, Rangnekar VM, Der CJ, Shields JM (2005) Ras-mediated loss of the pro-apoptotic response protein Par-4 is mediated by DNA hypermethylation through Raf-independent and Raf-dependent signaling cascades in epithelial cells. J Biol Chem 280:23363–23370
64. Burikhanov R, Sviripa VM, Hebbar N, Zhang W, Layton WJ, Hamza A, Zhan CG, Watt DS, Liu C, Rangnekar VM (2014) Arylquins target vimentin to trigger Par-4 secretion for tumor cell apoptosis. Nat Chem Biol 10:924–926
65. Shen Z, Qin X, Yan M, Li R, Chen G, Zhang J, Chen W (2017) Cancer-associated fibroblasts promote cancer cell growth through a miR-7-RASSF2-PAR-4 axis in the tumor microenvironment. Oncotarget 8:1290–1303

Regulation of Tumor Suppressor Par-4 by Ceramide

Sehamuddin Galadari, Anees Rahman Cheratta, and Faisal Thayyullathil

Abstract Prostate apoptosis response-4 (Par-4) is a pro-apoptotic protein that selectively induces apoptosis in cancer cells, thereby exerting prominent tumor-suppressing functions. Ceramides are important bioactive lipids belonging to the sphingolipid family. Both Par-4 and ceramide are critical regulators of cellular responses to acute stressors and has been clearly established as a key player in the execution of cell death. Though Par-4 and ceramide act as an apparent collaborator of cell-fate decisions, the relationship between these molecules is very complex, and mechanisms underlying their regulation are diverse and not fully characterized. In this chapter, we portray the role and mechanisms of action of Par-4 and ceramide in apoptosis and autophagy. Specifically, we are elaborating on the central role of Par-4 in ceramide-induced cell death signaling. Understanding the Par-4-ceramide connection will not only have profound importance to the understanding of programmed cell death pathways, but also will have an impact on interventions of cancer therapy and prevention.

Keywords Apoptosis · Autophagy · Caspases · Cell death · Ceramide · Mitochondria · NF-κB · Par-4 · ζPKC · PKA · Tumor suppression.

1 Introduction

Cells may die from accidental cell death or programmed cell death (PCD). PCD involves highly organized signaling cascades and molecularly defined effector mechanisms [3]. PCD plays a fundamental role in the biological process such as maintenance of tissue homeostasis, development of immune system, elimination of damaged cells, and sculpting structures during organogenesis [40]. Abnormal regulation of PCD is associated with human diseases, including developmental disorders, immunological disorders, cancer, and neurodegeneration [40, 69]. Both extrinsic signals and internal imbalances can lead to the activation of PCD. Studies

S. Galadari (✉) · A. R. Cheratta · F. Thayyullathil
Cell Death Signaling Laboratory, Division of Science (Biology), Experimental Research Building, New York University Abu Dhabi, Abu Dhabi, UAE
e-mail: sehamuddin@nuy.edu; arc10@nyu.edu; ftt1@nyu.edu

V. M. Rangnekar (ed.), *Tumor Suppressor Par-4*,
https://doi.org/10.1007/978-3-030-73572-2_10

over past decades have explored and elucidated several PCD pathways (apoptosis, autophagy, necroptosis, ferroptosis, etc.). Among them, apoptosis (type I PCD) and autophagy (type II PCD) are the best-studied forms of PCDs [56].

2 Apoptosis

Apoptosis is a form of PCD characterized by cell shrinkage, chromatin-cytoplasmic condensation, DNA fragmentation, and membrane blebbing [138]. In mammals, apoptosis occurs through two distinct, ultimately converging pathways: the intrinsic (mitochondrial) and the extrinsic (death-receptor) pathways [67]. A key player in the initiation and execution of both pathways is caspases, a family of specialized cysteine-dependent aspartate-directed proteases. Correspond to their positions in the apoptotic signaling cascade, caspases are classified into initiator (caspase-1, −2, −4, −5, −8, −9, −10, −11, −12) and effector (caspase-3, −6, −7) caspases. Initiator caspases are involved in interacting with the upstream adapter molecules, whereas effector caspases perform downstream execution of apoptosis [87].

The extrinsic pathway of apoptosis is mediated by the ligation of death-inducing ligands such as Fas ligand (FasL), TNF-α, and TNF-related apoptosis-inducing ligand (TRAIL) to their respective receptors, such as the Fas receptor (FasR), the TNF receptor (TNFR), and the death receptors (DR4 and DR5). The ligand-receptor interaction recruits adaptor protein (FADD and TRADD) and pro-caspase, leading to death-inducing signaling complex (DISC) formation and subsequent activation of caspase-8 and apoptosis (Fig. 1) [43–45]. On the contrary, intrinsic pathway is initiated by a disruption of mitochondria in response to various exogenous and endogenous stimuli (ionizing radiation, chemotherapy, ROS generation, DNA damage, unfolded protein response, and growth factor deprivation), leading to mitochondrial outer membrane permeabilization (MOMP) and release of apoptogenic factors such as cytochrome-c, HtrA2/Omi, second mitochondria-derived activator of caspase/direct inhibitor of apoptosis-binding protein with low pI (Smac/DIABLO), apoptosis-inducing factor (AIF), etc. Once released, these factors orchestrate the cell death process via caspase-dependent or -independent manner. For instance, cytochrome-c forms “apoptosome” complex with apoptotic protease activating factor 1 (Apaf1) which activates caspase-9 and downstream executioner caspases [74]. The Smac/DIABLO and HtrA2/Omi accelerate caspase activation by antagonizing inhibitors of apoptosis proteins (IAPs) family [81].

Apoptosis is tightly regulated by a variety of pro-apoptotic and anti-apoptotic proteins, viz. B-cell lymphoma-2 (Bcl-2) family proteins, IAPs, etc. Bcl-2 family composed of both pro-apoptotic (e.g., Bad, Bid, Bax, Bak, Bcl-Xs, Bim, Noxa, Puma, etc.) and anti-apoptotic (e.g., Bcl-2 itself, Bcl-xL, Bcl-W, Mcl-1, etc.) members [85]. In the case of intrinsic apoptotic pathway, accumulation of pro-apoptotic members causes mitochondrial membrane permeabilization and release of apoptogenic factors, while accumulation of anti-apoptotic members inhibits this process

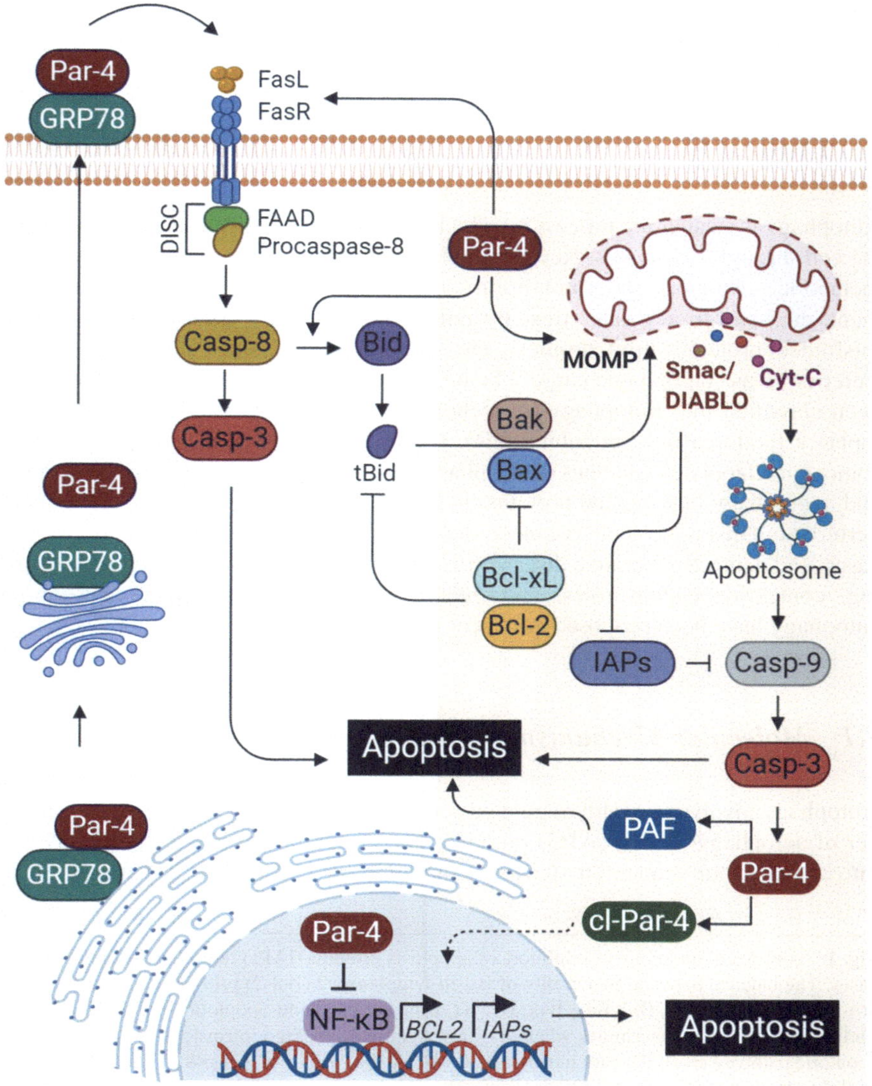

Fig. 1 Par-4-mediated apoptosis mechanisms. The extrinsic and intrinsic apoptotic pathway regulated by Par-4 is illustrated. The extrinsic pathway of apoptosis is activated upon ligation of receptors such as TNF receptor (TNFR), Fas receptor (FasR), or death receptors (DR4/DR5) by their respective cognate ligands TNF-α, FasL, or TRAIL. The ligation recruits adaptor proteins (FADD and TRADD) and pro-caspase, leading to the assembly of death-inducing signaling complex (DISC) and subsequent activation of caspase cascade and apoptosis. The intrinsic pathway of apoptosis involves mitochondrial membrane permeabilization (MOMP) and release of apoptogenic factors such as cytochrome-c (Cyt-C), second mitochondria-derived activator of caspase/direct inhibitor of apoptosis-binding protein with low pI (Smac/DIABLO), etc., into the cytosol. Cyt-C acts to form apoptosome and activate caspase cascade. Smac/DIABLO binds to and

(Fig. 1) [117]. On the other hand, IAPs (e.g., cIAP1, cIAP2, and XIAP) directly bind and inhibit caspases, and thus negatively regulate the apoptosis [90].

3 Autophagy

Autophagy is a catabolic process by which cellular components are degraded inside the cell through lysosomes. Key functions of autophagy are to provide energy and metabolic precursors under conditions of starvation, cell growth, organ development, etc., and to alleviate stress by removal of damaged organelles, pathogens, misfolded proteins, lipid droplets, etc. [46]. Based on the cellular component (referred to as "autophagic cargo") being delivered for degradation, autophagy has been classified into mitophagy (mitochondria), pexophagy (peroxisomes), reticulophagy (endoplasmic reticulum; ER), lysophagy (lysosomes), ribophagy (ribosomes), nucleophagy (nucleus), xenophagy (pathogens), lipophagy (lipid droplets), and aggrephagy (aggregated proteins) [47]. Interestingly, autophagy is often found perturbed in disorders such as cancer, diabetes, cardiovascular-liver-respiratory diseases, and neurodegenerative diseases [91, 94, 96]. In cancer, autophagy has opposing, context-dependent roles, and interventions to both stimulate and inhibit autophagy have been proposed as cancer therapies.

3.1 Molecular Mechanisms of Autophagy

Autophagy involves a highly regulated set of signaling events controlled by a number of autophagy-related (ATG) proteins. The events in autophagy can be divided into at least four sequential steps: (1) initiation, (2) nucleation and formation of

Fig. 1 (continued) antagonizes inhibitors of apoptosis proteins (IAPs) family, which inhibits caspases. The balance between two family of B-cell lymphoma-2 (Bcl-2) proteins composed of both pro-apoptotic (e.g., Bad, Bid, Bax, Bak, Bcl-Xs, Bim, etc.) and anti-apoptotic (e.g., Bcl-2, Bcl-xL, Mcl-1, etc.) members determines whether or not a cell will undergo intrinsic apoptosis. Extrinsic apoptotic pathway crosstalk with intrinsic pathway via the caspase-8 (Casp-8) induced conversion of Bid to truncated-Bid (tBid) leading to mitochondrial permeabilization. Par-4 induces FasL and FasR mobilization to the cell membrane facilitating DISC formation and caspase activation leading to the extrinsic apoptotic pathway. ER stress causes translocation of the Par-4-GRP78 complex from the endoplasmic reticulum (ER) to the plasma membrane. Secreted extracellular Par-4 interacts with cell surface GRP78 and facilitates the extrinsic pathway of apoptosis. Par-4 promotes MOMP and engagement of the intrinsic apoptotic pathway. Par-4 also promotes caspase-8-mediated Bid activation and its translocation to the mitochondria; resulting in an augmentation of apoptogenic factors release into the cytoplasm. Inside the nucleus, Par-4 inhibits the transcription factor NF-κB to promote apoptosis. Caspase-mediated cleavage of Par-4 generates PAF and cl-Par-4. The cl-Par-4 translocates into the nucleus and orchestrates its apoptotic function, while the PAF indirectly promotes apoptosis by rescuing Fbxo45-mediated degradation of full-length Par-4. Casp-8, caspase-8; Casp-9, caspase-9; Casp-3, caspase-3

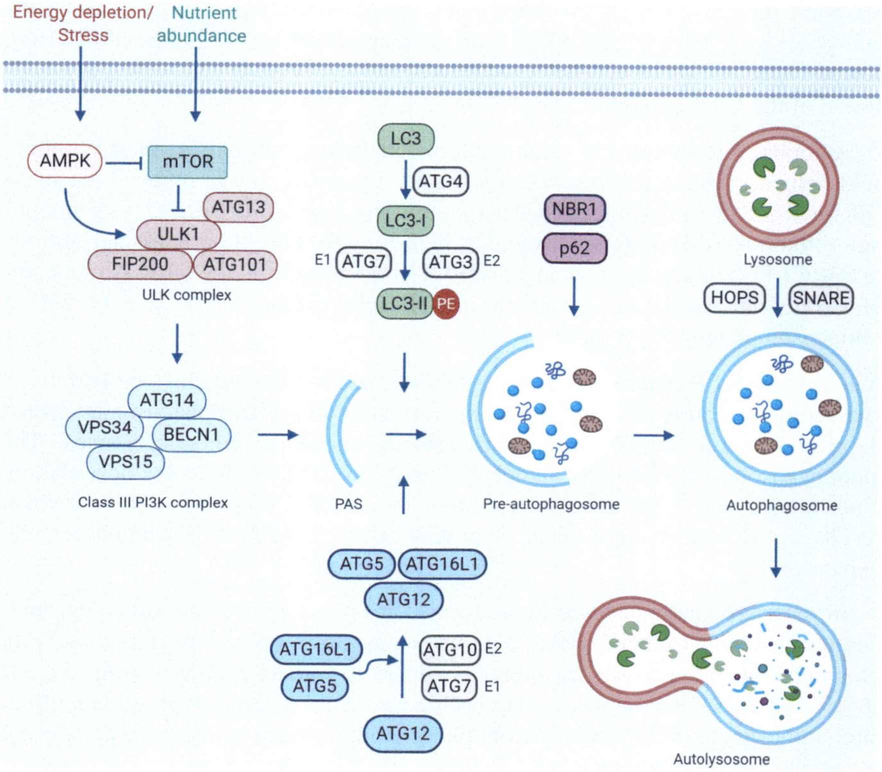

Fig. 2 Overview of the autophagy process. Nutrients or growth factors deprivation activates AMP-activated protein kinase (AMPK) and/or inhibition of mechanistic target of rapamycin (mTOR) leading to the dissociation of mTOR from Unc-51-like kinase (ULK) complex (consisting of ULK1, ATG13, FIP200, and ATG101), which triggers phagophore nucleation by phosphorylating class III PI3K complex (consisting of VPS34, VPS15, BECN1, ATG14). Multiple autophagy proteins such as ATG5 and ATG7 constitute two ubiquitin-like conjugation systems that catalyze the formation of phosphatidylethanolamine (PE)-conjugated microtubule-associated protein light chain 3 (LC3-II) and direct its proper incorporation into the phagophore membrane. The closure of an elongated phagophore marks the formation of a mature autophagosome, which eventually fuses with a lysosome, leading to cargo degradation and recycling of essential biomolecules

double-membrane pre-autophagosome, (3) elongation, cargo sequestration, and closure forming autophagosome, and (4) autophagosome-lysosome fusion, and degradation of autophagic cargo (Fig. 2) [86].

Initiation Autophagy initiation starts with the activation of the Unc-51-like kinase (ULK) complex. ULK complex consists of ULK family protein (mainly ULK1 or less characterized ULK2; orthologues of yeast ATG1), focal adhesion kinase interacting protein 200 kDa (FIP200; orthologue of yeast ATG17), ATG13, and ATG101 [4]; [115]. ULK complex activity is tightly regulated through nutrient and energy-sensing kinases mechanistic target of rapamycin complex 1 (mTORC1) and AMP-activated protein kinase (AMPK). Under normal condition, activated mTORC1 phosphorylates ULK1 and ATG13 to prevent autophagy initiation [48]; [68]. whereas inactivation of mTORC1 leads to dephosphorylation and subsequent activation of the ULK complex.

Autophagy is initiated by many stimuli, including nutrient deficiency (amino acids, glucose, etc.), growth factor deprivation (insulin), extra- or intracellular stress (ER stress, ROS, drugs, infection, etc.), energy shortage (AMP), etc. During autophagy initiation, AMPK gets activated, which directly phosphorylates and inhibits mTORC1 [37]. Once inactivated, mTORC1 dissociates from the ULK kinase complex, releasing the ULK1/2 available for phosphorylating ATG13 and FIP200 to initiate autophagy [95] (Fig. 2).

Nucleation and Formation of Double-Membrane Pre-Autophagosome Vesicle A characteristic feature of autophagy is the formation of pre-autophagosomes, which develop to form double-membraned vesicles known as autophagosomes. The autophagosomes delivers the autophagic cargo to the lysosomes for degradation. Nucleation refers to the process of mobilizing a small group of molecules, which results in the further recruitment of proteins that are needed for autophagosome expansion.

In yeasts, pre-autophagosomes are formed at a punctate structure called the "pre-autophagosomal structure" (PAS; also known as "'phagophore assembly site"). In mammals, autophagosomal membranes appear to initiate primarily from the ER [65]. Class III PI3K complex is responsible for the production of phosphatidylinositol 3-phosphate (PI3P) directly from phosphatidylinositol, an essential early event in autophagosome formation [93]. The class III PI3K complex consists of phosphatidylinositol 3-kinase catalytic subunit type-3 vacuolar protein sorting (VPS34), phosphoinositide-3-kinase regulatory subunit-4 (VPS15), ATG14L, and Beclin-1 (BECN1) (orthologue of yeast ATG6) [41]. ULK1 phosphorylates different members of class III PI3K complex, leading to their assembly at PAS and formation of PI3P [71]. This PI3P helps in the localization of several ATG proteins to the PAS to execute autophagosome formation (Fig. 2) [102].

Elongation, Cargo Sequestration, and Closure Forming Autophagosome The growing double membrane undergoes elongation to eventually form autophagosome: a process that is mediated by two ubiquitin-like conjugation systems [153]. The first ubiquitin-conjugation system is ATG12-ATG5-ATG16L [104]. ATG12 is activated in an E1-like reaction by ATG7 and transferred to ATG5 in an E2-like reaction by ATG10. The ATG5-ATG12 conjugate binds ATG16L to form ATG12-ATG5-ATG16L complex [104]. Microtubule-associated protein light chain 3 (LC3) (mammalian homolog of ATG8) is the second ubiquitin-like conjugation systems. LC3 undergoes a unique type of post-translational modification. LC3 is initially synthesized with a C-terminal extension that is proteolytically removed by the ATG4 cysteine protease, generating LC3-I. Using ATG7 as an E1 enzyme and ATG3 as an E2 enzyme, lipid phosphatidylethanolamine is then covalently attached to the LC3-I, generating LC3-II [72] (Fig. 2). In the autophagosome, LC3-II is distributed to both the exterior and the lumen of the vesicle, which is often used as a marker for the autophagosome and as an index of autophagy.

In addition to serving as an important structural component of autophagosomes, LC3 acts as a docking site for a growing list of autophagy cargo receptors that bring autophagic cargo to the autophagosomes. The best-characterized cargo receptor is p62 (sequestosome 1; SQSTM1). Other molecules, such as neighbor of BRCA1 (NBR1), while in yeast, Uth1p and Atg32 have been identified as proteins that promote mitophagy [50]. p62 contains a ubiquitin-associated UBA domain, a LC3-interacting region (LIR), and a multimerization PBI domain. These domains allow p62 to capture ubiquitinated cytosolic components and anchors them to LC3-II in the autophagosome. Because p62 can multimerize, a larger number of ubiquitinated proteins can be tethered to a single LC3-II.

The final stages of autophagosome formation include the closure to form a double-membrane vesicle. No factors that are specific for autophagosome closure have been identified yet. However, LC3-II has been reported to have properties that make it a candidate for membrane closure [101]. Soluble-N-ethylmaleimide-sensitive factor attachment protein receptors (SNARE) is another class of protein that is implicated in autophagosome formation and closure [73].

Autophagosome-Lysosome Fusion, and Degradation of Autophagic Cargo After the closure of the phagophore, the double-membrane autophagosome matures and fuses with lysosomes to degrade its contents. Initially, the autophagosome undergoes maturation, which involves gradual disassembly of ATG proteins from completed autophagosome, a step that presumably precedes fusion and recruitment of machinery responsible for lysosomal delivery and fusion (Fig. 2). In the first step of autophagosome-lysosome fusion, the outer membrane of autophagosome fuses with the lysosomal membrane. The inner membrane of autophagosome is then degraded by lysosomal hydrolases and exposes the cargo to the lumen of the lysosome. Once degraded, the resulting macromolecules are released back into the cytosol through various permeases [154].

The precise mechanism of autophagosome-lysosome fusion is still elusive. Three intracellular membrane-trafficking-associated protein families, RAS-related GTP-binding protein GTPases (Rab GTPases), membrane-tethering complex homotypic fusion and protein sorting (HOPS), and SNARE are known to be involved in this process [129]. In addition, evidence has shown that the microtubule system is essential for trafficking mature autophagosomes from random initiation sites toward the perinuclear area [39], where they fuse with lysosomes. Furthermore, the class III PI3K complex, in which ATG14L is replaced by UV irradiation resistance-associated gene (UVRAG), acts in regulating the trafficking of mature autophagosomes to lysosomes [7].

3.2 Autophagy as a Tumor Suppressor Mechanism

It is well-known that autophagy has opposing, context-dependent roles in cancer (i.e., as a promoter and a suppressor of tumors). In terms of promoting tumorigenesis, autophagy is a protective mechanism that allows premalignant cells to escape genotoxic stress, inflammation, nutrient deprivation, etc., and protects tumor cells from being killed. Pro-survival activity of autophagy also helps cancer cells to cope with therapy-induced stresses [2]. There is substantial evidence, moreover, that enforced overactivation of autophagy can lead to PCD. In such instances, the total area of autophagic vacuoles may roughly be equal to, or greater than, that of cytosol and organelles outside the autophagic vesicles. An autophagic activity of this magnitude destroys the majority of cytosolic components, finally leading to total collapse of cellular functions and irreversible commitment to PCD [54]. Unlike apoptosis, during autophagic cell death, caspases are not activated, and therefore DNA fragmentation is not pronounced, and the nucleus stays intact until the late phases of cell death. The demonstration of the autophagic vesicles by electron microscopy along with a substantial reduction in cellular viability is the "golden standard" for assessing the autophagic PCD. Autophagy may also be confirmed by monodansylcadaverine (MDC) staining and acridine orange (AO) staining, by measuring the LC3 processing and recruitment to autophagic membranes, or by inhibiting different stages of autophagic degradation pathways [54].

Autophagy was found to be reduced in cancer tissues compared with their normal counterparts, suggesting that autophagy is a tumor-suppressive process. In support of this premise, many core autophagy genes, including BECN1, UVRAG, Bif-1, and other ATGs were found to be mutated in many human cancers [79, 135]. Systemic mosaic deletion of Atg5 or liver-specific deletion of Atg7 in mice resulted in the spontaneous development of benign liver adenomas [136]. Autophagic cell death can be activated in cancers in response to various cytotoxic agents and drugs used in cancer treatment. For instance, treatment of malignant glioma cells with arsenic trioxide-induced G2/M arrest and autophagic cell death [80]. Similarly, tamoxifen treatment caused cell death with characteristics of autophagy in MCF7 breast carcinoma cells [20]. Furthermore, breast, prostate, and colon cancer cell lines responded to radiation treatment by increasing autophagy [106]. In addition to vital, tumor-suppressive role in established anticancer therapies, autophagy is explored in some of the most promising anticancer drugs that are currently undergoing rigorous investigation. Curcumin (polyphenolic compound in turmeric *Curcuma longa*), betulinic acid (pentacyclic triterpenoid in white birch trees *Betula papyrifera*), nimbolide (limonoid tetranortriterpenoid in *Azadirachta indica*), resveratrol (polyphenol in grape seeds), and sanguinarine (benzophenanthridine alkaloid in *Sanguinaria Canadensis*), etc., were reported to trigger autophagy-mediated tumor suppression in different cancer cells [51, 107, 110, 121, 140].

There are emerging evidences that induction of autophagy-mediated cell death is a promising tumor-suppressive strategy, especially when the apoptotic pathway is not functioning [128, 156]. For instance, apoptosis-resistant Bax/Bak double knockout mouse embryonic fibroblast (MEF) cells underwent to autophagic cell death when exposed to etoposide, a DNA damaging anticancer drug [128]. Autophagy-associated cytotoxicity has also been observed in caspase-8 inhibition-mediated death in mouse fibroblast L929 cells [156]. These studies indicate that autophagic cell death may be important to compensate when apoptosis is hindered or inactivated.

4 Prostate Apoptosis Response-4 (Par-4)

Par-4 (also known as WT1 regulator) is a therapeutically promising tumor suppressor protein encoded in humans by the *PAWR* gene [123]. Par-4 can be activated by diverse cellular stressors, including anticancer drugs [118], genotoxic stress [59], ROS generation [133, 140], bacterial infection [61], etc. Par-4 is unique, because it selectively induces apoptosis in cancer cells, but not in normal or immortalized cells. Despite years of research, the mechanism underlying Par-4's tumor-suppressor function has not been significantly defined and remains the focus of active investigation. Most of Par-4 functions have been considered in light of how Par-4 might induce apoptosis-mediated tumor suppression. However, other anticancer functions of Par-4 that might be as important as apoptosis are being uncovered. Indeed, Par-4 has recently been found to participate in other anticancer mechanism(s) such as metastasis, senescence, and autophagy [114, 130, 141]; however, it is still not clear which of these mechanisms is essential to its potent tumor-suppressor function and how these functions interact. The fate of the cell following Par-4 activation is determined by cell type, duration of exposure, and amplitude of stress signal. Understanding the tumor-suppressive functions of Par-4 will not only have profound importance to the understanding of PCD pathways, but also will have an impact on interventions of cancer therapy and prevention.

4.1 Domains of Par-4

Par-4 is a 38 kDa multi-domain protein comprising of 340 amino acids. Par-4 has several domains which appears to be critical for its regulation, localization, dimerization, and post translational modification. Par-4 has two putative nuclear localization sequences—NLS1 (amino acid residues 20–25) and NLS2 (a bipartite sequence comprising amino acid residues 137–153) (Fig. 3) [38]. Removal of NLS1 by deletion of the first 68 amino acids did not affect the apoptotic functions of Par-4 [124]. In contrast, deletion of NLS2 abrogates the ability of mutant to translocate to the nucleus and was also unable to induce apoptosis, indicating that NLS2 is critical for the apoptotic functions of Par-4. Both of these NLSs are located on the amino-terminal half of the molecule. Par-4 also has a nuclear export sequence (NES) at the carboxy-terminal half (Fig. 3), which has no known Par-4 regulatory functions

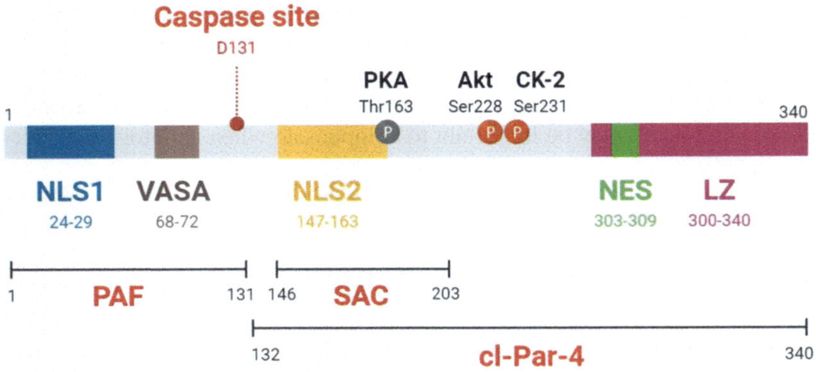

Fig. 3 Schematic representation of human Par-4. Functional domains of human Par-4 showing the two nuclear localization sequences (NLS1 and NLS2), the SAC domain, leucine zipper domain (LZ), nuclear export sequence (NES), VASA segment, Par-4 amino-terminal fragment (PAF), and cleaved Par-4 (cl-Par-4). The PKA phosphorylation site and NLS2 within the SAC domain, VASA segment within the PAF, and SAC domain within the cl-Par-4 are shown. Akt phosphorylation site, casein kinase 2 (CK-2) phosphorylation site, and caspase cleavage sites were also illustrated

identified thus far. Based on data from domain-based study, NES may be masked by protein-protein interaction, retaining the Par-4 in the nucleus to orchestrate its nuclear function [142]. At the carboxy-terminal end, Par-4 also has a leucine zipper (LZ) domain [124], which functions as the primary recognition and binding site for various Par-4 interacting proteins such as Wilms' tumor 1 (WT1) [76], protein kinase B (PKB/Akt) [52], Sequestosome 1 (SQSTM1/p62) [99], atypical protein kinase Cs (aPKC λ/ι and ζ) [99], death-associated protein kinase 3 (DAPK3/Dlk/ZIP Kinase) [105], topoisomerase 1 (TOP1) [53], apoptosis-antagonizing transcription factor (AATF/Che-1) (Q. [58]. Thanatos-associated domain-containing apoptosis-associated protein 1 (THAP1) [116], Amida [17], beta-secretase 1 (BACE1) (J. [152]. dopamine D2 receptor (D2DR) [108], and E2F transcription factor 1 (E2F1) [92]. The Par-4 deletion mutant which is lacking the LZ domain is still capable of inducing cell death. Therefore, it may be assumed that LZ domain is not indispensable for the cell death-inducing functions of Par-4 [38].

Par-4 also possesses phosphorylation sites for kinases such as protein kinase A (PKA), Akt, and casein kinase 2 (CK2) (Fig. 3) [35, 60]. PKA-mediated phosphorylation of Par-4 at Thr163 is critical for Par-4 nuclear translocation, an event implicated in the apoptotic function of Par-4 [60]. In contrast, Akt-mediated phosphorylation of Par-4 at Ser228 causes 14–3-3 chaperone-mediated sequestering of Par-4 in the cytoplasm and abrogation of apoptosis, even if it is PKA phosphorylated [52], CK-2 phosphorylates Par-4 at S231 and impairs its pro-apoptotic activity [33]. In addition to phosphorylation, Par-4 is also regulated by other post-translational mechanisms. For instance, Par-4 has a VASA domain which participate in its ubiquitylation and proteasomal degradation by Fbxo45 ubiquitin ligase [25, 66].

Par-4 contains a central core Selective for Apoptosis of Cancer cells domain (SAC domain) (amino acids 137–195), which is unique to Par-4 and is 100% conserved in human, mouse, and rat (Fig. 3). In fact, the pro-apoptotic function of Par-4 is localized in its SAC domain [38]. The SAC domain contains PKA phosphorylation site and NLS2, which are essential for Par-4 activation and nuclear entry, whereas it does not have Akt and CK-2 phosphorylation sites, which when phosphorylated, inhibits Par-4 activity [52, 60]. The SAC induces apoptosis in cancer cells which are even resistant to full-length Par-4-mediated apoptosis, yet does not induce apoptosis in normal cells [157].

Previously, we and others have discovered that Par-4 can be cleaved by caspases at D131 during apoptosis, resulting in the generation of two fragments: a 15 kDa amino-terminal fragment ("PAF" Par-4 Amino-terminal Fragment) and a 25 kDa C-terminal fragment, cleaved Par-4 (cl-Par-4) [23, 139]. The cl-Par-4 contains the NLS2, SAC, and LZ domains, while PAF contains the NLS1 and VASA domain. In addition to the above-mentioned domains, Par-4 also contains a phosphate-binding loop domain (P-loop domain), adenylyl cyclase domain, PKC phosphorylation site, additional CK-2 phosphorylation sites, and N-glycosylation site. The impact of these domains on par-4 regulation and function has not been resolved yet.

4.2 Establishment of Par-4 as Tumor Suppressor Protein

The realization of Par-4 as a bona fide tumor suppressor came from a number of important observations. Par-4 gene was first identified as a target gene upregulated in prostate cancer cells in response to apoptotic stimuli [123]. Later, it was shown that enforced expression of wild-type Par-4 in prostate (PC-3) and melanoma (A375-C6) cells confers super sensitivity to apoptotic stimuli [124].

Par-4 is located in chromosome 12q21, a region that is unstable and often deleted in pancreatic and gastric cancer [77]. Reinforcing its tumor-suppressive property, Par-4 is often depleted, mutated, or inactivated in a myriad of cancers such as cancers of blood [14], kidney [27], endometrium [98], brain [84], breast [100], prostate [52], and lungs [78]. Premature termination of Par-4 from a single base mutation at amino acid 189 (Arg [CGA] to Stop [TGA] at exon 3 or at the effector domain was noticed in human endometrial carcinoma [98]. The same study also demonstrated Par-4 promoter hypermethylation which is correlated with low levels of Par-4 in 32% of the analyzed tumors [98]. Diminished expression of Par-4 is also evinced in the presence of oncogenic Ras and restoration or overexpression of Par-4 induces apoptosis in these cells [6].

The tumor-suppressive property of Par-4 has been further examined and experimentally proven using Par-4 knockout (Par-4$^{-/-}$) mice. The Par-4$^{-/-}$ mice developed spontaneous tumors in various tissues such as endometrium, liver, and lung, and exhibited prostatic intraepithelial neoplasia. Interestingly, inactivation of Par-4 also markedly accelerated chemical- or hormone-driven tumors in the bladder and endometrium [49], while Par-4 or SAC transgenic mice were resistant to spontaneous tumors [19]. It has also been demonstrated that downregulation of the Par-4 is a key determinant in apoptosis evasion, which leads to breast cancer recurrence. Restoration of Par-4 function sensitizes such tumors to chemotherapy in vitro and in vivo, further advocating its tumor suppressor role [1].

4.3 Apoptosis-Inducing Functions of Par-4

Induction of apoptosis is the most extensively studied tumor-suppressive function of Par-4. Overexpression of Par-4 alone was sufficient to induce apoptosis in many cancers. Interestingly, Par-4 does not induce apoptosis in normal cells, immortalized cells, and hormone-responsive cancer cells. More importantly, the apoptotic action of Par-4 can overcome cell-protective mechanisms, such as the presence of Bcl-xL, Bcl-2, or the absence of wild-type p53 or PTEN function.

Par-4 is able to engage both extrinsic and intrinsic apoptotic pathways (Fig. 1). Many of the previously mentioned apoptosis regulators have been established as upstream regulators or downstream targets of Par-4-mediated apoptosis [16, 22, 83]. The canonical apoptotic functions of Par-4 require its Thr155 phosphorylation by PKA and nuclear translocation. Inside the nucleus, Par-4 inhibits NF-κB-dependent transcription activity and induces apoptosis [38] (Fig. 1). In the context of extrinsic apoptotic pathway, Par-4 activates FasL-Fas-FADD-caspase-8 pro-death pathway with concurrent inhibition of the NF-κB-mediated anti-apoptotic pathway [22]. Mechanistically, Par-4 activates mobilization of the FasL and the FasR to the cell membrane with a concomitant decrease in antiapoptotic cellular FLICE-like inhibitory protein (c-FLIP), leading to DISC formation and subsequent activation of caspase cascade [8].

Numerous studies have shown that Par-4 promotes the engagement of the intrinsic apoptotic pathway by enhancing the MOMP [14]. Boehrer et al. demonstrated that Par-4 expression in neoplastic lymphocytes increases cleavage of caspase-8, activation of pro-apoptotic Bid, and its translocation to the mitochondria; resulting in an augmentation of cytochrome-c and AIF release into cytoplasm [15]. One of the crucial mechanisms governing Par-4-mediated induction of intrinsic apoptosis is the inhibition of PKCζ-NF-κB signaling axis which regulate the transcription of pro-survival proteins such as Bcl-2, Bcl-xL, Bfl-1, and XIAP (Fig. 1) [26, 35]. Furthermore, Par-4 interacts with WT1 and promotes the transcriptional repression of Bcl-2 [24]. Par-4 can also inhibit Bcl-2 by upregulating miR-24a, a miRNA that represses Bcl-2 mRNA translation [148]. Par-4 was also shown to enhance Smac/DIABLO release by transcriptionally modulating its expression through an indirect interaction between Par-4 and *Smac* promoter [92].

In addition to directly regulating apoptosis, Par-4 also modulates functions of other pro-proliferative or pro-death proteins. For instance, Par-4 was shown to inhibit Akt activity through its ability to inhibit PKCζ [78]. On the other hand, Akt inhibition promotes Par-4-dependent apoptosis [30, 32, 126] (Fig. 1). Par-4 is also involved in an antagonistic relationship with oncogenic Ras [6, 35]. Oncogenic Ras induces Par-4 downregulation, while Par-4 expression inhibits Ras-inducible cellular transformation [143]. Phosphorylation of Par-4 at Thr155 was shown to be essential for DAPK3-mediated apoptosis [18]. In another study, Par-4 was shown to interact with THAP1 and potentiate both serum withdrawal- and TNF-induced apoptosis.

As mentioned earlier, caspase-mediated cleavage of Par-4 generates PAF and cl-Par-4 [23, 139]. The cl-Par-4 having NLS2 translocates into the nucleus and orchestrates its apoptotic function [23, 59]. while the PAF through its VASA segment rescues Fbxo45-mediated degradation of full-length Par-4 [66]. Specifically, PAF may be released from non-resistant cancer cells and translocates into the drug-resistant cancer cells (not the normal cells). Inside the cell, PAF competitively inhibits Fbxo45 and rescues full-length Par-4 from Fbxo45-mediated ubiquitination [66].

A breakthrough study by Burikhanov et al. provided an entirely new dimension to the apoptotic function of Par-4. The authors elegantly demonstrated that Par-4 is spontaneously secreted from mammalian cells via classical ER-Golgi pathway (Fig. 1) [19]. Interestingly, both normal and cancer cells are able to secrete Par-4; however, only cancer cells are susceptible to its pro-apoptotic effects [19]. Under conditions of ER stress, ER chaperone protein glucose-regulated protein-78 (GRP78) translocates from ER to plasma membrane, where it acts as a receptor for secreted Par-4 (also SAC) at the cell surface. Par-4/SAC binding to GRP78 facilitates extrinsic pathway of apoptosis via FADD-caspase-8-caspase-3 pathway (Fig. 1).

4.4 *Role of Par-4 in Autophagy*

Emerging evidences suggest that Par-4 also controls the autophagic cell death mechanism. Wang et al. observed that overexpression of Par-4 activates both apoptosis and autophagy in hypopharyngeal carcinoma (HPC) cells [150]. Par-4-induced cell death was significantly mitigated by autophagy inhibitor, while pan-caspase inhibitor had no apparent effect on cell death. Shortly thereafter, we have shown that curcumin induces ROS-dependent upregulation of Par-4 and autophagic death in human malignant glioma cells [140]. Extracellular treatment with H_2O_2 induced Par-4 accumulation and features of autophagic cell death, indicating that ROS generation is an important regulator of Par-4-mediated autophagy [140]. Interestingly, Par-4 overexpression alone was sufficient to induce autophagic cell death. In a subsequent follow-up study, we identified the tumor suppressor p53-dependent Bcl-2/adenovirus E1B 19 kDa interacting protein 3 (BNIP3) generation as a downstream target of Par-4-mediated autophagic cell death following curcumin treatment. Interestingly, the Par-4-p53-BNIP3 axis was also activated by other autophagy inducers such as ceramide and arsenic trioxide [141]. However, the precise signaling mechanism by which Par-4 regulates autophagic cell death is largely unresolved. Further studies are warranted to know: (1) *How Par-4 is activated during autophagy induced by various anticancer agents?* We have noticed that Par-4 is induced by classical autophagy inducers such as serum starvation [141] and rapamycin (unpublished data) treatment. Furthermore, Par-4 is induced during autophagic cell death induced by various anticancer drugs such as curcumin, ceramide, arsenic trioxide, etc. [141]. More importantly, Par-4 overexpression alone was sufficient to induce features autophagy, indicating that Par-4 may be playing a prominent role in the facilitation of autophagic cell death. However, the upstream signaling event responsible for Par-4 activation is not identified and warrants further investigation; (2) *what are the structural domains of Par-4 critical for its autophagic function?* Par-4 mutant lacking NLS1 and LZ domain was capable of inducing apoptosis. The SAC domain is sufficient for apoptosis induction in cancer cells. As mentioned earlier, apoptotic function of Par-4 is negatively regulated by Akt-mediated phosphorylation. Similarly, PKA-mediated phosphorylation of Par-4 leads to its nuclear entry and apoptosis. It is not known if these regions are critical for Par-4's autophagic function; (3) *what determines the type of cell death following Par-4 activation?* How Par-4 differentially activates apoptosis versus autophagic death remains poorly understood. It is not known whether it is cell-context-dependent or any specific regulatory mechanism is involved. Recent studies have suggested that under certain circumstances, induction of autophagy might facilitate the progression of apoptotic or other cell death pathways [57, 70]. In this view, Par-4 might coordinately activate autophagy with other tumor-suppressive functions.

5 Ceramide

Sphingolipids, a class of natural lipids comprised of a sphingoid base backbone, are important structural components of eukaryotic membranes. Although considered to be structural molecules, sphingolipids have recently been shown to act as signaling molecules in various cellular functions [21]. Now, these molecules are known to act on distinct targets, including kinases, lipases, phosphatases, membrane receptors, etc., and they exert distinct cellular functions. A plethora of physiological processes are critically modulated by bioactive sphingolipids, including signal transduction, growth regulation, cell-cell adhesion, cell migration, cell death regulation, and inflammatory responses [64]. At the organismal level, sphingolipids have been implicated in the control of pathobiological processes, such as neurodegenerative diseases, cancer, metabolic disorders, cardiovascular diseases, infectious diseases, inflammatory disorders, etc. [64]. Ceramides are central intermediates of sphingolipid metabolism and the most studied bioactive sphingolipid. Ceramides consist of a sphingoid base joined by an amide linkage to fatty acids of varying chain length and a varying degree of saturation giving rise to a variety of ceramide species. These differences in acyl chain length and saturation often dictate the biological activities of ceramide [21]. Topmost among these actions is the regulation of signaling cascades that lead to cell death or tumor suppression [44].

5.1 Ceramide Metabolism

The production of ceramide occurs via three major pathways, de novo synthesis, salvage pathway, and sphingomyelin hydrolysis (Fig. 4). De novo pathway begins with the condensation of amino acid serine onto a fatty acyl-CoA via serine palmitoyltransferase (SPT) forming 3-keto-dihydrosphingosine (3-keto-DHSph). 3-ketodihydrosphingosine reductase (KDSR) converts 3-keto-DHSph to dihydrosphingosine (DHSph). Different ceramide synthases (CS) add distinct fatty acyl chains to DHSph and forms dihydroceramides (DHCer). Finally, DHCer is desaturated by DHCer desaturases (DES) and produces ceramides.

Ceramide serves as a "metabolic hub" in the sphingolipid metabolic pathway for subsequent production of other sphingolipid signaling intermediates. Ceramidase (CDase) at different cellular organelles may hydrolyze ceramide to generate sphingosine (Sph), which seems to have one of the two fates. Sph may be phosphorylated to sphingosine-1-phosphate (S1P) by sphingosine kinase (SK) or may be recycled

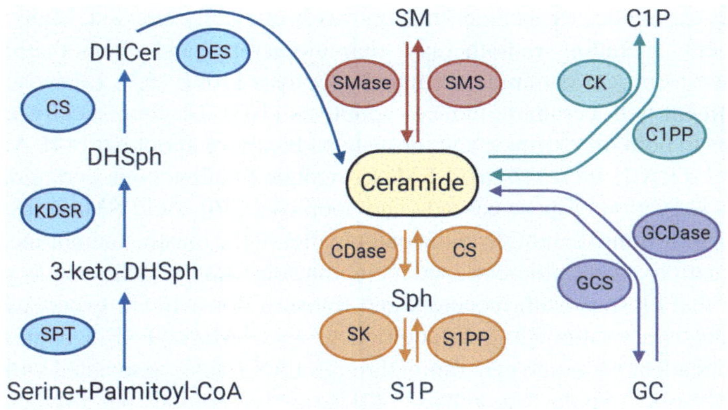

Fig. 4 Schematic representation of ceramide biosynthetic pathway. In the de novo ceramide synthesis, dihydrosphingosine (DHSph) is formed by condensation of serine and palmitoyl-CoA, catalyzed by the serine palmitoyltransferase (SPT), followed by reduction of the resultant 3-keto-dihydrosphingosine (3-keto-DHSph) by 3-ketodihydrosphingosine reductase (KDSR). DHSph is converted into dihydroceramide (DHCer) and then subsequently desaturated into ceramide by ceramide synthase (CS) and DHCer desaturases (DES), respectively. Catabolism of ceramides by ceramidases (CDase) yields sphingosine (Sph), which can be recycled back to ceramide by CS, or may be phosphorylated by sphingosine kinases (SK), yielding sphingosine-1-phosphate (S1P). The product S1P may be dephosphorylated back to Sph by S1P phosphatases (S1PP). More complex sphingolipids are formed from ceramide, including sphingomyelin (SM) and glucosylceramide (GC) by the action of SM synthase (SMS) and glucosylceramide synthase (GCS), respectively. SM and GC may be hydrolyzed back to ceramide by sphingomyelinase (SMase) and glucosyl ceramidase (GCDase), respectively. Functional headgroups are also transferred onto ceramide by ceramide kinase (CK) to form ceramide-1-phosphate (C1P), which may be hydrolyzed back by C1P phosphatase (C1PP)

back into ceramide by ceramide synthase (CS) [42]. The product S1P can be dephosphorylated to regenerate sphingosine through the action of specific intracellular S1P phosphatases (S1PP) and, possibly, non-specific extracellular lipid phosphate phosphatases [75] (Fig. 4). Simultaneously, ceramide may be phosphorylated by ceramide kinase (CK) to form ceramide-1-phosphate (C1P), which may be hydrolyzed back to ceramide by C1P phosphatase (C1PP) [151]. Alternatively, ceramide can be converted into sphingomyelin (SM) and glucosylceramide (GC) by the action of SM synthase (SMS) and glucosylceramide synthase (GCS), respectively [112]. SM and GC may be hydrolyzed back to ceramide by sphingomyelinase (SMase) and glucosyl ceramidase (GCDase), respectively [63] (Fig. 4).

5.2 Tumor-Suppressive Functions of Ceramide

Ceramides can activate a variety of tumor-suppressive programs, including senescence, apoptosis, autophagy, and necroptosis [44].

Apoptosis Ceramide engagement in apoptosis is clinically relevant. Many anticancer regimens, including radiotherapy, immunotherapy, and chemotherapy, have been shown to trigger ceramide-mediated apoptosis [109]; [62]. Likewise, exogenous treatment with ceramide induces apoptosis [103]. Mechanistically, ceramide contributes to both the extrinsic and intrinsic pathways of apoptosis [44]. Activation of Fas and TRAIL increases acid SMase-mediated endogenous ceramide levels, leading to activation caspase cascade and apoptosis [36]. Acid SMase also causes hydrolysis of SM into ceramide at the outer leaflet of the plasma membrane, leading to the formation of ceramide-enriched membrane domains, which serve as signaling platforms that cluster death receptors and transmit downstream events associated with apoptosis activation [55]. In addition to acid SMase, TNF activates neutral SMase-dependent ceramide generation through FAN (factor associated with neutral SMase activation) to induce apoptosis [122]. Notably, ceramide has also been shown to facilitate the extrinsic pathway of apoptosis by causing the downregulation of FLICE inhibitory protein which is a caspase-8 inhibitor [44].

Intrinsic apoptotic stimuli such as hypoxia, nutrient deprivation, radiation, heat shock, cellular stress, and cytotoxic drugs have been shown to increase endogenous ceramide through multiple mechanisms involving SM hydrolysis, de novo synthesis, and salvage pathway [63]. Ceramide can be transferred directly from the plasma membrane to the mitochondria, leading to mitochondrial ceramide accumulation and apoptosis [5]. Ceramide induces MOMP leading to the release of the pro-apoptotic molecule cytochrome-c, a key event in intrinsic signaling [12]. It has been demonstrated that SMase induction promotes Bax translocation and activation at the mitochondria, leading to cytochrome-c release and apoptosis [13]. Ceramide also indirectly elicits BAX translocation to mitochondria via p38 mitogen-activated protein kinase (MAPK) activation or Akt downregulation [82]. Ceramide-mediated inhibition of Akt also promotes cytochrome-c release by preventing the Akt-induced inactivation of pro-apoptotic Bcl-2 protein family members [44]. Other mechanism(s) by which ceramide enhances MOMP is by (1) activating and inducing the translocation of protein kinase Cδ to the mitochondria [132], (2) inducing oxidative stress [97], activating glycogen synthase kinase 3β (GSK3β) binding on voltage-dependent anion channel VDAC2 [28, 89], and by activating endolysosomal protease cathepsin D [119].

Autophagy Ceramide is activated by various stimuli that induce autophagic cell death, and multiple ceramide target proteins are identified to play crucial role in the autophagy regulation. Amino acid starvation is the most commonly used stimuli to induce autophagy in laboratory settings. Amino acid starvation activates ceramide generation via acid SMase and triggers ceramide-activated protein phosphatase (CAPP)-dependent inactivation of mTORC1 [137]. As mentioned earlier, mTOR is an important negative regulator of autophagy. Inhibition of the Akt/mTOR pathway in various cells can activate autophagic cell death [131]. Ceramide can inhibit Akt, leading to downregulation of mTOR activity and promotion of autophagy [120]. For instance, treatment of human cancer cells with exogenous C2-ceramide or GCS inhibitor D-threo-1-phenyl-2-decanoylamino-3-morpholino-l-propanol (PDMP) results in the inhibition of Akt-mTOR signaling and induction of autophagic cell death [120]; [127]. AMPK is another upstream regulator of mTOR. Recently, Yang et al. reported that deguelin increases the de novo ceramide production which activates AMPK-mediated phosphorylation of ULK1, eventually leading to autophagic cell death in human neck squamous cell carcinoma cells (HNSCC) (Y. L. [155]. Ceramide also blocks mTOR activity by inhibiting phospholipase D, which is an upstream regulator of mTOR activity [144].

Ceramide can also promote autophagy by targeting the 14–3-3ζ-VPS34 complex, which leads to VPS34 dissociation and activation, an early event in autophagy induction [111]. Ceramide also dissociates BECN1-Bcl-2 complex via c-Jun N-terminal kinase (JNK)-mediated phosphorylation of Bcl-2 (D. D. [88]. Previously, ceramide has been shown to transcriptionally activate BNIP-3, and induction of BNIP-3-dependent autophagy in human malignant glioma [29]. Recently, we have shown that ceramide induces autophagy via activation of Par-4-p53-BNIP3 signaling axis in malignant glioma cells [141]. Additionally, ceramide enhances the availability of LC3 by activating the JNK-mediated transcription of LC3 [134]. Exogenous C16-ceramide also triggers autophagy-dependent cell death that is accompanied by the interaction of phosphorylated emerin and LC3 [34]. Deroyer et al. hypothesized that phosphorylated emerin may exist in a complex with ceramide/LC3-II to promote LC3 lipidation for the induction of autophagic cell death [34]. Moreover, another study demonstrated that induction of ceramide stress using chemotherapeutic agents or exogenous ceramides increases the accumulation of ceramide synthase 1 (CS1)-generated C18-ceramide at the outer mitochondrial membrane, leading to the recruitment of autophagosomes to mitochondria via ceramide-LC3 binding and mitophagy in HNSCC [125]. Mitochondrial ceramide accumulation and lethal mitophagy was also reported in response to crenolanib treatment in acute myeloid leukemia [31].

6 Par-4 and Ceramide Connection

Though Par-4 and ceramide are made of entirely different entities, both of these molecules are important regulators and apparent collaborators of cell-fate decisions. The relationship between Par-4 and ceramide is very complex, and mechanisms underlying their coregulation are diverse and not fully characterized. Several reports suggest that Par-4 upregulation is an important downstream mediator of the ceramide response [10, 11]; [146]. whereas others have shown that ceramide and Par-4 are concomitantly upregulated in response to various cell stressors [9].

Par-4 in Ceramide-Induced Apoptosis Originally, Bieberich et al. had found that mouse embryonic stem cells expressing PAR-4 are rapidly induced to undergo apoptosis by the novel ceramide analog S18 [9]; [10]. Overexpression of Par-4 increases or antisense knockdown of Par-4 reduces the sensitivity of differentiating embryonic stem cells toward ceramide- or ceramide analog-inducible apoptosis [10]. In a follow-up study, the authors clearly demonstrated that S18 induces Par-4-ζPKC complex formation and inhibits ζPKC, triggering apoptosis in the embryoid body-derived cells [11]. It has been shown that long-chain ceramide induces apoptosis in primary cultured astrocytes from presenilin 1 knock-in mouse, a mouse model for early-onset familial Alzheimer's disease. In contrast, wild-type astrocytes appear to be resistant to ceramide. The authors noted that the sensitivity of presenilin 1 astrocytes is most likely due to the 9.5-fold elevated expression of Par-4 [147]. Wang and Bieberich provided in vitro and in vivo experimental evidence that neural crest-derived tissue culture and embryos are specifically sensitive to ethanol-induced apoptosis, which is associated with co-elevation of ceramide and Par-4 [145].

In an interesting study, Wang et al. demonstrated that treatment of astrocytes with amyloid peptide triggers the secretion of proapoptotic lipid vesicles that contains ceramide and Par-4. Antibodies against Par-4 and ceramide prevented amyloid-induced apoptosis in vitro and in vivo, suggesting that apoptosis was mediated by exogenous Par-4 and ceramide, potentially associated with secreted exosome [149] (Fig. 5). Moreover, amyloid peptide-induced apoptosis was not observed in astrocytes with deficient neutral SMase-2 (nSMase-2) indicating that ceramide generated by nSMase-2 and expression of Par-4 is critical for the apoptosis [149]. Furthermore, we have previously shown that H_2O_2 mediates sanguinarine-induced apoptosis in prostate cancer cells via simultaneous induction of ceramide generation and Par-4 cleavage [113].

Par-4 in Ceramide-Induced Autophagy In addition to apoptosis, Par-4-ceramide connection has implicated in autophagic cell death [150]. It has been shown that both ceramide and Par-4 were concomitantly upregulated during the process of autophagic cell death [140]. Recently, we have shown that exogenous ceramide induces Par-4-dependent autophagy via p53-dependent BNIP3 generation in human malignant glioma cells. Interestingly, inhibition of Par-4 protected malignant glioma cells from ceramide-induced autophagic cell death [141] (Fig. 5). These studies

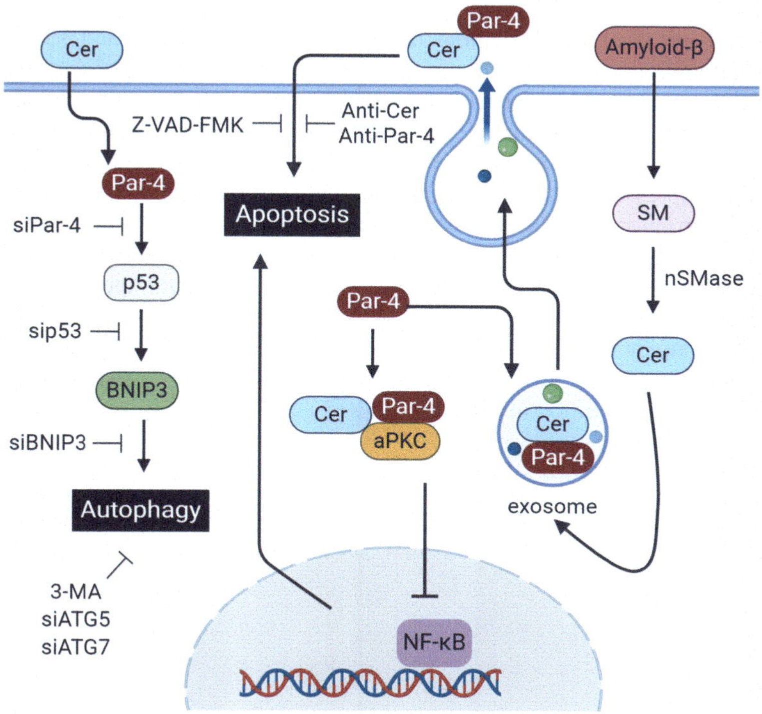

Fig. 5 Role of Par-4 in ceramide-mediated cell death. Ceramide (Cer) generated via neutral sphingomyelinase (nSMase)-mediated hydrolysis of sphingomyelin (SM) induces exosome formation and its association with Par-4, followed by its extracellular secretion. Secreted Par-4/Cer exosomes bind to astrocytes and induces apoptosis. Cer induces Par-4–PKCζ complex formation and inhibits PKCζ, triggering apoptosis via suppressing NF-κB survival pathway. Exogenous Cer activates Par-4, which leads to the activation of p53-BNIP3-mediated autophagic cell death pathway. 3-MA, 3-methyl adenine

shed light on the existence of a cross-connection between the ceramide and Par-4 with respect to the autophagic cell death.

7 Conclusion and Future Perspectives

Par-4 has recently turned 25. Intensive studies in these years have revealed details about protein structure, regulation, biological activities, and tumor-suppressor functions of Par-4. The Par-4 tumor-suppressor function is likely to be mediated through a number of collaborating effector functions rather than through a single pathway. Because of its unique cancer cell-specific apoptotic action, Par-4 has captured the fascination of cancer biologists. Par-4 is secreted, and can act both intracellularly and extracellularly to induce apoptosis specifically in cancer cells. In addition to apoptosis, Par-4 has recently demonstrated to mediate autophagy- and senescence-mediated tumor suppression. Ceramide has emerged as an important signaling entity in the early 1990s. Intensive research over the past 25 years has revealed significant information on ceramide's structural variability, interaction with other cellular components, and involvement in a vast array of biological functions.

Tumor-suppressor functions of ceramide include cell cycle arrest, senescence, apoptosis, ferroptosis. and autophagy. Numerous studies have indicated a connection between ceramide and Par-4 pathways. Ceramide can function as an upstream regulator of Par-4 induction. However, the role of Par-4 on ceramide regulation has never been explored. Further investigation will untangle the information regarding the Par-4-ceramide connection. As our knowledge of both the Par-4 and ceramide function expands, it will reveal the regulatory nodes that can be efficiently targeted for therapeutic intervention in cancers.

Acknowledgments The authors thank the members of the Sehamuddin Galadari laboratory for their helpful discussions. The authors apologize to those investigators whose publications were not mentioned in this review due to space limitations. This work is supported by Al Jalila Foundation for Medical Research grant RA 234 (AJF201631) and New York University Abu Dhabi Research grant AD 252.

References

1. Alvarez JV, Pan TC, Ruth J, Feng Y, Zhou A, Pant D et al (2013) Par-4 downregulation promotes breast cancer recurrence by preventing multinucleation following targeted therapy. Cancer Cell 24(1):30–44. https://doi.org/10.1016/j.ccr.2013.05.007
2. Apel A, Herr I, Schwarz H, Rodemann HP, Mayer A (2008) Blocked autophagy sensitizes resistant carcinoma cells to radiation therapy. Cancer Res 68(5):1485–1494. https://doi.org/10.1158/0008-5472.CAN-07-0562
3. Ashkenazi A, Salvesen G (2014) Regulated cell death: signaling and mechanisms. Annu Rev Cell Dev Biol 30:337–356. https://doi.org/10.1146/annurev-cellbio-100913-013226
4. Axe EL, Walker SA, Manifava M, Chandra P, Roderick HL, Habermann A et al (2008) Autophagosome formation from membrane compartments enriched in phosphatidylinositol 3-phosphate and dynamically connected to the endoplasmic reticulum. J Cell Biol 182(4):685–701. https://doi.org/10.1083/jcb.200803137
5. Babiychuk EB, Atanassoff AP, Monastyrskaya K, Brandenberger C, Studer D, Allemann C et al (2011) The targeting of plasmalemmal ceramide to mitochondria during apoptosis. PLoS One 6(8):e23706. https://doi.org/10.1371/journal.pone.0023706
6. Barradas M, Monjas A, Diaz-Meco MT, Serrano M, Moscat J (1999) The downregulation of the pro-apoptotic protein Par-4 is critical for Ras-induced survival and tumor progression. EMBO J 18(22):6362–6369. https://doi.org/10.1093/emboj/18.22.6362
7. Bento CF, Renna M, Ghislat G, Puri C, Ashkenazi A, Vicinanza M et al (2016) Mammalian autophagy: how does it work? Annu Rev Biochem 85:685–713. https://doi.org/10.1146/annurev-biochem-060815-014556
8. Bergmann M, Kukoc-Zivojnov N, Chow KU, Trepohl B, Hoelzer D, Weidmann E et al (2004) Prostate apoptosis response gene-4 sensitizes neoplastic lymphocytes to CD95-induced apoptosis. Ann Hematol 83(10):646–653. https://doi.org/10.1007/s00277-004-0922-3

9. Bieberich E, MacKinnon S, Silva J, Yu RK (2001) Regulation of apoptosis during neuronal differentiation by ceramide and b-series complex gangliosides. J Biol Chem 276(48):44396–44404. https://doi.org/10.1074/jbc.M107239200
10. Bieberich E, MacKinnon S, Silva J, Noggle S, Condie BG (2003) Regulation of cell death in mitotic neural progenitor cells by asymmetric distribution of prostate apoptosis response 4 (PAR-4) and simultaneous elevation of endogenous ceramide. J Cell Biol 162(3):469–479. https://doi.org/10.1083/jcb.200212067
11. Bieberich E, Silva J, Wang G, Krishnamurthy K, Condie BG (2004) Selective apoptosis of pluripotent mouse and human stem cells by novel ceramide analogues prevents teratoma formation and enriches for neural precursors in ES cell-derived neural transplants. J Cell Biol 167(4):723–734. https://doi.org/10.1083/jcb.200405144
12. Birbes H, El Bawab S, Hannun YA, Obeid LM (2001) Selective hydrolysis of a mitochondrial pool of sphingomyelin induces apoptosis. FASEB J 15(14):2669–2679. https://doi.org/10.1096/fj.01-0539com
13. Birbes H, Luberto C, Hsu YT, El Bawab S, Hannun YA, Obeid LM (2005) A mitochondrial pool of sphingomyelin is involved in TNFalpha-induced Bax translocation to mitochondria. Biochem J 386(Pt 3):445–451. https://doi.org/10.1042/BJ20041627
14. Boehrer S, Chow KU, Beske F, Kukoc-Zivojnov N, Puccetti E, Ruthardt M et al (2002) In lymphatic cells par-4 sensitizes to apoptosis by down-regulating bcl-2 and promoting disruption of mitochondrial membrane potential and caspase activation. Cancer Res 62(6):1768–1775
15. Boehrer S, Kukoc-Zivojnov N, Nowak D, Bergmann M, Baum C, Puccetti E et al (2004) Upon drug-induced apoptosis expression of prostate-apoptosis-response-gene-4 promotes cleavage of caspase-8, bid and mitochondrial release of cytochrome c. Hematology 9(5–6):425–431. https://doi.org/10.1080/10245330400010604
16. Boehrer S, Nowak D, Puccetti E, Ruthardt M, Sattler N, Trepohl B et al (2006) Prostate-apoptosis-response-gene-4 increases sensitivity to TRAIL-induced apoptosis. Leuk Res 30(5):597–605. https://doi.org/10.1016/j.leukres.2005.09.003
17. Boosen M, Vetterkind S, Koplin A, Illenberger S, Preuss U (2005) Par-4-mediated recruitment of Amida to the actin cytoskeleton leads to the induction of apoptosis. Exp Cell Res 311(2):177–191. https://doi.org/10.1016/j.yexcr.2005.09.010
18. Boosen M, Vetterkind S, Kubicek J, Scheidtmann KH, Illenberger S, Preuss U (2009) Par-4 is an essential downstream target of DAP-like kinase (Dlk) in Dlk/Par-4-mediated apoptosis. Mol Biol Cell 20(18):4010–4020. https://doi.org/10.1091/mbc.E09-02-0173
19. Burikhanov R, Zhao Y, Goswami A, Qiu S, Schwarze SR, Rangnekar VM (2009) The tumor suppressor Par-4 activates an extrinsic pathway for apoptosis. Cell 138(2):377–388. https://doi.org/10.1016/j.cell.2009.05.022
20. Bursch W, Ellinger A, Kienzl H, Torok L, Pandey S, Sikorska M et al (1996) Active cell death induced by the anti-estrogens tamoxifen and ICI 164 384 in human mammary carcinoma cells (MCF-7) in culture: the role of autophagy. Carcinogenesis 17(8):1595–1607. https://doi.org/10.1093/carcin/17.8.1595
21. Castro BM, Prieto M, Silva LC (2014) Ceramide: a simple sphingolipid with unique biophysical properties. Prog Lipid Res 54:53–67. https://doi.org/10.1016/j.plipres.2014.01.004
22. Chakraborty M, Qiu SG, Vasudevan KM, Rangnekar VM (2001) Par-4 drives trafficking and activation of Fas and Fasl to induce prostate cancer cell apoptosis and tumor regression. Cancer Res 61(19):7255–7263
23. Chaudhry P, Singh M, Parent S, Asselin E (2012) Prostate apoptosis response 4 (Par-4), a novel substrate of caspase-3 during apoptosis activation. Mol Cell Biol 32(4):826–839. https://doi.org/10.1128/MCB.06321-11
24. Cheema SK, Mishra SK, Rangnekar VM, Tari AM, Kumar R, Lopez-Berestein G (2003) Par-4 transcriptionally regulates Bcl-2 through a WT1-binding site on the bcl-2 promoter. J Biol Chem 278(22):19995–20005. https://doi.org/10.1074/jbc.M205865200
25. Chen X, Sahasrabuddhe AA, Szankasi P, Chung F, Basrur V, Rangnekar VM et al (2014) Fbxo45-mediated degradation of the tumor-suppressor Par-4 regulates cancer cell survival. Cell Death Differ 21(10):1535–1545. https://doi.org/10.1038/cdd.2014.92

26. Chendil D, Das A, Dey S, Mohiuddin M, Ahmed MM (2002) Par-4, a pro-apoptotic gene, inhibits radiation-induced NF kappa B activity and Bcl-2 expression leading to induction of radiosensitivity in human prostate cancer cells PC-3. Cancer Biol Ther 1(2):152–160. https://doi.org/10.4161/cbt.61
27. Cook J, Krishnan S, Ananth S, Sells SF, Shi Y, Walther MM et al (1999) Decreased expression of the pro-apoptotic protein Par-4 in renal cell carcinoma. Oncogene 18(5):1205–1208. https://doi.org/10.1038/sj.onc.1202416
28. Dadsena S, Bockelmann S, Mina JGM, Hassan DG, Korneev S, Razzera G et al (2019) Ceramides bind VDAC2 to trigger mitochondrial apoptosis. Nat Commun 10(1):1832. https://doi.org/10.1038/s41467-019-09654-4
29. Daido S, Kanzawa T, Yamamoto A, Takeuchi H, Kondo Y, Kondo S (2004) Pivotal role of the cell death factor BNIP3 in ceramide-induced autophagic cell death in malignant glioma cells. Cancer Res 64(12):4286–4293. https://doi.org/10.1158/0008-5472.CAN-03-3084
30. Damrauer JS, Phelps SN, Amuchastegui K, Lupo R, Mabe NW, Walens A et al (2018) Foxo-dependent Par-4 Upregulation prevents long-term survival of residual cells following PI3K-Akt inhibition. Mol Cancer Res 16(4):599–609. https://doi.org/10.1158/1541-7786.MCR-17-0492
31. Dany M, Gencer S, Nganga R, Thomas RJ, Oleinik N, Baron KD et al (2016) Targeting FLT3-ITD signaling mediates ceramide-dependent mitophagy and attenuates drug resistance in AML. Blood 128(15):1944–1958. https://doi.org/10.1182/blood-2016-04-708750
32. Das TP, Suman S, Alatassi H, Ankem MK, Damodaran C (2016) Inhibition of AKT promotes FOXO3a-dependent apoptosis in prostate cancer. Cell Death Dis 7:e2111. https://doi.org/10.1038/cddis.2015.403
33. de Thonel A, Hazoume A, Kochin V, Isoniemi K, Jego G, Fourmaux E et al (2014) Regulation of the proapoptotic functions of prostate apoptosis response-4 (Par-4) by casein kinase 2 in prostate cancer cells. Cell Death Dis 5:e1016. https://doi.org/10.1038/cddis.2013.532
34. Deroyer C, Renert AF, Merville MP, Fillet M (2014) New role for EMD (emerin), a key inner nuclear membrane protein, as an enhancer of autophagosome formation in the C16-ceramide autophagy pathway. Autophagy 10(7):1229–1240. https://doi.org/10.4161/auto.28777
35. Diaz-Meco MT, Lallena MJ, Monjas A, Frutos S, Moscat J (1999) Inactivation of the inhibitory kappaB protein kinase/nuclear factor kappaB pathway by Par-4 expression potentiates tumor necrosis factor alpha-induced apoptosis. J Biol Chem 274(28):19606–19612. https://doi.org/10.1074/jbc.274.28.19606
36. Dumitru CA, Gulbins E (2006) TRAIL activates acid sphingomyelinase via a redox mechanism and releases ceramide to trigger apoptosis. Oncogene 25(41):5612–5625. https://doi.org/10.1038/sj.onc.1209568
37. Egan DF, Shackelford DB, Mihaylova MM, Gelino S, Kohnz RA, Mair W et al (2011) Phosphorylation of ULK1 (hATG1) by AMP-activated protein kinase connects energy sensing to mitophagy. Science 331(6016):456–461. https://doi.org/10.1126/science.1196371
38. El-Guendy N, Zhao Y, Gurumurthy S, Burikhanov R, Rangnekar VM (2003) Identification of a unique core domain of par-4 sufficient for selective apoptosis induction in cancer cells. Mol Cell Biol 23(16):5516–5525. https://doi.org/10.1128/mcb.23.16.5516-5525.2003
39. Fass E, Shvets E, Degani I, Hirschberg K, Elazar Z (2006) Microtubules support production of starvation-induced autophagosomes but not their targeting and fusion with lysosomes. J Biol Chem 281(47):36303–36316. https://doi.org/10.1074/jbc.M607031200
40. Fuchs Y, Steller H (2011) Programmed cell death in animal development and disease. Cell 147(4):742–758. https://doi.org/10.1016/j.cell.2011.10.033
41. Funderburk SF, Wang QJ, Yue Z (2010) The Beclin 1-VPS34 complex--at the crossroads of autophagy and beyond. Trends Cell Biol 20(6):355–362. https://doi.org/10.1016/j.tcb.2010.03.002
42. Galadari S, Wu BX, Mao C, Roddy P, El Bawab S, Hannun YA (2006) Identification of a novel amidase motif in neutral ceramidase. Biochem J 393(Pt 3):687–695. https://doi.org/10.1042/BJ20050682

43. Galadari S, Rahman A, Pallichankandy S, Galadari A, Thayyullathil F (2013) Role of ceramide in diabetes mellitus: evidence and mechanisms. Lipids Health Dis 12:98. https://doi.org/10.1186/1476-511X-12-98
44. Galadari S, Rahman A, Pallichankandy S, Thayyullathil F (2015) Tumor suppressive functions of ceramide: evidence and mechanisms. Apoptosis 20(5):689–711. https://doi.org/10.1007/s10495-015-1109-1
45. Galadari S, Rahman A, Pallichankandy S, Thayyullathil F (2017) Reactive oxygen species and cancer paradox: to promote or to suppress? Free Radic Biol Med 104:144–164. https://doi.org/10.1016/j.freeradbiomed.2017.01.004
46. Galluzzi L, Vicencio JM, Kepp O, Tasdemir E, Maiuri MC, Kroemer G (2008) To die or not to die: that is the autophagic question. Curr Mol Med 8(2):78–91. https://doi.org/10.2174/156652408783769616
47. Galluzzi L, Baehrecke EH, Ballabio A, Boya P, Bravo-San Pedro JM, Cecconi F et al (2017) Molecular definitions of autophagy and related processes. EMBO J 36(13):1811–1836. https://doi.org/10.15252/embj.201796697
48. Ganley, I. G., Lam du, H., Wang, J., Ding, X., Chen, S., & Jiang, X. (2009). ULK1.ATG13. FIP200 complex mediates mTOR signaling and is essential for autophagy. J Biol Chem, 284(18), 12297–12305, doi:https://doi.org/10.1074/jbc.M900573200
49. Garcia-Cao I, Duran A, Collado M, Carrascosa MJ, Martin-Caballero J, Flores JM et al (2005) Tumour-suppression activity of the proapoptotic regulator Par4. EMBO Rep 6(6):577–583. https://doi.org/10.1038/sj.embor.7400421
50. Glick D, Barth S, Macleod KF (2010) Autophagy: cellular and molecular mechanisms. J Pathol 221(1):3–12. https://doi.org/10.1002/path.2697
51. Gonzalez P, Mader I, Tchoghandjian A, Enzenmuller S, Cristofanon S, Basit F et al (2012) Impairment of lysosomal integrity by B10, a glycosylated derivative of betulinic acid, leads to lysosomal cell death and converts autophagy into a detrimental process. Cell Death Differ 19(8):1337–1346. https://doi.org/10.1038/cdd.2012.10
52. Goswami A, Burikhanov R, de Thonel A, Fujita N, Goswami M, Zhao Y et al (2005) Binding and phosphorylation of par-4 by akt is essential for cancer cell survival. Mol Cell 20(1):33–44. https://doi.org/10.1016/j.molcel.2005.08.016
53. Goswami A, Qiu S, Dexheimer TS, Ranganathan P, Burikhanov R, Pommier Y et al (2008) Par-4 binds to topoisomerase 1 and attenuates its DNA relaxation activity. Cancer Res 68(15):6190–6198. https://doi.org/10.1158/0008-5472.CAN-08-0831
54. Gozuacik D, Kimchi A (2004) Autophagy as a cell death and tumor suppressor mechanism. Oncogene 23(16):2891–2906. https://doi.org/10.1038/sj.onc.1207521
55. Grassme H, Cremesti A, Kolesnick R, Gulbins E (2003) Ceramide-mediated clustering is required for CD95-DISC formation. Oncogene 22(35):5457–5470. https://doi.org/10.1038/sj.onc.1206540
56. Green DR (2019) The coming decade of cell death research: five riddles. Cell 177(5):1094–1107. https://doi.org/10.1016/j.cell.2019.04.024
57. Gump JM, Staskiewicz L, Morgan MJ, Bamberg A, Riches DW, Thorburn A (2014) Autophagy variation within a cell population determines cell fate through selective degradation of Fap-1. Nat Cell Biol 16(1):47–54. https://doi.org/10.1038/ncb2886
58. Guo Q, Xie J (2004) AATF inhibits aberrant production of amyloid beta peptide 1-42 by interacting directly with Par-4. J Biol Chem 279(6):4596–4603. https://doi.org/10.1074/jbc.M309811200
59. Guo H, Treude F, Kramer OH, Luscher B, Hartkamp J (2019) PAR-4 overcomes chemo-resistance in breast cancer cells by antagonizing cIAP1. Sci Rep 9(1):8755. https://doi.org/10.1038/s41598-019-45209-9
60. Gurumurthy S, Goswami A, Vasudevan KM, Rangnekar VM (2005) Phosphorylation of Par-4 by protein kinase a is critical for apoptosis. Mol Cell Biol 25(3):1146–1161. https://doi.org/10.1128/MCB.25.3.1146-1161.2005

61. Han JY, Lim YJ, Choi JA, Lee JH, Jo SH, Oh SM et al (2016) The role of prostate apoptosis Response-4 (Par-4) in Mycobacterium tuberculosis infected macrophages. Sci Rep 6:32079. https://doi.org/10.1038/srep32079
62. Hannun YA, Luberto C (2000) Ceramide in the eukaryotic stress response. Trends Cell Biol 10(2):73–80. https://doi.org/10.1016/s0962-8924(99)01694-3
63. Hannun YA, Obeid LM (2008) Principles of bioactive lipid signalling: lessons from sphingolipids. Nat Rev Mol Cell Biol 9(2):139–150. https://doi.org/10.1038/nrm2329
64. Hannun YA, Obeid LM (2018) Author correction: sphingolipids and their metabolism in physiology and disease. Nat Rev Mol Cell Biol 19(10):673. https://doi.org/10.1038/s41580-018-0046-6
65. Hayashi-Nishino M, Fujita N, Noda T, Yamaguchi A, Yoshimori T, Yamamoto A (2009) A subdomain of the endoplasmic reticulum forms a cradle for autophagosome formation. Nat Cell Biol 11(12):1433–1437. https://doi.org/10.1038/ncb1991
66. Hebbar N, Burikhanov R, Shukla N, Qiu S, Zhao Y, Elenitoba-Johnson KSJ et al (2017) A naturally generated decoy of the prostate apoptosis Response-4 protein overcomes therapy resistance in Tumors. Cancer Res 77(15):4039–4050. https://doi.org/10.1158/0008-5472.CAN-16-1970
67. Hengartner MO (2000) The biochemistry of apoptosis. Nature 407(6805):770–776. https://doi.org/10.1038/35037710
68. Hosokawa N, Hara T, Kaizuka T, Kishi C, Takamura A, Miura Y et al (2009) Nutrient-dependent mTORC1 association with the ULK1-Atg13-FIP200 complex required for autophagy. Mol Biol Cell 20(7):1981–1991. https://doi.org/10.1091/mbc.E08-12-1248
69. Hotchkiss RS, Strasser A, McDunn JE, Swanson PE (2009) Cell death. N Engl J Med 361(16):1570–1583. https://doi.org/10.1056/NEJMra0901217
70. Hou W, Xie Y, Song X, Sun X, Lotze MT, Zeh HJ 3rd et al (2016) Autophagy promotes ferroptosis by degradation of ferritin. Autophagy 12(8):1425–1428. https://doi.org/10.1080/15548627.2016.1187366
71. Hurley JH, Young LN (2017) Mechanisms of autophagy initiation. Annu Rev Biochem 86:225–244. https://doi.org/10.1146/annurev-biochem-061516-044820
72. Ichimura Y, Kirisako T, Takao T, Satomi Y, Shimonishi Y, Ishihara N et al (2000) A ubiquitin-like system mediates protein lipidation. Nature 408(6811):488–492. https://doi.org/10.1038/35044114
73. Ishihara N, Hamasaki M, Yokota S, Suzuki K, Kamada Y, Kihara A et al (2001) Autophagosome requires specific early sec proteins for its formation and NSF/SNARE for vacuolar fusion. Mol Biol Cell 12(11):3690–3702. https://doi.org/10.1091/mbc.12.11.3690
74. Jiang X, Wang X (2000) Cytochrome c promotes caspase-9 activation by inducing nucleotide binding to Apaf-1. J Biol Chem 275(40):31199–31203. https://doi.org/10.1074/jbc.C000405200
75. Johnson KR, Johnson KY, Becker KP, Bielawski J, Mao C, Obeid LM (2003) Role of human sphingosine-1-phosphate phosphatase 1 in the regulation of intra- and extracellular sphingosine-1-phosphate levels and cell viability. J Biol Chem 278(36):34541–34547. https://doi.org/10.1074/jbc.M301741200
76. Johnstone RW, See RH, Sells SF, Wang J, Muthukkumar S, Englert C et al (1996) A novel repressor, par-4, modulates transcription and growth suppression functions of the Wilms' tumor suppressor WT1. Mol Cell Biol 16(12):6945–6956. https://doi.org/10.1128/mcb.16.12.6945
77. Johnstone RW, Tommerup N, Hansen C, Vissing H, Shi Y (1998) Mapping of the human PAWR (par-4) gene to chromosome 12q21. Genomics 53(2):241–243. https://doi.org/10.1006/geno.1998.5494
78. Joshi J, Fernandez-Marcos PJ, Galvez A, Amanchy R, Linares JF, Duran A et al (2008) Par-4 inhibits Akt and suppresses Ras-induced lung tumorigenesis. EMBO J 27(16):2181–2193. https://doi.org/10.1038/emboj.2008.149

79. Kang MR, Kim MS, Oh JE, Kim YR, Song SY, Kim SS et al (2009) Frameshift mutations of autophagy-related genes ATG2B, ATG5, ATG9B and ATG12 in gastric and colorectal cancers with microsatellite instability. J Pathol 217(5):702–706. https://doi.org/10.1002/path.2509
80. Kanzawa T, Kondo Y, Ito H, Kondo S, Germano I (2003) Induction of autophagic cell death in malignant glioma cells by arsenic trioxide. Cancer Res 63(9):2103–2108
81. Kilbride SM, Prehn JH (2013) Central roles of apoptotic proteins in mitochondrial function. Oncogene 32(22):2703–2711. https://doi.org/10.1038/onc.2012.348
82. Kim HJ, Oh JE, Kim SW, Chun YJ, Kim MY (2008) Ceramide induces p38 MAPK-dependent apoptosis and Bax translocation via inhibition of Akt in HL-60 cells. Cancer Lett 260(1–2):88–95. https://doi.org/10.1016/j.canlet.2007.10.030
83. Kline CL, Shanmugavelandy SS, Kester M, Irby RB (2009) Delivery of PAR-4 plasmid in vivo via nanoliposomes sensitizes colon tumor cells subcutaneously implanted into nude mice to 5-FU. Cancer Biol Ther 8(19):1831–1837. https://doi.org/10.4161/cbt.8.19.9592
84. Kogel D, Reimertz C, Mech P, Poppe M, Fruhwald MC, Engemann H et al (2001) Dlk/ZIP kinase-induced apoptosis in human medulloblastoma cells: requirement of the mitochondrial apoptosis pathway. Br J Cancer 85(11):1801–1808. https://doi.org/10.1054/bjoc.2001.2158
85. Kroemer G (1997) The proto-oncogene Bcl-2 and its role in regulating apoptosis. Nat Med 3(6):614–620. https://doi.org/10.1038/nm0697-614
86. Levine B, Klionsky DJ (2004) Development by self-digestion: molecular mechanisms and biological functions of autophagy. Dev Cell 6(4):463–477. https://doi.org/10.1016/s1534-5807(04)00099-1
87. Li J, Yuan J (2008) Caspases in apoptosis and beyond. Oncogene 27(48):6194–6206. https://doi.org/10.1038/onc.2008.297
88. Li DD, Wang LL, Deng R, Tang J, Shen Y, Guo JF et al (2009) The pivotal role of c-Jun NH2-terminal kinase-mediated Beclin 1 expression during anticancer agents-induced autophagy in cancer cells. Oncogene 28(6):886–898. https://doi.org/10.1038/onc.2008.441
89. Lin CF, Chen CL, Chiang CW, Jan MS, Huang WC, Lin YS (2007) GSK-3beta acts downstream of PP2A and the PI 3-kinase-Akt pathway, and upstream of caspase-2 in ceramide-induced mitochondrial apoptosis. J Cell Sci 120(Pt 16):2935–2943. https://doi.org/10.1242/jcs.03473
90. Liston P, Fong WG, Korneluk RG (2003) The inhibitors of apoptosis: there is more to life than Bcl2. Oncogene 22(53):8568–8580. https://doi.org/10.1038/sj.onc.1207101
91. Liu J, Debnath J (2016) The evolving, multifaceted roles of autophagy in Cancer. Adv Cancer Res 130:1–53. https://doi.org/10.1016/bs.acr.2016.01.005
92. Lu C, Chen JQ, Zhou GP, Wu SH, Guan YF, Yuan CS (2008) Multimolecular complex of Par-4 and E2F1 binding to Smac promoter contributes to glutamate-induced apoptosis in human- bone mesenchymal stem cells. Nucleic Acids Res 36(15):5021–5032. https://doi.org/10.1093/nar/gkn426
93. Marino G, Lopez-Otin C (2004) Autophagy: molecular mechanisms, physiological functions and relevance in human pathology. Cell Mol Life Sci 61(12):1439–1454. https://doi.org/10.1007/s00018-004-4012-4
94. Mathew R, Karp CM, Beaudoin B, Vuong N, Chen G, Chen HY et al (2009) Autophagy suppresses tumorigenesis through elimination of p62. Cell 137(6):1062–1075. https://doi.org/10.1016/j.cell.2009.03.048
95. Mizushima N (2010) The role of the Atg1/ULK1 complex in autophagy regulation. Curr Opin Cell Biol 22(2):132–139. https://doi.org/10.1016/j.ceb.2009.12.004
96. Mizushima N, Komatsu M (2011) Autophagy: renovation of cells and tissues. Cell 147(4):728–741. https://doi.org/10.1016/j.cell.2011.10.026
97. Morales MC, Perez-Yarza G, Rementeria NN, Boyano MD, Apraiz A, Gomez-Munoz A et al (2007) 4-HPR-mediated leukemia cell cytotoxicity is triggered by ceramide-induced mitochondrial oxidative stress and is regulated downstream by Bcl-2. Free Radic Res 41(5):591–601. https://doi.org/10.1080/10715760701218558

98. Moreno-Bueno G, Fernandez-Marcos PJ, Collado M, Tendero MJ, Rodriguez-Pinilla SM, Garcia-Cao I et al (2007) Inactivation of the candidate tumor suppressor par-4 in endometrial cancer. Cancer Res 67(5):1927–1934. https://doi.org/10.1158/0008-5472.CAN-06-2687
99. Moscat J, Diaz-Meco MT, Wooten MW (2009) Of the atypical PKCs, Par-4 and p62: recent understandings of the biology and pathology of a PB1-dominated complex. Cell Death Differ 16(11):1426–1437. https://doi.org/10.1038/cdd.2009.119
100. Nagai MA, Gerhard R, Salaorni S, Fregnani JH, Nonogaki S, Netto MM et al (2010) Down-regulation of the candidate tumor suppressor gene PAR-4 is associated with poor prognosis in breast cancer. Int J Oncol 37(1):41–49. https://doi.org/10.3892/ijo_00000651
101. Nakatogawa H, Ichimura Y, Ohsumi Y (2007) Atg8, a ubiquitin-like protein required for autophagosome formation, mediates membrane tethering and hemifusion. Cell 130(1):165–178. https://doi.org/10.1016/j.cell.2007.05.021
102. Obara K, Sekito T, Niimi K, Ohsumi Y (2008) The Atg18-Atg2 complex is recruited to autophagic membranes via phosphatidylinositol 3-phosphate and exerts an essential function. J Biol Chem 283(35):23972–23980. https://doi.org/10.1074/jbc.M803180200
103. Obeid LM, Linardic CM, Karolak LA, Hannun YA (1993) Programmed cell death induced by ceramide. Science 259(5102):1769–1771. https://doi.org/10.1126/science.8456305
104. Ohsumi Y (2001) Molecular dissection of autophagy: two ubiquitin-like systems. Nat Rev Mol Cell Biol 2(3):211–216. https://doi.org/10.1038/35056522
105. Page G, Kogel D, Rangnekar V, Scheidtmann KH (1999) Interaction partners of Dlk/ZIP kinase: co-expression of Dlk/ZIP kinase and Par-4 results in cytoplasmic retention and apoptosis. Oncogene 18(51):7265–7273. https://doi.org/10.1038/sj.onc.1203170
106. Paglin S, Hollister T, Delohery T, Hackett N, McMahill M, Sphicas E et al (2001) A novel response of cancer cells to radiation involves autophagy and formation of acidic vesicles. Cancer Res 61(2):439–444
107. Pallichankandy S, Rahman A, Thayyullathil F, Galadari S (2015) ROS-dependent activation of autophagy is a critical mechanism for the induction of anti-glioma effect of sanguinarine. Free Radic Biol Med 89:708–720. https://doi.org/10.1016/j.freeradbiomed.2015.10.404
108. Park SK, Nguyen MD, Fischer A, Luke MP, Affarel B, Dieffenbach PB et al (2005) Par-4 links dopamine signaling and depression. Cell 122(2):275–287. https://doi.org/10.1016/j.cell.2005.05.031
109. Pettus BJ, Chalfant CE, Hannun YA (2002) Ceramide in apoptosis: an overview and current perspectives. Biochim Biophys Acta 1585(2–3):114–125. https://doi.org/10.1016/s1388-1981(02)00331-1
110. Pooladanda V, Bandi S, Mondi SR, Gottumukkala KM, Godugu C (2018) Nimbolide epigenetically regulates autophagy and apoptosis in breast cancer. Toxicol In Vitro 51:114–128. https://doi.org/10.1016/j.tiv.2018.05.010
111. Pozuelo-Rubio M (2011) Regulation of autophagic activity by 14-3-3zeta proteins associated with class III phosphatidylinositol-3-kinase. Cell Death Differ 18(3):479–492. https://doi.org/10.1038/cdd.2010.118
112. Raas-Rothschild A, Pankova-Kholmyansky I, Kacher Y, Futerman AH (2004) Glycosphingolipidoses: beyond the enzymatic defect. Glycoconj J 21(6):295–304. https://doi.org/10.1023/B:GLYC.0000046272.38480.ef
113. Rahman A, Pallichankandy S, Thayyullathil F, Galadari S (2019) Critical role of H2O2 in mediating sanguinarine-induced apoptosis in prostate cancer cells via facilitating ceramide generation, ERK1/2 phosphorylation, and Par-4 cleavage. Free Radic Biol Med 134:527–544. https://doi.org/10.1016/j.freeradbiomed.2019.01.039
114. Rasool RU, Nayak D, Chakraborty S, Katoch A, Faheem MM, Amin H et al (2016) A journey beyond apoptosis: new enigma of controlling metastasis by pro-apoptotic Par-4. Clin Exp Metastasis 33(8):757–764. https://doi.org/10.1007/s10585-016-9819-5

115. Ravikumar B, Moreau K, Jahreiss L, Puri C, Rubinsztein DC (2010) Plasma membrane contributes to the formation of pre-autophagosomal structures. Nat Cell Biol 12(8):747–757. https://doi.org/10.1038/ncb2078
116. Roussigne M, Cayrol C, Clouaire T, Amalric F, Girard JP (2003) THAP1 is a nuclear pro-apoptotic factor that links prostate-apoptosis-response-4 (Par-4) to PML nuclear bodies. Oncogene 22(16):2432–2442. https://doi.org/10.1038/sj.onc.1206271
117. Saelens X, Festjens N, Vande Walle L, van Gurp M, van Loo G, Vandenabeele P (2004) Toxic proteins released from mitochondria in cell death. Oncogene 23(16):2861–2874. https://doi.org/10.1038/sj.onc.1207523
118. Santos RVC, de Sena WLB, Dos Santos FA, da Silva Filho AF, da Rocha Pitta MG, da Rocha Pitta MG et al (2019) Potential therapeutic agents against Par-4 target for Cancer treatment: where are we going? Curr Drug Targets 20(6):635–654. https://doi.org/10.2174/1389450120666181126122440
119. Sanvicens N, Cotter TG (2006) Ceramide is the key mediator of oxidative stress-induced apoptosis in retinal photoreceptor cells. J Neurochem 98(5):1432–1444. https://doi.org/10.1111/j.1471-4159.2006.03977.x
120. Scarlatti F, Bauvy C, Ventruti A, Sala G, Cluzeaud F, Vandewalle A et al (2004) Ceramide-mediated macroautophagy involves inhibition of protein kinase B and up-regulation of beclin 1. J Biol Chem 279(18):18384–18391. https://doi.org/10.1074/jbc.M313561200
121. Scarlatti F, Maffei R, Beau I, Codogno P, Ghidoni R (2008) Role of non-canonical Beclin 1-independent autophagy in cell death induced by resveratrol in human breast cancer cells. Cell Death Differ 15(8):1318–1329. https://doi.org/10.1038/cdd.2008.51
122. Segui B, Cuvillier O, Adam-Klages S, Garcia V, Malagarie-Cazenave S, Leveque S et al (2001) Involvement of FAN in TNF-induced apoptosis. J Clin Invest 108(1):143–151. https://doi.org/10.1172/JCI11498
123. Sells SF, Wood DP Jr, Joshi-Barve SS, Muthukumar S, Jacob RJ, Crist SA et al (1994) Commonality of the gene programs induced by effectors of apoptosis in androgen-dependent and -independent prostate cells. Cell Growth Differ 5(4):457–466
124. Sells SF, Han SS, Muthukkumar S, Maddiwar N, Johnstone R, Boghaert E et al (1997) Expression and function of the leucine zipper protein Par-4 in apoptosis. Mol Cell Biol 17(7):3823–3832. https://doi.org/10.1128/mcb.17.7.3823
125. Sentelle RD, Senkal CE, Jiang W, Ponnusamy S, Gencer S, Selvam SP et al (2012) Ceramide targets autophagosomes to mitochondria and induces lethal mitophagy. Nat Chem Biol 8(10):831–838. https://doi.org/10.1038/nchembio.1059
126. Sharma AK, Kline CL, Berg A, Amin S, Irby RB (2011) The Akt inhibitor ISC-4 activates prostate apoptosis response protein-4 and reduces colon tumor growth in a nude mouse model. Clin Cancer Res 17(13):4474–4483. https://doi.org/10.1158/1078-0432.CCR-10-2370
127. Shen W, Henry AG, Paumier KL, Li L, Mou K, Dunlop J et al (2014) Inhibition of glucosylceramide synthase stimulates autophagy flux in neurons. J Neurochem 129(5):884–894. https://doi.org/10.1111/jnc.12672
128. Shimizu S, Kanaseki T, Mizushima N, Mizuta T, Arakawa-Kobayashi S, Thompson CB et al (2004) Role of Bcl-2 family proteins in a non-apoptotic programmed cell death dependent on autophagy genes. Nat Cell Biol 6(12):1221–1228. https://doi.org/10.1038/ncb1192
129. Stolz A, Ernst A, Dikic I (2014) Cargo recognition and trafficking in selective autophagy. Nat Cell Biol 16(6):495–501. https://doi.org/10.1038/ncb2979
130. Subburayan K, Thayyullathil F, Pallichankandy S, Rahman A, Galadari S (2018) Par-4-dependent p53 up-regulation plays a critical role in thymoquinone-induced cellular senescence in human malignant glioma cells. Cancer Lett 426:80–97. https://doi.org/10.1016/j.canlet.2018.04.009
131. Sui Y, Yao H, Li S, Jin L, Shi P, Li Z et al (2017) Delicaflavone induces autophagic cell death in lung cancer via Akt/mTOR/p70S6K signaling pathway. J Mol Med (Berl) 95(3):311–322. https://doi.org/10.1007/s00109-016-1487-z

132. Sumitomo M, Ohba M, Asakuma J, Asano T, Kuroki T, Asano T et al (2002) Protein kinase Cdelta amplifies ceramide formation via mitochondrial signaling in prostate cancer cells. J Clin Invest 109(6):827–836. https://doi.org/10.1172/JCI14146
133. Sun B, Lu C, Zhou GP, Xing CY (2011a) Suppression of Par-4 protects human renal proximal tubule cells from apoptosis induced by oxidative stress. Nephron Exp Nephrol 117(3):e53–e61. doi:10.1159/000320593 10.1159/000320594
134. Sun T, Li D, Wang L, Xia L, Ma J, Guan Z et al (2011b) c-Jun NH2-terminal kinase activation is essential for up-regulation of LC3 during ceramide-induced autophagy in human nasopharyngeal carcinoma cells. J Transl Med 9:161. https://doi.org/10.1186/1479-5876-9-161
135. Takahashi Y, Coppola D, Matsushita N, Cualing HD, Sun M, Sato Y et al (2007) Bif-1 interacts with Beclin 1 through UVRAG and regulates autophagy and tumorigenesis. Nat Cell Biol 9(10):1142–1151. https://doi.org/10.1038/ncb1634
136. Takamura A, Komatsu M, Hara T, Sakamoto A, Kishi C, Waguri S et al (2011) Autophagy-deficient mice develop multiple liver tumors. Genes Dev 25(8):795–800. https://doi.org/10.1101/gad.2016211
137. Taniguchi M, Kitatani K, Kondo T, Hashimoto-Nishimura M, Asano S, Hayashi A et al (2012) Regulation of autophagy and its associated cell death by "sphingolipid rheostat": reciprocal role of ceramide and sphingosine 1-phosphate in the mammalian target of rapamycin pathway. J Biol Chem 287(47):39898–39910. https://doi.org/10.1074/jbc.M112.416552
138. Taylor RC, Cullen SP, Martin SJ (2008) Apoptosis: controlled demolition at the cellular level. Nat Rev Mol Cell Biol 9(3):231–241. https://doi.org/10.1038/nrm2312
139. Thayyullathil F, Pallichankandy S, Rahman A, Kizhakkayil J, Chathoth S, Patel M et al (2013) Caspase-3 mediated release of SAC domain containing fragment from Par-4 is necessary for the sphingosine-induced apoptosis in Jurkat cells. J Mol Signal 8(1):2. https://doi.org/10.1186/1750-2187-8-2
140. Thayyullathil F, Rahman A, Pallichankandy S, Patel M, Galadari S (2014) ROS-dependent prostate apoptosis response-4 (Par-4) up-regulation and ceramide generation are the prime signaling events associated with curcumin-induced autophagic cell death in human malignant glioma. FEBS Open Bio 4:763–776. https://doi.org/10.1016/j.fob.2014.08.005
141. Thayyullathil F, Cheratta AR, Pallichankandy S, Subburayan K, Tariq S, Rangnekar VM et al (2020) Par-4 regulates autophagic cell death in human cancer cells via upregulating p53 and BNIP3. Biochim Biophys Acta, Mol Cell Res 1867(7):118692. https://doi.org/10.1016/j.bbamcr.2020.118692
142. Tiruttani Subhramanyam UK, Kubicek J, Eidhoff UB, Labahn J (2017) Structural basis for the regulatory interactions of proapoptotic Par-4. Cell Death Differ 24(9):1540–1547. https://doi.org/10.1038/cdd.2017.76
143. Vasudevan KM, Ranganathan P, Rangnekar VM (2006) Regulation of Par-4 by oncogenic Ras. Methods Enzymol 407:422–442. https://doi.org/10.1016/S0076-6879(05)07035-7
144. Venable ME, Bielawska A, Obeid LM (1996) Ceramide inhibits phospholipase D in a cell-free system. J Biol Chem 271(40):24800–24805. https://doi.org/10.1074/jbc.271.40.24800
145. Wang G, Bieberich E (2010) Prenatal alcohol exposure triggers ceramide-induced apoptosis in neural crest-derived tissues concurrent with defective cranial development. Cell Death Dis 1:e46. https://doi.org/10.1038/cddis.2010.22
146. Wang G, Silva J, Krishnamurthy K, Tran E, Condie BG, Bieberich E (2005) Direct binding to ceramide activates protein kinase Czeta before the formation of a pro-apoptotic complex with PAR-4 in differentiating stem cells. J Biol Chem 280(28):26415–26424. https://doi.org/10.1074/jbc.M501492200
147. Wang G, Silva J, Dasgupta S, Bieberich E (2008) Long-chain ceramide is elevated in presenilin 1 (PS1M146V) mouse brain and induces apoptosis in PS1 astrocytes. Glia 56(4):449–456. https://doi.org/10.1002/glia.20626
148. Wang BD, Kline CL, Pastor DM, Olson TL, Frank B, Luu T et al (2010) Prostate apoptosis response protein 4 sensitizes human colon cancer cells to chemotherapeutic 5-FU through

mediation of an NF kappaB and microRNA network. Mol Cancer 9:98. https://doi.org/10.1186/1476-4598-9-98
149. Wang G, Dinkins M, He Q, Zhu G, Poirier C, Campbell A et al (2012) Astrocytes secrete exosomes enriched with proapoptotic ceramide and prostate apoptosis response 4 (PAR-4): potential mechanism of apoptosis induction in Alzheimer disease (AD). J Biol Chem 287(25):21384–21395. https://doi.org/10.1074/jbc.M112.340513
150. Wang LJ, Chen PR, Hsu LP, Hsu WL, Liu DW, Chang CH et al (2014) Concomitant induction of apoptosis and autophagy by prostate apoptosis response-4 in hypopharyngeal carcinoma cells. Am J Pathol 184(2):418–430. https://doi.org/10.1016/j.ajpath.2013.10.012
151. Wijesinghe DS, Massiello A, Subramanian P, Szulc Z, Bielawska A, Chalfant CE (2005) Substrate specificity of human ceramide kinase. J Lipid Res 46(12):2706–2716. https://doi.org/10.1194/jlr.M500313-JLR200
152. Xie J, Guo Q (2005) PAR-4 is involved in regulation of beta-secretase cleavage of the Alzheimer amyloid precursor protein. J Biol Chem 280(14):13824–13832. https://doi.org/10.1074/jbc.M411933200
153. Xie Z, Klionsky DJ (2007) Autophagosome formation: core machinery and adaptations. Nat Cell Biol 9(10):1102–1109. https://doi.org/10.1038/ncb1007-1102
154. Yang Z, Huang J, Geng J, Nair U, Klionsky DJ (2006) Atg22 recycles amino acids to link the degradative and recycling functions of autophagy. Mol Biol Cell 17(12):5094–5104. https://doi.org/10.1091/mbc.e06-06-0479
155. Yang YL, Ji C, Bi ZG, Lu CC, Wang R, Gu B et al (2013) Deguelin induces both apoptosis and autophagy in cultured head and neck squamous cell carcinoma cells. PLoS One 8(1):e54736. https://doi.org/10.1371/journal.pone.0054736
156. Yu L, Alva A, Su H, Dutt P, Freundt E, Welsh S et al (2004) Regulation of an ATG7-beclin 1 program of autophagic cell death by caspase-8. Science 304(5676):1500–1502. https://doi.org/10.1126/science.1096645
157. Zhao Y, Rangnekar VM (2008) Apoptosis and tumor resistance conferred by Par-4. Cancer Biol Ther 7(12):1867–1874. https://doi.org/10.4161/cbt.7.12.6945

PAWR as a Direct SRC-1/HOXC11 Suppression Target

Damir Varešlija and Leonie Young

Abstract Regulation of cell fate is controlled through complex and highly regulated transcription of key genes including growth factors, pluripotency, cell differentiation and apoptotic genes. Dysregulation of this process in cancer is central to tumour cell plasticity, the ability of the cancer to resist drug treatment and adaptation to new metastatic environments. The p160 family of coactivator proteins, including SRC-1, are known to partner with nuclear receptor proteins to promote gene transcription and enhance cell growth. More recently, SRC-1 has also been shown to associate with other transcription factors including the homeobox protein HOXC11 to activate pluripotency and development genes. The HOXC11/SRC-1 partnership can also repress gene expression. SRC-1 has been shown to utilise the DNA methylation machinery to silence cancer cell differentiation genes including CD24, NTRK2, NR2F2, CTDP1, SETBP1, and POU3F2. Prostate apoptosis response-4 (Par-4, PAWR) is an apoptotic gene associated with prolonged disease-free survival in breast cancer patients. HOXC11-SRC-1 can utilise the histone demethylase, jumonji domain containing 2C (JMJD2C) to recruit the histone deacetylating protein HDAC1 and histone H3 lysine 27 trimethylation to silence PAWR and promote cancer cells survival in the advanced setting. Master transcriptional regulatory machinery can cooperate to aberrantly silence PAWR, to inhibit mechanisms of controlled cell death and provide a survival advantage for cancer cells.

Keywords Par-4 · PAWR · HOXC11 · SRC-1 DNA methylation · Breast cancer · Histone demethylase · Jumonji domain · Histone deacetylating protein HDAC1 · Histone H3 lysine 27 trimethylation

D. Varešlija · L. Young (✉)
Endocrine Oncology Research Group, Department of Surgery, Royal College of Surgeons in Ireland, Dublin, Ireland
e-mail: lyoung@rcsi.ie

V. M. Rangnekar (ed.), *Tumor Suppressor Par-4*,
https://doi.org/10.1007/978-3-030-73572-2_11

1 Steroid Receptor Co-activator 1 (SRC-1) and HOXC11 in Cancer

SRC-1 regulates disease progression through a complex series of transcriptional networks [1, 3]. HOXC11, a member of the homeobox superfamily of developmental proteins, was identified as a non-steroidal functional binding partner for the master co-regulator SRC-1 in models of advanced breast cancer [4]. This novel interaction between two proteins from two separate super-families raises the concept of a new regulatory complex which may determine cell fate, differentiation programmes and molecular switching with disease progression.

1.1 SRC-1

SRC-1, a member of the p160 family of steroid coactivator proteins, is a master transcriptional regulator. It has been identified as the gene with the strongest selective pressure among ethnic populations analysed by the International HapMap Project [5], illustrating the importance of the coactivator protein in human evolutionary adaptation.

SRC-1 has been shown to be central to the ability of oestrogen receptor (ER) positive tumours to adapt and facilitate metastatic disease progression [6, 7]. Typically, SRC-1 binds to and coactivates nuclear receptors such as ER to regulate a network of proliferation- and differentiation-associated genes critical to breast cancer. More recently, a steroid-independent role for SRC-1 has been recognised and the coactivator has been shown to interact with other transcription factors including members of the Ets family, Ets2 and PEA3 [8] [9] and the developmental protein HOXC11 [4].

There is substantial evidence that SRC-1 is central to the ability of endocrine tumours to adapt and overcome targeted therapy facilitating the development of distant metastatic disease [4, 9, 10, 11, 12, 17, 18, 22]. Though other members of the p160 family including SRC-2 (GRIP1/TIF2/NCoA2) and SRC-3 (AIB1/ACTR/RAC-3/pCIP/TRAM-1/NCoA3) have been investigated in endocrine-related cancer [7], specific roles for both SRC-1 and SRC-3 in the development of metastasis have also been described [1, 3, 6, 11]. Global analysis of SRC-1 targets in breast cancer cells has identified pathways involved in growth factor signalling and cellular adaptability [3]. Furthermore, molecular and in vivo studies have provided evidence that SRC-1 promotes disease progression and the development of metastasis, through the transcriptional activation of tumorigenic genes including integrin-alpha5, ADAM22 and Myc [9, 12, 13]. Modulations of these endocrine-resistant pathways can be driven by SRC-1 through both transcriptomic and epigenetic mechanisms [1, 3].

1.2 The HOMEOBOX Family

The HOX genes encode a family of transcription factors that play key roles in the determination and maintenance of cell fate and cell identity; they are 'master regulators' of normal development and control processes such as proliferation, apoptosis, migration and invasion [14, 15]. In humans, 39 HOX genes have been identified and are found organised contiguously among four clusters. HOX genes within different clusters are arranged along the chromosomes 7, 12 and 2 as 13 paralogous groups [16]. The hallmark of the homeobox family of genes is that they share a common nucleotide sequence motif which encodes a 61 amino acid homeodomain. The homeodomain is a helix-turn-helix DNA-binding domain which mediates binding to a TAAT consensus site, thus allowing these proteins to act as either activators or repressors of transcription [17]. HOX proteins can function as monomers or homodimers to directly drive the transcription of downstream targets. It has also been suggested that HOX proteins can sequester other proteins to enhance or repress gene expression [18].

1.3 The Role of HOX Genes in the Normal Mammary Gland

The first demonstrated role for homeobox genes in the control of mammary function came from the targeted deletion of three paralogous *HOX* genes: *HOXa9, HOXb9,* and *HOXd9.* In triple mutant adult female mice, the mammary glands appeared normal; however, their development was severely impaired during pregnancy, leading to a failure in lactation [19]. Single mutant lines showed only a small decrease in newborn survival suggesting that functional cooperativity is at play between the three gene products during lobulo-alveolar expansion. In contrast, *HOXc6*-deficient mammary glands appeared underdeveloped at birth and revealed reduced ductal branching in the adult gland [20].

Five HOX genes were also identified in CID-9 mouse mammary epithelial cells. When these cells were cultured on a reconstituted basement membrane in the presence of lactogenic hormones, they formed alveolar-like hollow spheres and differentiated to express milk proteins. *HOXa1* and *HOXb7* gene expression were downregulated by exogenous basement membrane suggesting that homeobox gene expression may be modulated in vivo by interactions between epithelial cells and components of the basement membrane. Such interactions have been shown to be critical for mammary differentiation and function and tend to be altered on neoplastic progression [14].

1.4 The Role of HOX Proteins in Breast Cancer

The accurate spatial and temporal regulation of HOX gene expression is imperative during embryogenesis to acquire the proper differentiated state and cell identity. There is an accumulating body of evidence which suggests that developmental genes are often aberrantly expressed in human cancers and that this can impact neoplastic disease through the re-initiation of developmental programmes [21].

Aberrant expression of homeobox genes can lead to failure of differentiation or loss of the differentiated state, activation of anti-apoptotic pathways or adoption of an alternative cellular identity. This disrupted HOX gene expression can impact on many cell types within the body, causing affected cells to incur inappropriate characteristics such as loss of cell cycle, decreased apoptosis, altered cell-cell adhesions, altered hormone and growth factor responses or increased protease expression, all of which could easily contribute to cancer initiation or progression [17]. The expression of the HOX genes and their targets could be useful as a crude measure of the identity or functional state of a given cell thus HOX gene expression has become an area of increased interest in cancer biology.

HOXB7 has been identified as an ERα-responsive gene and overexpression of HOXB7 has been shown to render endocrine sensitive MCF-7 cells resistant to tamoxifen via cross-talk between receptor tyrosine kinases (RTK) and ERα signalling. Extended treatment of MCF-7 cells with tamoxifen resulted in progressively increasing levels of HOXB7 expression, along with EGFR and EGFR ligands. The study also showed upregulation of EGFR occurs through direct binding of HOXB7 to the EGFR promoter, thus enhancing transcriptional activity. Finally, higher expression levels of HOXB7 in the tumour significantly correlated with poor disease-free survival in ERα-positive breast cancer patients on adjuvant endocrine monotherapy [22].

A large study by Ma et al. used microarray analyses of ERα-positive tumours from patients with disease recurrences despite receiving ongoing tamoxifen treatment and found that HOXB13 expression was markedly upregulated compared to tumour samples from patients without recurrences [23]. Later work showed that these patients who overexpressed HOXB13 were also at risk of distant metastasis at recurrence [24]. Other overexpression studies using HOXB13 demonstrated that HOXB13 could suppress the ERα-dependent transcription and expression of components of the ERα pathway such as PgR and cyclin D1. In such cases, no mutation or amplification of HOXB13 of the gene itself was observed, suggesting that the mutation may lie within the gene regulating HOXB13 or in a gene that leads to the activation of alternative growth factor pathways [18].

It is also suggested that aberrant HOX expression could impact on apoptosis to promote tumourigenesis. Loss of HOXA5 expression and methylation of the HOXA5 promoter were shown to correlate with reduced expression of the tumour suppressor p53, in both primary breast tumours and breast cancer cell lines [25]. Further, in vitro studies identified a HOXA5 binding site on the p53 promoter whereby upregulation of HOXA5 induced p53 expression and increased the rates of apoptosis in the MCF-7 breast cancer cell line. HOXA10 was shown to have a similar effect on p53 in ERα-positive breast cancer cell lines [26].

Whilst there is a general consensus as to the overall function of the HOX gene family, there is still limited information on the specific normal and aberrant functions of each individual family member. Often the effects of one HOX gene can vary due to their tissue-specific functions; the majority of HOX genes seem to lose expression during the initiation or progression of neoplasia; however, there are also a significant few whose overexpression remains strongly associated with tumour progression. Previous observations suggest that the consequences of a given HOX gene mutation cannot be predicted definitively, but will depend, at least in part, on the nature of the mutation and on the battery of downstream genes deregulated as a result.

1.5 SRC-1 and HOXC11 in Transcriptional Activation

HOXC11 has been previously associated with the differentiation of neuroblastoma GOTO cells to Schwann cells [27]. The study conducted by Zhang et al. was the first to identify the Schwann cell marker; S100β as a direct target gene of HOXC11. In terms of endocrine-resistant breast cancer, mass spectrometry analysis identified HOXC11 as an interacting partner of SRC-1 [4]. Moreover, immunohistochemical analysis of a tamoxifen-treated patient population showed increased localisation and expression of HOXC11 in the nuclei of tumour cells. Strong associations were also observed between HOXC11 and poor disease survival in breast cancer patients. Moreover, coexpression of HOXC11 with SRC-1 was noted as a superior predictor of poor clinical prognosis than any of the other classic parameters of disease progression [4]. Of interest, S100β has been recognised as the coactivated transcriptional target for the SRC-1/HOXC11 complex in both breast cancer and melanoma [4, 28] and acts as a secreted biomarker to predict poor response to treatment in an ER-positive breast cancer patient population [29].

2 Gene Regulation Through Epigenetic Modification

Epigenetics is defined as the heritable changes in gene expression that drive phenotypic expression, without making direct changes to the genetic code. Initially characterised by Conrad Waddington [30], it now defines DNA methylation and histone modifications which drive functional processes such as X chromosome inactivation, embryogenesis, tissue differentiation and carcinogenesis.

2.1 DNA Methylation

DNA methylation is amongst the best-studied epigenetic modifications. DNA methylation events occur at replication with the addition of a methyl (CH3) group to the fifth position of a cytosine's pyrimidine ring of its CpG dinucleotide sequence. Within the genome most CpG sites are methylated; however, clusters of CpG dinucleotides called CpG islands remain un-methylated. CpG islands are often located at the 5′ position of a gene promoter. In a normal cell, methylation occurs to repress predominantly repetitive regions including long interspersed transposable elements, short interspersed transposable elements and endogenous retroviruses which make up 90% of the human genome.

The establishment and maintenance of a cell's methylation pattern is mediated by DNA methyltransferases (DNMTs) [31]. The addition of methyl groups is catalysed by DNMTs which utilise S-adenosyl-l-methionine as the methyl group donor [32]. The methyltransferase family includes DNMT1, DNMT2, DNMT3A, DNMT3B and DNMT3L, all of which have a methyltransferase catalytic domain. DNMT1 is the maintenance methyltransferase which has a preference for hemi-methylated DNA, allowing it to add methyl groups during replication [33, 34]. DNMT1 is highly expressed in differentiated cells [35]. DNMT2 is an RNA methyltransferase [35]. DNMT3A and DNMT3B are preferentially expressed in stem cells, with low expression levels in differentiated cells. They function as de novo methyltransferases which establish novel patterns of methylation [36].

When a site is methylated it becomes a candidate region for a regulative chromatin state to be established. This is achieved by recruitment of methyl-CpG binding domain (MBD) proteins, which in turn recruit HDACs and chromatin re-modellers. These MBD proteins form a platform linking DNA methylation, polycomb/trithorax protein complexes and the other chromatin modifiers which subsequently drive epigenetic regulation. There are five MBD family members known, MBD1, MBD2, MBD3, MBD4 and MeCP2, with each possessing a common methyl CpG binding domain. MBD1 associates with the Suv39-HP1 complex to drive transcriptional repression [37] as well as associating with SETDB1 to drive H3K9 histone methylation [38]. MBD2 binds methylated CpG's and drives repression through its association with HDAC1 and HDAC2 often in the nucleosome remodelling and deacetylase (NuRD) chromatin remodelling complex [38]. MBD3 is also part of the NuRD complex, however, evidence suggests these functions in complexes independently to one another [38]. MBD4 is a DNA repair protein which specifically targets sites of cytosine deamination [39]. The MBD protein MeCP2 is the most studied of the MBD proteins [40]. It is involved in an array of cellular functions including transcriptional repression, gene activation, nuclear organisation and splicing [41]. The full suppressive ability of MeCP2 depends on its interaction with HDAC2 [42].

2.2 Post-Translational Modifications

DNA is packaged into chromatin which is a highly organised structure [43]. The open and closed formation of chromatin is determined by histone modifications that determine chromatin regulation. The nucleosome is an octamer of four core histones including an H3/H4 tetramer and two H2A/H2B dimers, which allow for 146 bp of DNA to coil [44]. Histone structure is controlled by several post-translational modifications including methylation, acetylation, phosphorylation, ubiquitination, sumoylation, ADP ribosylation and deamination. The specific combination of histone modifications dictates the regulatory capacity of a genomic region, i.e. the 'histone code'. For a gene to become activated, the chromatin must be unwound allowing transcriptional machinery to bind to the DNA thus aiding transcriptional activation [45]. Histone acetylation is the most studied modification involving the addition of an amino group on the lysine residues [46]. Histone acetyltransferases enable the addition of amino groups to the N-terminal of lysine, whilst histone deacetylases remove these groups. Lysine methylation of H3 or H4 can be associated with either activation or repression depending on the lysine residue affected and whether that site is mono, di or tri-methylated. H3K4me1 and H3K4me3 are both marks of activation found at enhancers and gene promoters respectively, whereas, K9 and K27 tri-methylation on histone H3 are repressive marks [47]. Often histone methylation and histone acetylation mediate regulation simultaneously. Trimethylation of sites such as H3K4, H3K36 and H3K79 is an example of such regulation, which creates an active site associated with elevated levels of acetylation [48].

DNA methylation and histone modifications are carried out by different reactions; however, the two mechanisms appear to bi-directionally affect each other [48–50]. It is now known that these processes are involved in a complex interplay which alters their regulative patterns, as well as target gene expression. DNA methylation can control histone edits either by direct interplay between HDACs and DNMTs (51, Roth, 2001 #53) or through MBD proteins which subsequently recruit histone-modifying proteins [42, 50, 51].

2.3 Epigenetic Regulation of Cancer

Due to the importance of these epigenetic processes, it is not surprising that dysregulation of such mechanisms drives an array of human diseases including cancer. In addition to genetic mutations in carcinogenesis, there are vast epigenetic changes which occur in a developing tumour cell and these changes can alter the genetic landscape of a cell, promoting oncogenesis. In 1983, cancer was one of the first diseases to be linked to epigenetic dysregulation. Genes in colorectal cancer were found to be substantially hypo-methylated compared to normal tissue [52]. Furthermore, studies using DNA methyltransferase inhibitors (DNMTi) or DNA methyltransferase-deficient mice described an important role for hypo-methylation in cancer [53].

After this initial discovery, the field of epigenetics shifted focus to DNA hyper-methylation, investigating repression of tumour suppressor genes. These hyper-methylated tumour suppressors were demonstrated to have functions in cell cycle regulation, cell invasion, DNA damage repair, chromatin remodelling, cell signalling, transcription and apoptosis [54]. In studies looking at primary and metastatic breast tumours, global elevations in DNA hyper-methylation have been observed with disease progression (Fig. 1) [1]. Furthermore, pathway analysis suggests that

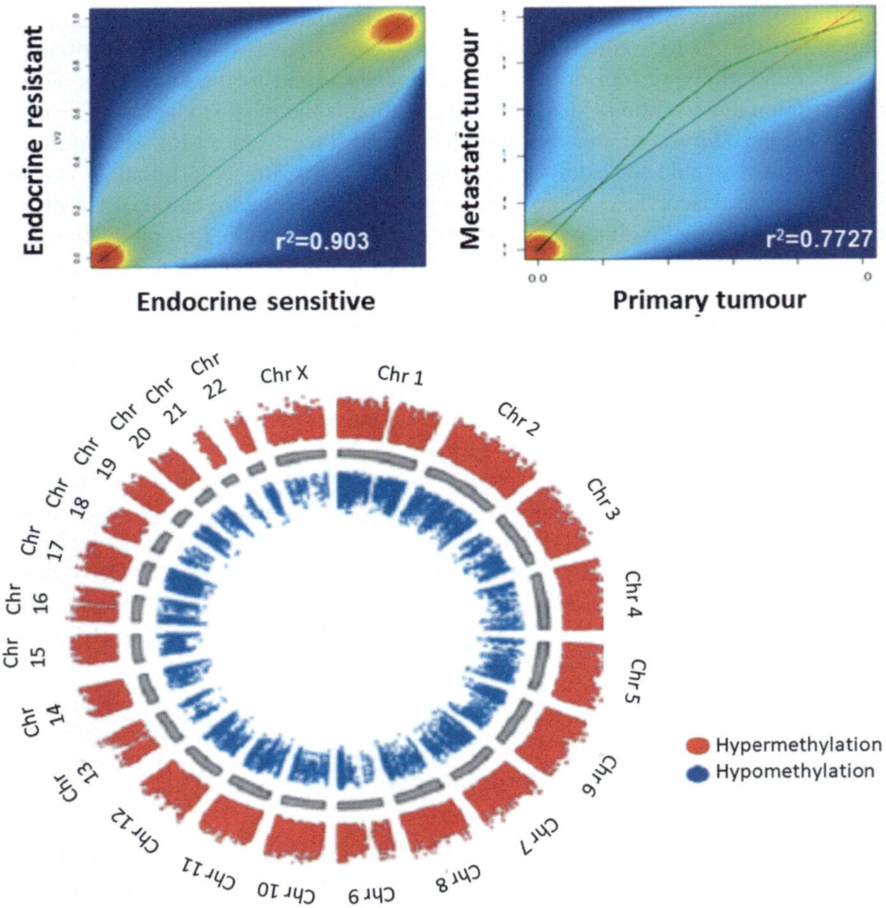

Fig. 1 DNA hypermethylation with disease progression. Colourimetric density plot of the correlation between the global methylation profile of endocrine sensitive and endocrine-resistant cell models and patient primary breast tumour with a matched brain metastatic tumour using Roche SeqCap Epi-targeted bisulphite methylation sequencing. The correlation plot demonstrates an asymmetric density distribution: Endocrine-resistant and the brain metastatic tumour have increased hyper-methylation relative to the sensitive cells and matched primary tumour sample, respectively. Circos plot of differentially methylated CpGs detected in brain metastasis compared to the primary tumour. Adapted from [1]

aberrant methylation events are associated with the suppression of genes important in tumour cell differentiation [1].

2.4 *SRC-1 and Transcriptional Repression*

All of the p160 family members, including SRC-1, have been described as transcriptional coactivators, partnering with transcription factors including nuclear receptors to activate transcription of target genes. However, SRC-2 has a somewhat amphipathic role in transcriptional regulation, working as both a transactivator and repressor of the glucocorticoid receptor (GR) depending on the transcriptional context [55, 56]. This opposing regulatory role for SRC-2 opens up the possibility that SRC-1 could also suppress transcriptional activity to fulfil its oncogenic potential. Recently a dual role for SRC-1 has also been described, where SRC-1 the classic steroid receptor coactivator protein, has been shown to transcriptionally repress both pro-survival and differentiation genes [1, 2]. Integrated multi-omics studies have uncovered specific global SRC-1 dependent hypermethylation, corresponding to regions of transcriptional repression of direct SRC-1 target differential genes including NTRK2, NR2F2, CTDP1, SETBP1, and POU3F2 [1]. SRC-1 can coordinate the repression of these differential genes through the recruitment of members of the methyl binding domain (MBD) protein family, MBD2 and MECP2, which deciphers the DNA methylation code and the histone deacetylator, HDAC2, a known MECP2 binding partner [1].

Collectively these data revealed an epigenetic reprograming pathway, whereby concerted differential DNA methylation is potentiated by SRC-1 in the endocrine-resistant setting.

3 SRC-1 and HOX Transcriptional Repression of PAWR

As previously described HOXC11 can act as a transcription factor partner for SRC-1 which can activate genes to mediate increased tumour plasticity and adaptation to the therapeutic environment [4]. SRC-1 can however also partner with HOXC11 to induce transcriptional repression.

The differentiation marker CD24 and the apoptotic protein PAWR have both been identified as direct SRC1/HOXC11 suppression targets [2]. In endocrine-resistant cell models silencing of SRC-1 and HOXC11 with siRNA resulted in enhanced levels of PAWR. Furthermore, in SRC-1 knockout mice (SRC-$1^{-/-}$/PyMT mice) levels of PAWR transcript and protein were found to be elevated in comparison to wild-type mice ((WT/PyMT mice) (Fig. 2) [2].

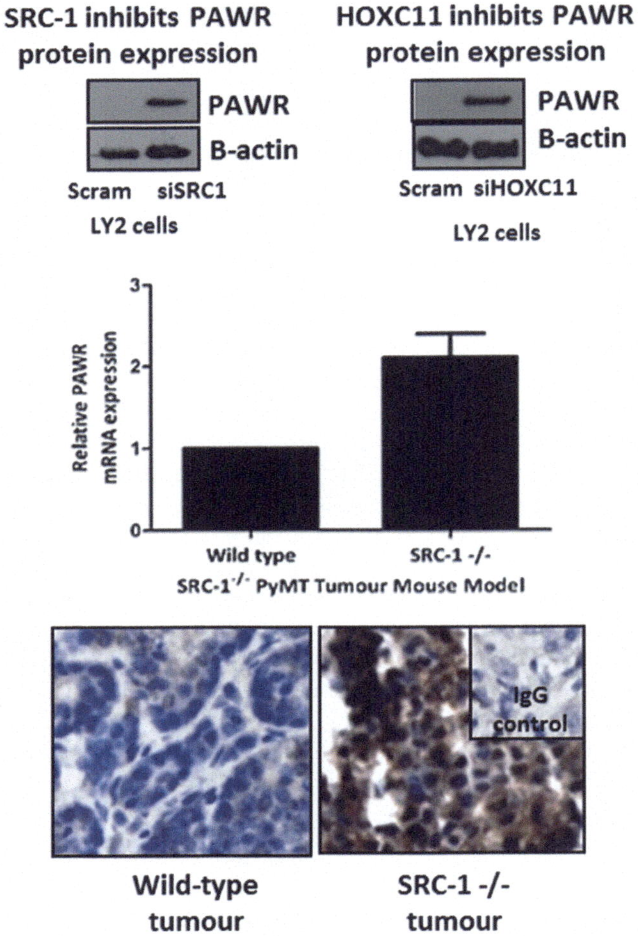

Fig. 2 PAWR expression is repressed by HOXC11 and SRC-1. PAWR protein levels are enhanced following knockdown of SRC-1 and HOXC11 with siRNA in LY2 cells in endocrine-resistant breast cancer cell models
Elevated levels of CD24 and PAWR transcript and protein as determined by q-RT-PCR and immunohistochemistry for PAWR were detected in mammary tumours from SRC-$1^{-/-}$/PyMT mice in comparison to wild-type mice (WT/PyMT mice). Adapted from [2]

PAWR has a documented role as an endocrine tumour suppressor [57]. As a well-known pro-apoptotic protein, PAWR initiates apoptosis in tumour cells by inactivating several pro-survival pathways including NF-κB, Akt and ζPKC [58]. Establishing the functional consequences of SRC-1 target gene expression, PAWR has been reported to be inversely associated with SRC-1 expression in breast cancer patients

[2]. Similarly, in HER2-driven breast cancer, PAWR was found to be downregulated in recurrent tumours following either HER2 or chemotherapy treatment. It was reported that PAWR silencing was directly responsible for promoting residual cancer survival in both murine models and patient association studies [59]. Furthermore, where SRC-1 associates with poor disease-free survival in patients [8, 12], PAWR is a marker of good prognosis in endocrine-resistant ER-positive breast cancers (Fig. 3) [2]. That SRC-1 can silence this apoptotic protein is of prime importance when considering the role of SRC-1 as a master regulator of disease progression in cancer. Not only does SRC-1 promote expression of classic oncogenes such as Myc, it can simultaneously work to suppress expression of genes which are striving to maintain controlled cell death.

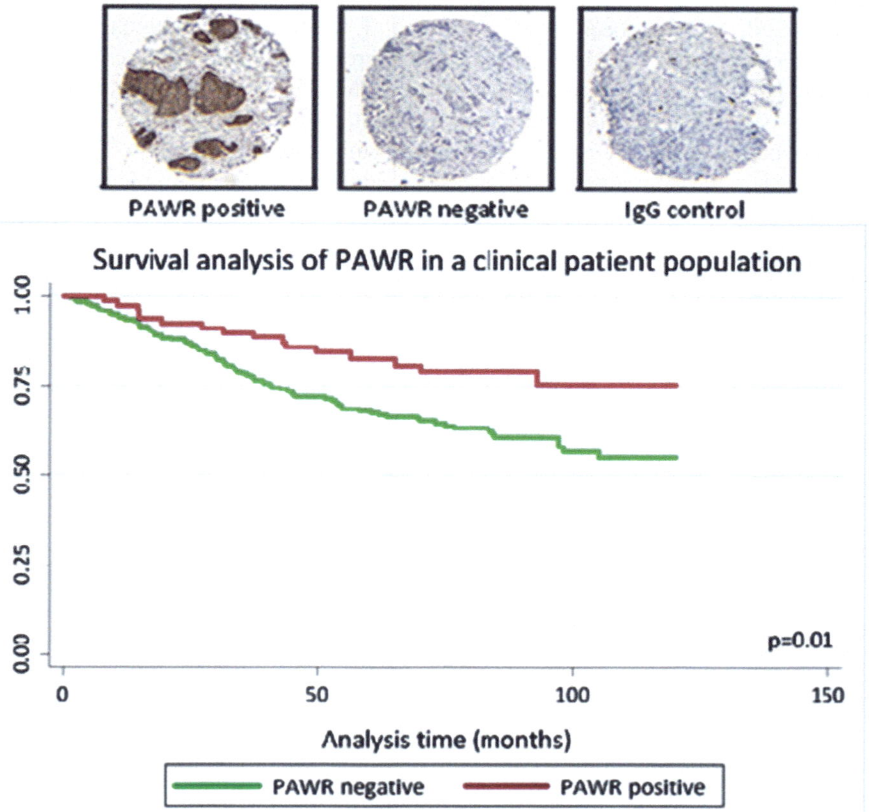

Fig. 3 PAWR associates with prolonged disease-free survival in breast cancer patients Immunolocalisation of PAWR in breast cancer patients (n = 560). PAWR-positive patients have significantly improved 5 year survival compared to PAWR-negative patients (p = 0.01). Adapted from [2]

As there is now evidence that the SRC-1/HOXC11 transcriptional complex can both promote and repress gene expression, the mechanism of how some genes are marked for activation and others repression remains to be fully elucidated. For other co-regulatory protein/transcription factor pairing such as SRC-2-GR, SRC-2 can serve as a GR coactivator at palindromic GREs; however, when SRC-2 is tethered to GR-AP1 sites and GR-NF-kappaB sites it mediates transcriptional repression [60, 61].

From motif analysis of the SRC-1 cistrome, AP-1 sites are reported to be over-represented in the promoters of SRC-1 repressed genes in comparison to activated genes [2]. Furthermore, treatment-induced interactions between the AP-1 interaction partner jun and SRC-1, as well as recruitment of jun to the promoter of PAWR, have been described [2], suggesting that SRC-1 can regulate gene expression in a promoter-specific manner similar to that of SRC-2 [62].

A number of chromatin re-modelling proteins have been identified as SRC-1 interacting partners in luminal B breast cancer sub-type [2, 3]. Several transcriptional regulators have been identified including the histone demethylase, jumonji domain containing 2C (JMJD2C) [2]. JMJD2 proteins are demethylases that target histone H3 on lysines 9 and 3. Both AP1 and JMJD2C proteins have been shown previously to induce transcriptional repression through recruitment of the histone deacetylating protein HDAC1 [63, 64]. Molecular studies in models of advanced breast cancer have demonstrated that SRC-1 can mediate transcriptional repression of PAWR, at least in part, by facilitating histone deacetylation [2]. JMJD2C has a well-described role in modulating histone methylation, consistent with this SRC-1 can utilise histone H3 lysine 27 trimethylation, the transcriptional repression mark, to silence PAWR in cancer cells (Fig. 4) [2]. Attempts to reverse this phenotype and re-express PAWR expression levels have been investigated [65]. There is further

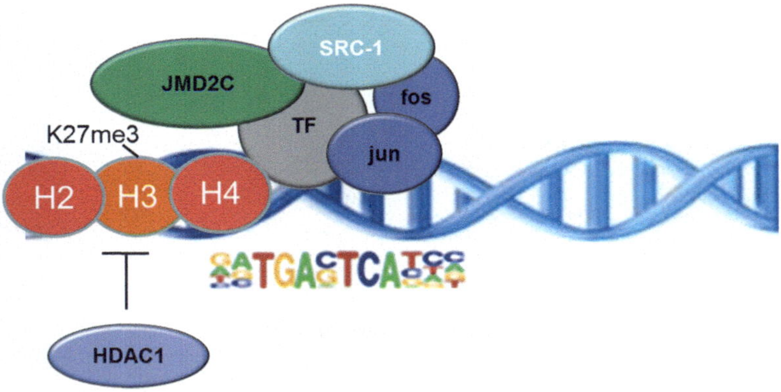

Fig. 4 SRC-1 transcriptional repression of PAWR. Cartoon illustrating the putative SRC-1 interactome regulating transcriptional repression of target genes. Adapted from [2]

evidence of epigenetic silencing of PAWR in breast cancer and indeed studies have found that using epigenome editing to reverse PAWR silencing can be a tool to resensitise tumour cells to overcome treatment resistance [66].

4 Conclusions

Dysregulation of the transcriptional process in cancer can lead to the aberrant expression of pluripotency genes and silencing of tumour suppressor and apoptotic genes. Master regulators HOXC11 and SRC-1 are associated with enhanced transcriptional activity. The transcription factor/co-regulatory complex can also induce transcriptional repression in advanced cancer. Using the DNA methylation machinery, HOXC11/SRC-1 can suppress expression of the apoptotic protein PAWR.

As a tumour suppressor gene and a cancer-specific apoptotic gene, PAWR not only offers promise as a prognostic marker and indicator of response to therapy in cancer, but also as a potential therapeutic target. Strategies to inhibit DNA methylation and re-express tumour suppressor and differentiation genes is the subject of several clinical trials over a range of cancers. Tailoring these therapies to specific tumour suppression pathways has the potential to open up a new management strategy for advanced cancer. Specifically, mechanisms of epigenome editing suitable for the clinic to induce re-expression of PAWR may reverse treatment resistance and offer a new therapeutic modality to treat metastatic disease.

References

1. Ward E, Varešlija D, Charmsaz S, Fagan A, Browne AL, Cosgrove N et al (2018) Epigenome-wide SRC-1-mediated gene silencing represses cellular differentiation in advanced breast Cancer. Clin Cancer Res Off J Am Assoc Cancer Res 24(15):3692–3703
2. Walsh CA, Bolger JC, Byrne C, Cocchiglia S, Hao Y, Fagan A et al (2014) Global gene repression by the steroid receptor coactivator SRC-1 promotes oncogenesis. Cancer Res 74(9):2533–2544
3. Browne AL, Charmsaz S, Varešlija D, Fagan A, Cosgrove N, Cocchiglia S et al (2018) Network analysis of SRC-1 reveals a novel transcription factor hub which regulates endocrine resistant breast cancer. Oncogene 37(15):2008–2021
4. McIlroy M, McCartan D, Early S (2010) P OG, Pennington S, Hill AD, et al. interaction of developmental transcription factor HOXC11 with steroid receptor coactivator SRC-1 mediates resistance to endocrine therapy in breast cancer [corrected]. Cancer Res 70(4):1585–1594
5. Voight BF, Kudaravalli S, Wen X, Pritchard JK (2006) A map of recent positive selection in the human genome. PLoS Biol 4(3):e72
6. Xu J, Qiu Y, DeMayo FJ, Tsai SY, Tsai M-J, O'Malley BW (1998) Partial hormone resistance in mice with disruption of the steroid receptor Coactivator-1 (SRC-1) gene. Science 279(5358):1922–1925
7. Xu J, Wu RC, O'Malley BW (2009) Normal and cancer-related functions of the p160 steroid receptor co-activator (SRC) family. Nat Rev Cancer 9(9):615–630

8. Myers E, Hill ADK, Kelly G, McDermott EW, O'Higgins NJ, Buggy Y et al (2005) Associations and interactions between Ets-1 and Ets-2 and coregulatory proteins, SRC-1, AIB1, and NCoR in breast Cancer. Clin Cancer Res 11(6):2111–2122
9. Al-azawi D, Ilroy MM, Kelly G, Redmond AM, Bane FT, Cocchiglia S et al (2008) Ets-2 and p160 proteins collaborate to regulate c-Myc in endocrine resistant breast cancer. Oncogene 27(21):3021–3031
10. Zhao S, Geybels MS, Leonardson A, Rubicz R, Kolb S, Yan Q et al (2017) Epigenome-wide tumor DNA methylation profiling identifies novel prognostic biomarkers of metastatic-lethal progression in men diagnosed with clinically localized prostate Cancer. Clin Cancer Res 23(1):311–319
11. Varešlija D, Ward E, Purcell SP, Cosgrove NS, Cocchiglia S, O'Halloran PJ et al (2021) Comparative analysis of the AIB1 interactome in breast cancer reveals MTA2 as a repressive partner which silences E-Cadherin to promote EMT and associates with a pro-metastatic phenotype. Oncogene
12. McBryan J, Theissen SM, Byrne C, Hughes E, Cocchiglia S, Sande S et al (2012) Metastatic progression with resistance to aromatase inhibitors is driven by the steroid receptor coactivator SRC-1. Cancer Res 72(2):548–559
13. McCartan D, Bolger JC, Fagan A, Byrne C, Hao Y, Qin L et al (2012) Global characterization of the SRC-1 transcriptome identifies ADAM22 as an ER-independent mediator of endocrine-resistant breast Cancer. Cancer Res 72(1):220–229
14. Lewis MT (2000) Homeobox genes in mammary gland development and neoplasia. Breast Cancer Res BCR 2(3):158–169
15. McCoy EL, Iwanaga R, Jedlicka P, Abbey NS, Chodosh LA, Heichman KA et al (2009) Six1 expands the mouse mammary epithelial stem/progenitor cell pool and induces mammary tumors that undergo epithelial-mesenchymal transition. J Clin Invest 119(9):2663–2677
16. Pineault KM, Wellik DM (2014) Hox genes and limb musculoskeletal development. Curr Osteoporos Rep 12(4):420–427
17. Visvader JE, Lindeman GJ (2003) Transcriptional regulators in mammary gland development and cancer. Int J Biochem Cell Biol 35(7):1034–1051
18. Shah N, Sukumar S (2010) The Hox genes and their roles in oncogenesis. Nat Rev Cancer 10(5):361–371
19. Chen F, Capecchi MR (1999) Paralogous mouse Hox genes, Hoxa9, Hoxb9, and Hoxd9, function together to control development of the mammary gland in response to pregnancy. Proc Natl Acad Sci U S A 96(2):541–546
20. Garcia-Gasca A, Spyropoulos DD (2000) Differential mammary morphogenesis along the anteroposterior axis in Hoxc6 gene targeted mice. Dev DynamicsOff Pub Am Assoc Anatomists 219(2):261–276
21. Pearson JC, Lemons D, McGinnis W (2005) Modulating Hox gene functions during animal body patterning. Nat Rev Genet 6(12):893–904
22. Jin K, Kong X, Shah T, Penet MF, Wildes F, Sgroi DC et al (2012) The HOXB7 protein renders breast cancer cells resistant to tamoxifen through activation of the EGFR pathway. Proc Natl Acad Sci U S A 109(8):2736–2741
23. Ma XJ, Wang Z, Ryan PD, Isakoff SJ, Barmettler A, Fuller A et al (2004) A two-gene expression ratio predicts clinical outcome in breast cancer patients treated with tamoxifen. Cancer Cell 5(6):607–616
24. Jerevall PL, Brommesson S, Strand C, Gruvberger-Saal S, Malmström P, Nordenskjöld B et al (2008) Exploring the two-gene ratio in breast cancer--independent roles for HOXB13 and IL17BR in prediction of clinical outcome. Breast Cancer Res Treat 107(2):225–234
25. Raman V, Martensen SA, Reisman D, Evron E, Odenwald WF, Jaffee E et al (2000) Compromised HOXA5 function can limit p53 expression in human breast tumours. Nature 405(6789):974–978
26. Chu MC, Selam FB, Taylor HS (2004) HOXA10 regulates p53 expression and matrigel invasion in human breast cancer cells. Cancer Biol Ther 3(6):568–572

27. Zhang X, Hamada J, Nishimoto A, Takahashi Y, Murai T, Tada M et al (2007) HOXC6 and HOXC11 increase transcription of S100β gene in BrdU-induced in vitro differentiation of GOTO neuroblastoma cells into Schwannian cells. J Cell Mol Med 11(2):299–306
28. deBlacam C, Byrne C, Hughes E, McIlroy M, Bane F, Hill AD et al (2011) HOXC11-SRC-1 regulation of S100β in cutaneous melanoma: new targets for the kinase inhibitor dasatinib. Br J Cancer 105(1):118–123
29. Charmsaz S, Hughes É, Bane FT, Tibbitts P, McIlroy M, Byrne C et al (2017) S100β as a serum marker in endocrine resistant breast cancer. BMC Med 15(1):79
30. Waddington CH (1959) Canalization of development and genetic assimilation of acquired characters. Nature 183(4676):1654–1655
31. Lande-Diner L, Zhang J, Ben-Porath I, Amariglio N, Keshet I, Hecht M et al (2007) Role of DNA methylation in stable gene repression. J Biol Chem 282(16):12194–12200
32. Bird AP (1996) The relationship of DNA methylation to cancer. Cancer Surv 28:87–101
33. Cheng X, Blumenthal RM (2008) Mammalian DNA methyltransferases: a structural perspective. Structure (London, England : 1993) 16(3):341–350
34. Dahl C, Grønbæk K, Guldberg P (2011) Advances in DNA methylation: 5-hydroxymethylcytosine revisited. Clinica chimica acta; international journal of clinical chemistry 412(11–12):831–83e
35. Schaefer M, Pollex T, Hanna K, Tuorto F, Meusburger M, Helm M et al (2010) RNA methylation by Dnmt2 protects transfer RNAs against stress-induced cleavage. Genes Dev 24(15):1590–1595
36. Bestor TH (2000) The DNA methyltransferases of mammals. Hum Mol Genet 9(16):2395–2402
37. Fujita N, Watanabe S, Ichimura T, Tsuruzoe S, Shinkai Y, Tachibana M et al (2003) Methyl-CpG binding domain 1 (MBD1) interacts with the Suv39h1-HP1 heterochromatic complex for DNA methylation-based transcriptional repression. J Biol Chem 278(26):24132–24138
38. Bogdanović O, Veenstra GJ (2009) DNA methylation and methyl-CpG binding proteins: developmental requirements and function. Chromosoma 118(5):549–565
39. Hendrich B, Hardeland U, Ng HH, Jiricny J, Bird A (1999) The thymine glycosylase MBD4 can bind to the product of deamination at methylated CpG sites. Nature 401(6750):301–304
40. Lewis JD, Meehan RR, Henzel WJ, Maurer-Fogy I, Jeppesen P, Klein F et al (1992) Purification, sequence, and cellular localization of a novel chromosomal protein that binds to methylated DNA. Cell 69(6):905–914
41. Chahrour M, Jung SY, Shaw C, Zhou X, Wong ST, Qin J et al (2008) MeCP2, a key contributor to neurological disease, activates and represses transcription. Science 320(5880):1224–1229
42. Nan X, Ng HH, Johnson CA, Laherty CD, Turner BM, Eisenman RN et al (1998) Transcriptional repression by the methyl-CpG-binding protein MeCP2 involves a histone deacetylase complex. Nature 393(6683):386–389
43. Kornberg RD (1974) Chromatin structure: a repeating unit of histones and DNA. Science 184(4139):868–871
44. Strahl BD, Allis CD (2000) The language of covalent histone modifications. Nature 403(6765):41–45
45. Ito K, Adcock IM (2002) Histone acetylation and histone deacetylation. Mol Biotechnol 20(1):99–106
46. Grunstein M (1997) Histone acetylation in chromatin structure and transcription. Nature 389(6649):349–352
47. Greer EL, Shi Y (2012) Histone methylation: a dynamic mark in health, disease and inheritance. Nat Rev Genet 13(5):343–357
48. Cedar H, Bergman Y (2009) Linking DNA methylation and histone modification: patterns and paradigms. Nat Rev Genet 10(5):295–304
49. Turner BM (2000) Histone acetylation and an epigenetic code. BioEssays: News Rev Mol Cell Dev Biol 22(9):836–845
50. Roth SY, Denu JM, Allis CD (2001) Histone acetyltransferases. Annu Rev Biochem 70:81–120

51. Bachman KE, Rountree MR, Baylin SB (2001) Dnmt3a and Dnmt3b are transcriptional repressors that exhibit unique localization properties to heterochromatin. J Biol Chem 276(34):32282–32287
52. Feinberg AP, Ohlsson R, Henikoff S (2006) The epigenetic progenitor origin of human cancer. Nat Rev Genet 7(1):21–33
53. Carr BI, Reilly JG, Smith SS, Winberg C, Riggs A (1984) The tumorigenicity of 5-azacytidine in the male Fischer rat. Carcinogenesis 5(12):1583–1590
54. Baylin SB, Herman JG (2000) DNA hypermethylation in tumorigenesis: epigenetics joins genetics. Trends Genetics : TIG 16(4):168–174
55. Qin L, Chen X, Wu Y, Feng Z, He T, Wang L et al (2011) Steroid receptor Coactivator-1 upregulates integrin α_5 expression to promote breast Cancer cell adhesion and migration. Cancer Res 71(5):1742–1751
56. Dobrovolna J, Chinenov Y, Kennedy MA, Liu B, Rogatsky I (2012) Glucocorticoid-dependent phosphorylation of the transcriptional Coregulator GRIP1. Mol Cell Biol 32(4):730–739
57. Chakraborty M, Qiu SG, Vasudevan KM, Rangnekar VM (2001) Par-4 drives trafficking and activation of Fas and FasL to induce prostate Cancer cell apoptosis and tumor regression. Cancer Res 61(19):7255–7263
58. Hebbar N, Wang C, Rangnekar VM (2012) Mechanisms of apoptosis by the tumor suppressor Par-4. J Cell Physiol 227(12):3715–3721
59. Alvarez JV, Pan TC, Ruth J, Feng Y, Zhou A, Pant D et al (2013) Par-4 downregulation promotes breast cancer recurrence by preventing multinucleation following targeted therapy. Cancer Cell 24(1):30–44
60. Rogatsky I, Luecke HF, Leitman DC, Yamamoto KR (2002) Alternate surfaces of transcriptional coregulator GRIP1 function in different glucocorticoid receptor activation and repression contexts. Proc Natl Acad Sci 99(26):16701–16706
61. Rogatsky I, Zarember KA, Yamamoto KR (2001) Factor recruitment and TIF2/GRIP1 corepressor activity at a collagenase-3 response element that mediates regulation by phorbol esters and hormones. EMBO J 20(21):6071–6083
62. Uhlenhaut NH, Barish Grant D, Yu Ruth T, Downes M, Karunasiri M, Liddle C et al (2013) Insights into negative regulation by the glucocorticoid receptor from genome-wide profiling of inflammatory Cistromes. Mol Cell 49(1):158–171
63. Mittelstadt ML, Patel RC (2012) AP-1 mediated transcriptional repression of matrix metalloproteinase-9 by recruitment of histone deacetylase 1 in response to interferon β. PLoS One 7(8):e42152
64. Lizcano F, Romero C, Vargas D (2011) Regulation of adipogenesis by nuclear receptor PPARγ is modulated by the histone demethylase JMJD2C. Genet Mol Biol 34(1):19–24
65. Santos RVC, de Sena WLB, Dos Santos FA, da Silva Filho AF, da Rocha Pitta MG, da Rocha Pitta MG et al (2019) Potential therapeutic agents against Par-4 target for Cancer treatment: where are we going? Curr Drug Targets 20(6):635–654
66. Mabe NW, Fox DB, Lupo R, Decker AE, Phelps SN, Thompson JW et al (2018) Epigenetic silencing of tumor suppressor Par-4 promotes chemoresistance in recurrent breast cancer. J Clin Invest 128(10):4413–4428

Index

V. M. Rangnekar (ed.), *Tumor Suppressor Par-4*,
https://doi.org/10.1007/978-3-030-73572-2

The manufacturer's authorised representative in the EU is Springer Nature Customer Service Centre GmbH, Europaplatz 3, 69115 Heidelberg, Germany. If you have any concerns regarding our products, please contact ProductSafety@springernature.com

Printed and bound by CPI Group (UK) Ltd, Croydon, CR0 4YY
30/09/2026
02225687-0005